Brewing
Second Edition

Michael J. Lewis, PhD, FIBrew
Professor Emeritus of Brewing Science
University of California, Davis
Academic Director of Brewing Programs
University Extension
Davis, California

Tom W. Young, PhD, FIBrew, ILTM
Senior Lecturer in Biochemistry
School of Biosciences
The University of Birmingham
Birmingham, United Kingdom

Kluwer Academic / Plenum Publishers
New York, Boston, Dordrecht, London, Moscow
2002

Library of Congress Cataloging-in-Publication Data

Lewis, Michael, 1936–
Brewing / Michael J. Lewis, Tom W. Young.—2nd ed.
p. cm.
Includes bibliographical references and index.
ISBN 0-8342-1851-8
1. Brewing. I. Young, Tom W. II. Title.

TP570.L475 2001
663′.3—dc21

© KLUWER ACADEMIC / PLENUM PUBLISHERS
233 Spring Street, New York, New York 10013
http://www.wkap.com

Library of Congress Catalog Card Number: TP570.L475 2002
ISBN: 0-306-47274-0

Printed in the United States of America

1 2 3 4 5

Contents

Preface

The second edition of this little book on brewing arises because many readers told us they found the first edition useful. We were pleased and encouraged and even flattered by that, as any authors would be. This second edition gives us the chance to correct some things that needed correcting and to add two chapters that should have been in the first edition but somehow got overlooked—a chapter devoted to water and another to analytical methods and statistical process control (Chapters 4 and 7).

We hope we have heeded the advice not to turn a useful little book into a big clumsy one, but nevertheless we have also added some material on engineering subjects. One of these additions is a stand-alone chapter on heat transfer and refrigeration (Chapter 3) and we have added sections to Chapters 17 (fluids), Chapter 19 (carbonation) and a little to Chapter 20 (packaging).

The engineering material deliberately falls far short of a standard course on brewery engineering (which is a different book from this one and one that we will not write!). The objective here is to demystify the subject, as far as possible, by showing how engineers do what they do, and to expose our non-engineer readers to some principles of engineering that (whether they know it or not) they observe and work with every day in the brewery. We hope that through an understanding, or at least an appreciation, of these principles, solving everyday problems may prove easier and elementary mistakes might be avoided.

We have some colleagues to thank. We are most grateful to professor emeritus of food engineering R. Larry Merson whose impeccable attention to the new engineering material was immensely helpful; to the Anheuser-Busch endowed professor of brewing science Charles W. Bamforth who made wise and perceptive comments on the malting and brewing chapters and to Ms. Candace Wallin for her Trojan work on the section on brewing calculations. These are all colleagues at the University of California Davis, Department of Food Science and Technology. We are delighted, incidentally, to say that the research and teaching program in malting and brewing science continues there under the new and tireless and charismatic leadership of Charlie Bamforth.

Tom Young remains active in research and teaching at the University of Birmingham, UK, as Senior Lecturer of Biochemistry. Since the first edition was written one of us (MJL) has retired from the University of California

Davis and now enjoys the honorific professor emeritus of brewing science. However, as Academic Director and Lead Instructor, he continues to be responsible for the University Extension program in brewing science of which the Masterbrewers program and the related Professional Brewers Certificate Program are the flagships, and he is active in the profession as a speaker and writer. His is a non-retirement.

We are indebted to our students, at the university and in shorter courses, including those from unusual and even exotic backgrounds who, coming to the subject fresh have often provided us with new insights into brewing. These students are highly enthusiastic and dedicated women and men many of whom wish to enter the world of corporate brewing and others who are determined to build, own and run their own microbrewing enterprises. Their demands for knowledge of brewing has helped us to sharpen our focus on what to teach and how to best teach it, and accounts for some of our original editorial choices and some of the changes in this edition of the book.

We are also grateful to the many brewing scientists and production brewers in various parts of the world who, over the years, have discussed brewing with us and generously provided of their time to show us their brewing operations. Equally, we are indebted to those experienced practitioners of the brewers art, who, while attending short courses for example, have often illuminated our science with their practice, and from which we have learned. Despite the help of all our colleagues and students the approach we have taken to the subject, the selection (and omission!) of topics and the view we have expressed are entirely our own and any errors are solely our responsibility.

It is our intention and our hope that this book will provide a useful and practical grounding in the fundamentals of brewing science and the practice of brewing to all those who wish to acquire it. For those who will go further we have provided a reference list to texts, journals and proceedings as a starting point.

Finally and foremostly, we thank our wives Sheila and Dorothy, to whom we dedicate this book, for without their support we would never have completed the task.

Michael J. Lewis
Department of Food Science and Technology
University of California
Davis
California. USA

Tom W. Young
School of Biosciences
University of Birmingham
Birmingham. UK

November 2000

Units and conversion tables

SPECIFIC GRAVITY CONVERSION CHART

Values are at 20°C and rounded up.
°Plato=Degrees Brix=Cane sugar % w/w.

Specific gravity	°Plato	Specific gravity	°Plato
1.003	0.6	1.047	11.7
1.005	1.3	1.050	12.3
1.007	1.9	1.062	15.3
1.010	2.6	1.065	15.8
1.012	3.2	1.067	16.4
1.015	3.8	1.070	17.0
1.017	4.4	1.087	21.0
1.020	5.1	1.090	21.5
1.022	5.7	1.092	22.0
1.025	6.3	1.095	22.6
1.027	6.9	1.097	23.1
1.030	7.5	1.100	23.7
1.032	8.1	1.102	24.2
1.035	8.7	1.105	24.8
1.037	9.4	1.107	25.3
1.040	10.0	1.110	25.8
1.042	10.6	1.112	26.4
1.045	11.2	1.115	27.0

TEMPERATURE CONVERSION CHART

°Celsius	°Fahrenheit	°Celsius	°Fahrenheit	°Celsius	°Fahrenheit
−6	21	40	104	70	158
−4	25	42	108	71	160
−2	28	44	111	72	162
−1	30	46	115	73	163
0	32	48	118	74	165
2	36	50	122	76	169
4	39	52	126	78	172
8	46	54	129	80	176
10	50	55	131	82	180
12	54	56	133	84	183
14	57	57	135	86	187
16	61	58	136	88	190
18	64	59	138	90	194
20	68	60	140	92	198
22	72	61	142	94	201
24	75	62	144	96	205
26	79	63	145	98	208
28	82	64	147	100	212
30	86	65	149	105	221
32	90	66	151	110	230
34	93	67	153	115	239
36	97	68	154	120	248
38	100	69	156	121	250

BEER VOLUME CONVERSION TABLE

To convert the units in the left column to units in one of the right 5 columns **multiply** by the value in the table.

Example—to convert Hectolitres to U.S. barrels **multiply** by 0.852
1000 hl = 1000 × 0.852 = 852 U.S. barrels.

	Hectolitres	U.S. barrels	U.K. barrels	U.S. gallons	U.K. gallons
Hectolitres		0.852	0.611	26.412	21.997
U.S. barrels	1.174		0.717	31.000	25.813
U.K. barrels	1.634	1.395		43.234	36.000
U.S. gallons					1.201
U.K. gallons				0.833	

UNITS AND CONVERSION TITLES

Volumes

1 U.S. barrel = 0.717 U.K. barrel = 1.174 hl
1 U.K. barrel = 1.395 U.S. barrel = 1.634 hl
1 hl = 100 l
1 U.S. barrel of water weighs 258 lb.
1 U.K. barrel of water weighs 360 lb.
1 hl of water weighs 220.5 lb.

Weights

1 metric ton (tonne) = 1000 kg = 2204.6 lb.
1 U.K. ton (ton) = 2240 lb.
1 Zentner = 50 kg
1 kg = 2.205 lb.
1 lb. = 0.4536 kg

English traditional cask volumes

1 Butt = 3 barrels = 108 gallons
1 Hogshead = 54 gallons
1 Kilderkin = 0.5 barrel = 18 gallons
1 Firkin = 0.25 barrel = 9 gallons
1 Pin = 0.125 barrel = 4.5 gallons

Yeast pitching rate

1 lb./U.K. barrel of pressed yeast = 0.3kg/hl = about 10 million cells/ml
Pressed yeast is about 20% solids
Yeast slurry is about half the solids content of pressed yeast

Carbon dioxide in beer

1 g carbon dioxide in 100 ml beer = 5.06 vol/vol
1 vol carbon dioxide/vol beer = 0.198g/100ml = 0.198% w/v

Alcohol (Ethanol) in beer

Alcohol % by weight = 0.79 × alcohol % by volume.
Alcohol % by volume = 1.266 × alcohol % by weight

BARLEY AND MALT MEASURES

Country	Pounds (lb)	Kilograms (kg)
U.K. and South Africa		
Barley quarter	448	203.21
Barley bushel	56	25.40
Malt quarter	336	152.41
Malt bushel	42	19.05
Australia and New Zealand		
Barley bushel	50	22.68
Malt bushel	40	18.14
U.S. and Canada		
Barley bushel	48	21.78
Malt bushel	34	15.42

Components

Components

CHAPTER 1

Overview of the brewing process

I f beer were invented today, the complex processes described in this book would probably be considered a most inappropriate technology. We would almost certainly make beer directly from barley or other grain using enzymes, we would not heat and cool and dry and wet the process as much as we do today; we would shorten fermentation and aging and we would not dilute the bulk product and fill it into tiny, yet heavy packages for distribution! The process would be a good deal more rational.

But beer was not invented today. Many of our processes and practices have come down to us through centuries and even millennia of brewing and from generations of brewers who have gone before us. These ideas form the traditional background (the brewer's art) against which the rational or scientific approach to brewing is thrown into relief. Brewing science is the approach of this book, but we cannot ignore, or indeed avoid, the influence that previous generations of brewers have had on the way we think today. Nor should we wish to. We do not throw over the traditional ways of making beer simply because it is technically possible to do so; we are, after all, brewers. It is the essence of brewers (and perhaps other artisans or craftsmen) to hold to certain traditions while working firmly in the present. Brewers have always done that; they have been at the forefront of scientific inquiry and technological advance while remaining true to their roots. To become a brewer means that we must develop a feel for the interplay of the old with the new and the art with the science and be comfortable with it. It's not easy. Where we have arrived at today is determined by yesterday. In a flight of fancy, we would imagine that a brewer from ancient Egypt, though awed by the extraordinary scale and technology of a modern brewery, would quickly identify it as a brewery.

What drives brewing change? Internal to the industry and its suppliers, we might point to the drive for quality, the drive for process control and efficiency, the drive for market share, and the drive for economy and a decent profit. Externally, we can recognize the demands of the marketplace and consumer preference, the pressure of the neo-prohibitionists, laws about drinking and driving and their enforcement, government policy about alcohol consumption, and, especially, the pressure of taxation.

There is little doubt that in former times, regulations and the imposition of a tax on malt had much to do with the type of malting processes employed. The malts produced, to a large extent, governed the technologies of the brewhouse, which effects we can still observe and recognize today. Brewers in Japan have recently invented very low-malt beers to avoid heavy taxation on regular beers. Are such beers a way of the future? In former years (and it is no different today), profitability and success went to those brewers who could understand and control their processes, minimize losses, maximize yields, and produce beers of consistent high quality and preferred flavor. The introduction of the thermometer and saccharometer and the development of biochemistry and microbiology gave a rational basis for regulating brewing processes and, therefore, it is no accident that many of today's most successful brewing companies have their origins in the late 19th century. Those technically forward-looking brewers prospered, and consumers benefited from quality and consistency.

Today in the world's major brewing companies, the advent of new technologies and increased scientific understanding has led to a reappraisal of many traditional processes. This, in turn, has produced significant changes in thinking and fostered internationalization of brewing practices as ideas, developments, and new inventions become available to all brewers unimpeded (or at least less impeded than they once were) by local traditions. It is therefore feasible to distill, from many traditions and regions, a common worldwide view of the basic science of brewing operations, because brewers and breweries around the world are much more similar today than formerly (even in comparatively recent history). Most brewers, therefore, operate processes that follow the basic outlines briefly reviewed in this opening chapter.

1.1 RAW MATERIALS AND PROCESS

We begin with a brief description of the raw materials of brewing, which might also serve as a brief dictionary of useful brewing terms. We provide a flow chart of the brewing process (Figure 1–1) and also present an overview of the key stages of brewing, their description, purposes, temperatures and approximate time scales condensed into Table 1–1.

Traditionally the raw materials of brewing are **water**, **malt**, **hops** and **yeast**. Many brewers use in addition **adjuncts** and/or various **processing aids**.

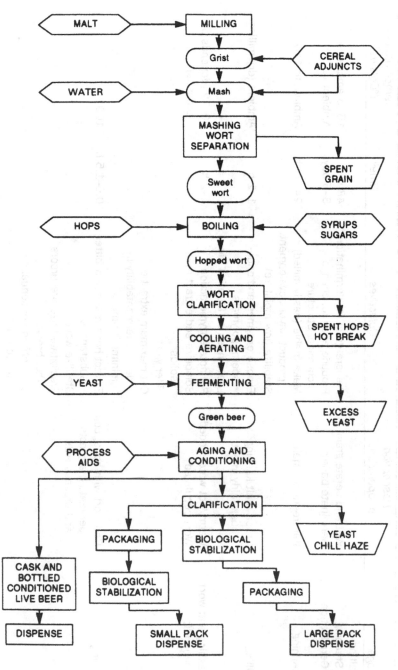

Figure 1–1 An overview of brewing, from malt milling to beer dispense.

Table 1-1 Stages, operations, and objectives of the brewing process

Stage	Description of operations	Objectives	Time[a]	Temperature (°Celsius)
Malting				
Steeping	Wet and aerate grain	Prepare grain for germination	48 h	15 to ambient
Germination	Germinate barley	Produce enzyme, change chemical structure	3–5 days	Ambient to 22
Kilning	Kiln resultant malt	Make friable (easily milled) product flavor development. Stabilize (low moisture)	24–48 h	Ambient to 110
Milling	Crush malt keeping husk fairly intact	Expose enzymes and endo-sperm polymers. Leave husk as filtration aid	1–2 h	Ambient (dry milling)
Mashing and wort separation	Warm/hot water added. Wort circulated	Promote enzyme action. Solubilize and extract malt solids. Filter wort. Get maximum extract of desired fermentability in minimum time	1–2 h	30–72 (depends on process)
Boiling	Boil wort with hops, copper (kettle) adjuncts and process aids	Extract hops make new bitter substances. Sterilize wort. Precipitate haze precursors (hot break). Drive off hop essential oil. Increase color	0.5–1.5 h	100+
Wort clarification	Straining, settling, or centrifugation	Remove spent hops and hot break (trub), clarify wort	<1 h	100–80

continued

	Action	Purpose	Time	Temperature
Cooling and aerating	Pass through heat exchanger Inject oxygen	Create conditions for yeast growth in fermentation	<1 h	12–18
Fermenting	Add yeast Control desired specific gravity Remove yeast	Produce green beer from wort Obtain yeast for further fermentations Collect carbon dioxide	2–7 days	12–22 Cool to 4–15
Aging and conditioning	Pass beer to oxygen-free tanks Chill beer Make additions of process aids Transfer live beers to package	Mature beer Modify flavor Adjust carbon dioxide level Settle out yeast and cold break Stabilize the beer	7–21 days	−1–0
Clarification	Centrifuge, filter mature beer	Remove yeast and suspended solids (cold break) Produce bright beer	1–2 h	−1–0
Biological stabilization	Pasteurize or sterile filter	Kill or remove any microbes	1–2 h	62–72 (pasteurization) −1–0 (sterile filtration)
Packaging	Bottle, can, keg or cask Pasteurize small volumes in the package	Produced packaged beer to desired specifications	0.5–1.5 h	−1 to ambient
Dispense	Pour, pump or displace with carbon dioxide	Present beer of desired specifications to consumer	Seconds	4 to ambient

aTimings are based on a beer production volume of 500 hl

1.1.1 Water

To brew one volume of beer, breweries use 4 to 12 volumes of water. It is an increasingly expensive commodity and may be in short supply. Large breweries use primarily mains (municipal, town) water, although many also use some bore-hole water (well water). Water for brewing must be potable and is often modified by filtration and inclusion of salts to change its hardness and sometimes acids to change its alkalinity.

1.1.2 Malts, special malts, and malt substitutes

Malt

This is derived from special varieties of barley by controlled germination of the grain in the malting process. Germination changes the physical and chemical structure of the barley, so that malt is a friable (easily ground) product which provides a source of enzymes able to break down the polymers (starches and proteins) of malt. Ground malt is extracted with warm water in the brewhouse of the brewery to give "sweet wort" or "extract." Arguably, malt is the most important brewing raw material.

Special malts

These are colored malts with increased flavor made by heating normal malt at high heat intensity. They are used in small amounts to give flavor and color to certain beers. A cheaper alternative in some beers is roasted barley.

Malt substitutes

Malt extract or barley syrup (barley converted with microbial enzymes) may be used to compensate for a lack of malt extraction capacity or to enhance the gravity (density) of a wort.

1.1.3 Adjuncts

These are additions which are used to supplement extract from malt. They include cereals (such as rice, corn (maize), barley and wheat products), syrups, and sugars. Typical examples are flaked cereals (similar in appearance to cornflakes), flours (usually wheat flour) and grits (coarse starchy endosperm particles of maize or rice).

Syrups are derived from maize, barley, or wheat and are produced by acid or acid/enzyme treatments to hydrolyze the starch, followed by concentration. Sugars, usually sucrose (from cane rather than beet) or invert sugar (hydrolyzed sucrose) can be used for special purposes. When used in the copper (kettle), sugars are referred to as "copper sugar"; when added to beer they are called "priming sugar."

1.1.4 Hops

Whole hops

Whole hops are kiln-dried cones which are the fruit of the hop vine. Other forms of hop include powdered hops (hammer-milled whole hops), hop pellets (powdered hops compressed as pellets), or hop extracts (hops extracted with liquid carbon dioxide). These are used in the kettle of the brewhouse. Isomerized extracts (usually extracts that are heated to convert α-acids to the iso α-acids) are used after the fermentation stage to provide "post-fermentation bitterness." Reduced hop extracts are light stable.

1.1.5 Yeast

Yeast is a living organism and therefore, properly, it should not be considered a "raw material." Yeast converts the fermentable sugar of wort to alcohol, carbon dioxide, and a host of necessary and desirable beer flavor compounds. It is, thus, more than a mere constituent. Although formal taxonomy classifies all brewers yeasts as *Saccharomyces cerevisiae*, brewers commonly identify two types of yeast:

- Ale yeast (*Saccharomyces cerevisiae*)—this is usually used for fermentation at temperatures of 15 to 17°C (59 to 63°F) or higher. It may be used for ale or lager production.
- Lager yeast (*Saccharomyces carlbergensis* or *Saccharomyces uvarum*)—this is used at a fermentation temperatures of less than 15°C (and often much lower), primarily for making lager beers.

1.1.6 Other additions

- **Caramels** can be used to provide color and flavor.
- **Hop oil** is distilled essential oil of hops. It is used to give "hoppy" aroma.
- **Anti-microbials**—only sulfur dioxide is permitted (the concentration allowed, 10 to 50 ppm, depends on beer type and local regulations. SO_2 is also an effective anti-oxidant and stabilizes beer flavor. If used in the malt kiln or as a kettle addition, it helps conserve pale color.
- **Carbon dioxide** is used as a beer conditioner (to boost carbonation).

- **Head (foam) promoters**, usually alginate esters, can be added to enhance beer foam stability, mainly by protecting against foam-collapsing compounds such as oils or grease on beer glasses.
- **Reducing agents** including ascorbic acid (vitamin C), isoascorbate, or sulfur dioxide (SO_2) can be added to protect beer against the ravages of oxygen that otherwise catalyzes haze formation and induces undesirable flavor changes.
- **Nitrogen** gas can be added to relatively low-CO_2 beers, including packaged beers, to bring up a creamy foam similar to that on beers dispensed with a traditional beer engine.

1.1.7 Processing aids

- **Yeast foods** are used to improve the nutritional status of wort and aid efficient fermentation.
- **Clarifying agents** such as copper (kettle) finings are used to aid separation of proteins in the copper (wort kettle). Auxiliary finings are similar preparations to copper finings and are used to complement the action of isinglass finings. Isinglass finings are collagen preparations (from fish swim bladders) dissolved in acid and used to precipitate proteins and yeast from beer to render it clear and more stable against haze formation
- **Stabilizing agents** are added to beer in the aging tank to prevent protein/polyphenol haze. Polyphenol adsorbents, such as polyvinyl pyrrolidone (PVPP), which is insoluble in beer, binds and removes haze-forming polyphenols. Protein adsorbents, such as silica hydrogels, also insoluble in beer, bind and remove haze-forming proteins. These contemporary materials are added to the beer in the conditioning tank and are removed by filtration.

CHAPTER 2

Basic chemistry for brewing science

2.1 INTRODUCTION

The events of brewing can only be explained and understood in a scientific context. Scientific terms, e.g., proteins, enzymes, molecules, bonds, pH, heat transfer, conductivity, etc., provide a necessary shorthand by which brewing events can be explained and not merely described. Absent a firm grasp of the scientific basis of brewing, the profession reverts to an unbridled art. Although all brewers cherish the arts and mysteries of what they do and the long history of human endeavor of which they are the present embodiment, few indeed could imagine modern brewing without modern science and engineering. The following sections therefore deal with some scientific concepts that might make the chapters on brewing science more intelligible.

2.2 CHEMICAL BONDS AND WATER

All life requires water because the curious collection of coordinated chemical reactions we call life can take place only in an aqueous environment. A brief study of the chemistry of this remarkable substance repays the effort, not merely because beer is an aqueous product, but because those who understand water understand much of the basic chemistry that governs brewing processes.

The water molecule comprises two hydrogen atoms joined to an oxygen atom by covalent bonds. A **covalent bond** is the sharing of electrons among the atoms involved in the bond to reach the most stable state. This, it so happens, requires eight electrons, an **octet** (except for hydrogen, in which case it is two electrons). Most biological substances are made up of only six elements: hydrogen (H), nitrogen (N), oxygen (O), carbon (C), sulfur (S), and phosphorus (P), joined, for the most part, covalently. The outer, or valence (bond-forming) electrons of these atoms, can be represented in

11

the Lewis structure by dots (Figure 2–1a). Thus, water creates its stable molecular form because the hydrogen atoms share one each of the oxygen valence electrons and the oxygen shares the two hydrogen electrons to capture eight in its orbit (Figure 2–1b). Oxygen has a valency of two.

Similarly, carbon, with four valence electrons of its own, needs to find four electrons to share to make up its stable complement of eight, as in methane (CH_4) (Figure 2–1b). Carbon has a valency of four. The four valencies of carbon can be satisfied by two double valencies, for example, with oxygen to form carbon dioxide (CO_2) (Figure 2–1b). This arrangement creates a double bond. These ideas may be extended for example to ethanol, the alcohol in fermented beverages (Figure 2–1c).

In short, covalent bonds are **shared electron pairs** which achieve a stable octet for each atom. Double bonds (and triple bonds) occur when the molecule forms with two (or three) shared electron pairs to form the octet. In double or triple bonds, the flexibility about the bond is reduced and a double bond is shorter, stronger and more energetic than a single bond.

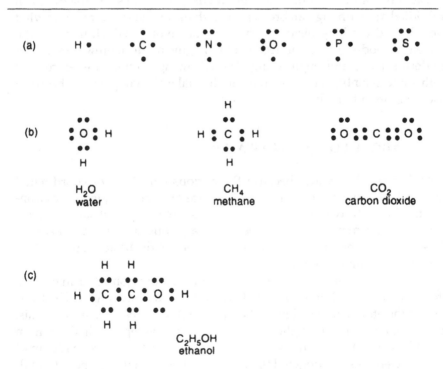

Figure 2–1 The Lewis structure of some atoms and molecules in which the valence electrons are shown as dots.

The shared electrons, or covalent bonds, among atoms in molecules are shown as connecting lines among the atoms (see Figures 2–8 and 2–15). The most common structures in organic molecules important to brewing are carboxylic acids R-COOH (or R-COO⁻), aldehydes R-CHO, and alcohols R-CH$_2$OH. The R group represents the rest of the molecule which might be small (e.g., H or CH$_3$, or progressively larger). But, however large the molecule, the properties of it are governed by whether it is an acid, an aldehyde, or an alcohol. These reactive groups are fairly easily interconverted, as shown in many biological pathways (Figure 18–4a, for example). Pathways are simply roadmaps and are more easily followed if the reader keeps count of the carbon atoms and recognizes the interconversion of these main reactive groups.

The four electron pairs around an atom in a molecule repel each other and so the bonds they form are most stable when farthest apart. In other words, when all four pairs of electrons are shared, they disperse around the atom uniformly and therefore point to the four corners of a tetrahedron. Around the carbon atom in methane (for example) the four valencies or bonds point towards the corners of a tetrahedron (Figure 2–2a) and the angle between the bonds is 109.5°. This gives the environment around a carbon atom a distinct three-dimensional character (Figure 2–3). This quality becomes crucial in some biological structures and shows up for example in the L- and D-forms of amino acids and sugars.

Other structures are linear as in the CO$_2$ molecule because there are two groups of two electron pairs (double bonds) around the C-atom. However, in water and many other molecules, there are pairs of **unshared elec-**

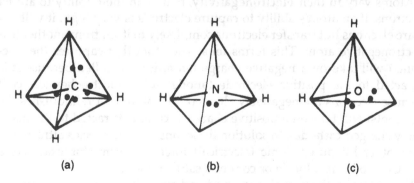

(a) (b) (c)

Figure 2–2 Carbon with four valencies occupies the center of a tetrahedron: (a) as in methane (CH$_4$). The ammonia NH$_3$ (b) and water H$_2$O (c) molecules are nearly tetrahedral in which the extra electron pair(s) (●●) can be thought of as directed toward one or two corners of the tetrahedron.

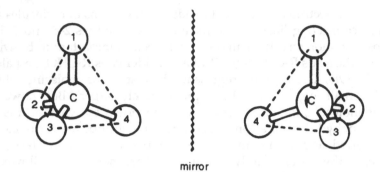

mirror

Figure 2–3 Because the carbon atom (C) in the tetrahedron is linked to four different groups (1, 2, 3, 4), two different compounds can be formed. These are called **chiral** molecules and are related to each other structurally just as a left and right hand are related, i.e., as mirror images. They are enatiomers of each other or an enantiomeric pair. Nature chooses specific chiral structures.

---

trons that affect the molecular geometry or shape. These unshared electron pairs behave in space just as if they were valencies or bonds and so a water molecule (or an ammonia molecule) occupies a tetrahedron-shaped space just as the methane molecule does. But, since oxygen in a water molecule has only two H atoms (not four as in methane, and nitrogen in an ammonia molecule has three H atoms, not four) the water and ammonia molecules are "bent" (Figure 2–2 b,c). This means that one part of the tetrahedron they occupy is different from another. This explains dipole moment.

Atoms vary in their **electronegativity**, that is, in their ability to attract electrons. If an atom's ability to capture electrons is very high, it will not share electrons but transfer electrons completely to itself from another less electronegative atom. This forms **ions**. The atom that captures the electron(s) will take on a negative charge (an **anion**, so called because it is attracted to the positive electrode (**anode**) in an electrochemical cell) because electrons are negatively charged. The atom that relinquishes the electron(s) will take on a positive charge (a **cation**, attracted to the negatively charged **cathode**). In solution these ions are free species attracted to each other by purely ionic (electrical) forces to form, for example, a Na^+Cl^- (sodium chloride or common salt) solution.

If, in contrast, the participants in a bond are rather equally electronegative, the electrons will be shared in a covalent bond and there will be little or no electric charge or ionic character. However, in the case of water, oxygen is somewhat more electronegative than hydrogen and so, in a water molecule, the shared electrons tend to orbit a bit closer to oxygen than to the hydrogen atoms. The oxygen atom therefore takes on a distinct though small negative charge, and the hydrogen atoms to the same extent take on a slight positive charge. Such a bond is dipolar and a molecule containing such bonds is a **dipole**, and the molecule is **polar**. The direction of the charge determines the dipole moment influenced by the "bent" or V-shape of the molecule as discussed above. Polar molecules are **hydrophilic** (water loving) and water soluble. Polar molecules also tend to line up or organize themselves in crystals and in solution with the positive part of one molecule juxtaposed to the negative part of another. This electrostatic attraction is called a dipole force and is as remarkable and as effective as the attraction of the north and south poles of magnets.

If two atoms are about equally electronegative, no dipole develops as a result of the bond. Such a bond is said to be **non-polar**; for example, C—H bonds are non-polar, because carbon and hydrogen are almost equally electronegative. Non-polar molecules are **hydrophobic** or water insoluble, as for example, oils or hydrocarbons.

The **hydrogen bond** is an unusually strong type of dipole force that is common and centrally important in biological systems. Hydrogen bonds arise in or among molecules in which a hydrogen atom is linked to oxygen or to nitrogen (as in water or ammonia or organic molecules). Because these molecules are "bent," and the difference in electronegativity of the atoms is significant, the dipole moment (direction of charge) is quite strong. Also, the small atomic radius of hydrogen, oxygen, and nitrogen allows those parts of molecules containing these atoms to approach closely to each other to form the relatively strong hydrogen bond.

Though many other factors influence hydrogen bond formation in particular circumstances, the presence of hydrogen bonds within molecules, among molecules and between organic molecules and water is a crucially important condition in biology. The ability to form hydrogen bonds significantly modifies the properties and behavior of compounds. In water itself, for example, an extensive network of hydrogen bonds among the water molecules dominates its properties (Figure 2–4). For example, water boils at **plus** 100°C rather than at **minus** 150°C or so which its molecular weight predicts. In fact to vaporize water (i.e., turn it into a gas or create steam) requires much more energy than for any other liquid because the strength of the H-bonds that form liquid water must be overcome. In practice, this explains why steam is such a convenient way of moving heat around a brewery (Chapter 18). Similarly, much heat can be dissipated by vaporization of water, an important consideration in malt kilning for example. The high specific heat, latent heat of vaporization (evaporation),

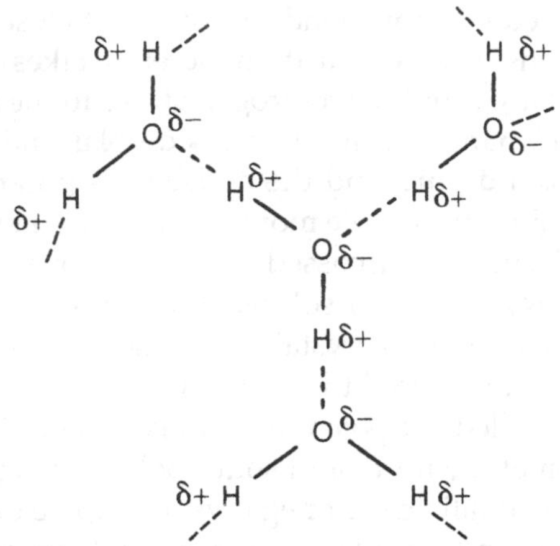

Figure 2–4 Hydrogen bonding (- - - -) among water molecules is almost complete in ice, and short chains occur in water. The energy of these bonds accounts for the extraordinary properties of ice, water, and steam including, e.g., their high specific heat. $\delta+$ and $\delta-$ indicate the slight positive and negative charges present in water and responsible for the extensive H-bonding present.

and heat of fusion (melting) of water have important biological consequences, and arise because of hydrogen bonding. Hydrogen bonds dramatically affect the properties of all materials in which they arise including proteins and polysaccharides. Of course, ammonia is also an extensively hydrogen-bonded material with unusual properties that find particular application in refrigeration.

Besides hydrogen bonds, the idea of an electrical dipole helps to explain another weak force that is important in holding together large molecules. This is the **dispersion force**, more commonly called the van der Waals force or interaction. This force works in either polar or non-polar molecules or parts of molecules. The hydrogen bond depends on a permanent dipole, but van der Waals' forces depend on temporary dipoles that arise as the electron cloud around atoms becomes momentarily concentrated at one end of the molecule. This induces temporary dipoles elsewhere and in other molecules that match the first and so attract. These forces are more effective in large molecules than in small ones because electrons polarize more easily in such molecules.

A third weak force that stabilizes large molecules is the **hydrophobic force** or interaction. This is really the absence of bonding. A polar substance easily dissolves in water because, through formation of hydrogen bonds, it easily inserts itself into the extensively hydrogen-bonded net-work of the water. A non-polar substance will not dissolve because it can-not force itself into this network. If, however, it should be carried into the water network by a molecule that is otherwise water soluble, the water will build a cage around the non-polar part by electrical repulsion. Moreover, if several non-polar parts on the same or different molecules approach closely to each other, they will tend to be herded into the same cage or forced together, rather as tiny oil droplets in water emulsion coa-lesce into large ones. This is the nature of the hydrophobic bond.

A fourth force that holds organic molecules together—**ionic forces**—has already been described. It depends upon the simple electrical attrac-tion of positive and negative charges. In biological systems these are pri-marily associated with carboxylic acids (as —COO⁻ not —COOH) and amino groups (as —NH₃⁺ not —NH₂). These charges arise because of a deficit or excess of one electron caused by (partial) dissociation to lose a proton (H⁺) or association to gain one.

2.3 pH AND BUFFERS

Although water is a covalently-bonded molecular material, it can dis-sociate to form a tiny proportion of ions, represented as H^+ and OH^- (the H^+ reacts again with H_2O to form H_3O^+ but this does not alter the argu-ment that follows) and

$$K_w = [H^+] \times [OH^-] = 1.0 \times 10^{-14} \text{ M (ion product at 25°C)} \qquad [2.1]$$

(where K_w is the equilibrium constant of water). In pure water or neu-tral solution, the $[H^+]$ always equals the $[OH^-]$ and so in this case

$$[H^+] = 1.0 \times 10^{-7} \text{ M}$$
$$\text{(from } [H^+] = K_w/[OH^-] \text{ or square root of } K_w) \qquad [2.2]$$

But solutions are not always neutral. In such a case, since the constant K_w cannot alter, if one ion increases, the other must decrease proportion-ally. Therefore, the concentration of one ion can always be calculated from the concentration of the other through equation [2.1]. Concentrations of H^+ greater than 10^{-7} M (e.g. 10^{-6} or 10^{-3}) are acidic; below this concen-tration (e.g. 10^{-8} or 10^{-10}) are alkaline or basic. Complementarily, a con-centration of OH^- above 10^{-7} M is alkaline or basic, below is acidic. The maximum and minimum concentration of H^+ or OH^- approaches 10^0 M (i.e. 1 M) and 10^{-14} M respectively, and if both ions are at 10^{-7} M, the solu-tion is neutral (equation [2.2]).

The pH scale is merely a more convenient way of expressing these ideas. pH is defined as the negative $\log_{10}$ of the hydrogen ion concentration or

$$pH = -\log_{10}[H^+]$$

Then if the H^+ concentration is 10^{-4} or 10^{-9} M the pH is 4.0 or 9.0, and so on. Consequently, the full range of the pH scale runs from pH = 0 to pH = 14. What is important to remember is that the pH scale is logarithmic and pH 3 contains ten times **more** H^+ and ten times *less* OH^- than a solution at pH 4. Similarly, a wort at pH 5.2 contains nearly four times as much H^+ as wort at pH 5.8.

By definition, an acid releases a proton (an H^+) and a base combines with one. Strong acids such as hydrochloric acid, dissociate completely in water into two ions H^+ and Cl^-, because of the great difference in electronegativity of the atoms. A weak acid (represented by HA), on the other hand, only partially dissociates in water to release H^+ and its negatively charged anion A^- leaving some HA; this dissociation is reversible, and the lost proton can be regained (by association) thus:

$$HA \rightleftharpoons H^+ + A^-$$

and the equilibrium constant K_a for this reaction will be

$$K_a = [H^+][A^-]/[HA]$$

By converting everything to logarithms, this can be derived:

$$pH \rightleftharpoons pK_a + \log_{10}[A^-]/[HA] \qquad [2.3]$$

If A^- is equal to HA (i.e. the acid is half dissociated and $\log_{10}[A^-]/[HA]$ is zero), then equation [2.3] says in plain terms that the pK_a (the dissociation constant) of any acid is equal to the pH at which half the acid molecules are dissociated and half are not.

If a solution of base be slowly added to a weak acid, such as acetic acid, the pH does not rise steadily with added base, but passes through a distinct plateau (Figure 2–5). At the middle of this plateau the pH is equal to the pK_a, as predicted from equation [2.3]. The plateau is the region in which the weak acid is said to "buffer"; that is, the pH does not change much with the addition or subtraction of H^+ or OH^-, i.e., addition of acid or base.

Most **buffers** (weak acids or weak bases in the presence of their salts) are effective between one pH unit above and one below their pK_a. Resistance to pH change is an important quality of biological systems and depends on the concentration of the buffering compounds present. In brewing, phosphate ions and amino acids and even proteins exercise strong buffering effects in wort. Buffering explains why the pH of wort is

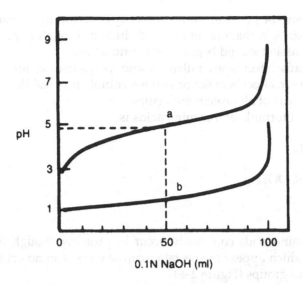

Figure 2–5 (a) Titration curve of 0.1N acetic acid (weak acid, 100 ml) with 0.1N NaOH. (b) Titration curve of 0.1N HCl (strong acid, 100 ml) with 0.1N NaOH. The buffering capacity of the mixed weak acid and base is best when the acid is half-neutralized (50 ml NaOH added). This is the pK_a (dotted lines), which is pH 4.73 for acetic acid. HCl does not buffer.

hard to change by adding acid and why the pH of brewing water (as opposed to the composition and concentration of salts dissolved in it) is, of itself, inconsequential to mash pH.

2.4 AMINO ACIDS AND PROTEINS

Proteins perform many useful functions in nature as skin, hair, blood, muscle, tendons, hormones and enzymes. In foods, proteins affect (among other things) nutrition, foams, gels and emulsions, and water-binding capacity. When cooked, the reaction of proteins to heat influences food appearance, flavor, and texture. In beer and brewing, the list of important functions of proteins is also impressive: proteins affect beer clarity and beer foam; as enzymes proteins affect extract quality and dictate the behavior of yeast; as a source of **amino acids**, proteins affect yeast performance and beer flavor, and through the Maillard reaction, they affect malt and beer color and flavor. The list could be lengthened. Suffice it to say that brewers and maltsters need to have some understanding of proteins and protein structure to fully appreciate how brewing processes affect beer quality,

because important practical events often operate through protein chemistry. A protein is a chain of amino acids linked through **peptide bonds**,
and this discussion should begin with amino acids.

Amino acids affect yeast nutrition and performance, influence malt
color and flavor, affect beer flavor and microbiological stability, and make
up the main part of all protein molecules.

The general formula of all amino acids is:

$$
\begin{array}{c}
NH_2 \\
| \\
R\!-\!C\!-\!COOH \\
| \\
H
\end{array}
$$

Twenty amino acids commonly occur in proteins, though there are a
few others which appear in special circumstances. Amino acids fall into
one of several groups (Figure 2–6)

- Group 1 amino acids contain one amino and one acidic group; examples are alanine and valine
- Group 2A amino acids contain the same as Group 1, plus sulfur; an
  example is cysteine
- Group 2B amino acids contain the same as Group 1, plus a benzene
  ring; an example is tyrosine
- Group 3 amino acids contain one amino and two acidic groups;
  examples are glutamic and aspartic acid
- Group 4 amino acids contain two amino groups and one acidic
  group; an example is lysine
- Group 5 amino acids are the heterocyclic amino acids, such as tryptophan and proline.

The amino acids have two features of particular note. First, they have
an asymmetric carbon atom or chiral center (called the α-carbon, i.e., a
carbon atom attached through its four valencies to four different substituents R–, H–, –NH_2, and –COOH). This means, because of the
three-dimensional or tetrahedral structure around any carbon atom, that
two different versions of the same amino acid structure can be built. These
structures are mirror images of each other and cannot be superimposed
on each other (Figure 2–3). Chemically, they are called **enantiomorphs**;
they are enantiomers of each other in the same way that one's left hand is
an enantiomer of one's right hand. The two forms are called L- and D- by
convention in comparison to the structures of L- and D-lactic acid; these

Group 1

Notes

glycine

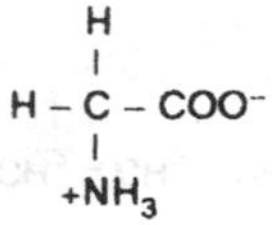

$$H - \underset{\underset{+NH_3}{|}}{\overset{\overset{H}{|}}{C}} - COO^-$$

alanine

$$CH_3 - \underset{\underset{+NH_3}{|}}{\overset{\overset{H}{|}}{C}} - COO^-$$

serine — R-group is polar

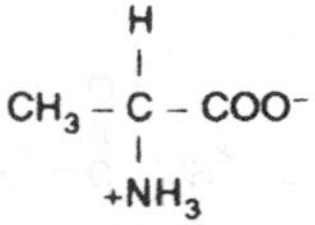

$$HO - CH_2 - \underset{\underset{+NH_3}{|}}{\overset{\overset{H}{|}}{C}} - COO^-$$

threonine — R-group is polar

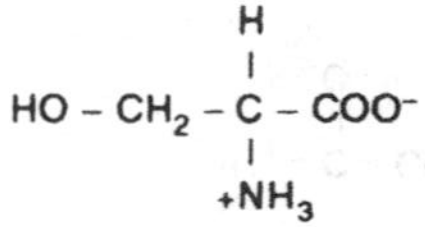

$$\underset{CH_3}{\overset{OH}{\diagdown}} CH - \underset{\underset{+NH_3}{|}}{\overset{\overset{H}{|}}{C}} - COO^-$$

valine

$$\underset{CH_3}{\overset{CH_3}{\diagdown}} CH - \underset{\underset{+NH_3}{|}}{\overset{\overset{H}{|}}{C}} - COO^-$$

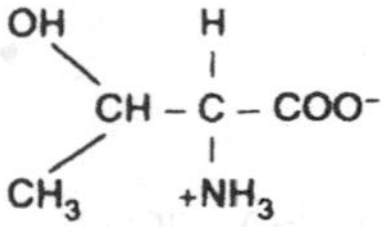

leucine

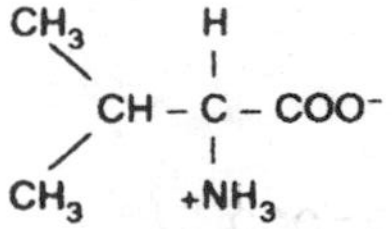

$$\underset{CH_3}{\overset{CH_3}{\diagdown}} CH - CH_2 - \underset{\underset{+NH_3}{|}}{\overset{\overset{H}{|}}{C}} - COO^-$$

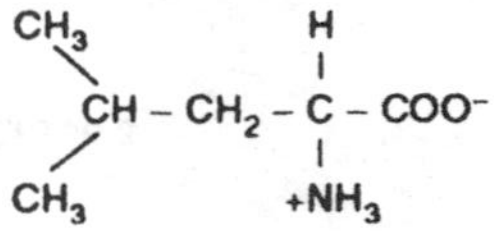

Notes

iso-leucine

$$CH_3 - CH_2 - \underset{CH_3}{CH} - \underset{\underset{+NH_3}{|}}{\overset{\overset{H}{|}}{C}} - COO^-$$

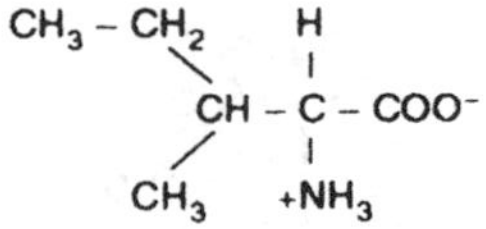

Group 2A

cysteine — R-group is polar

$$HS - CH_2 - \underset{\underset{+NH_3}{|}}{\overset{\overset{H}{|}}{C}} - COO^-$$

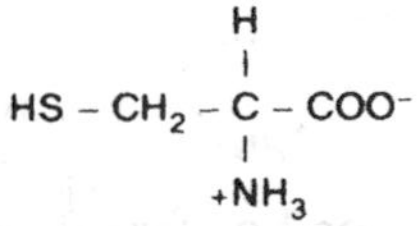

methionine

$$CH_3 - S - CH_2 - CH_2 - \underset{\underset{+NH_3}{|}}{\overset{\overset{H}{|}}{C}} - COO^-$$

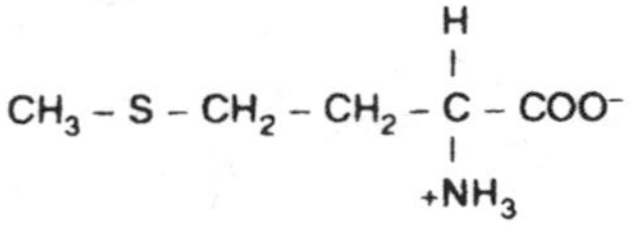

Group 2B

phenylalanine

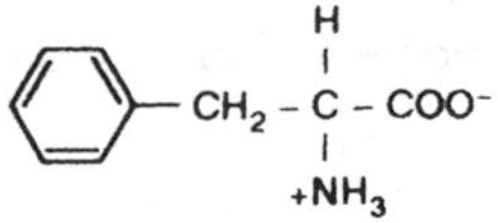

$$C_6H_5 - CH_2 - \underset{\underset{+NH_3}{|}}{\overset{\overset{H}{|}}{C}} - COO^-$$

tyrosine — R-group is polar

$$HO - C_6H_4 - CH_2 - \underset{\underset{+NH_3}{|}}{\overset{\overset{H}{|}}{C}} - COO^-$$

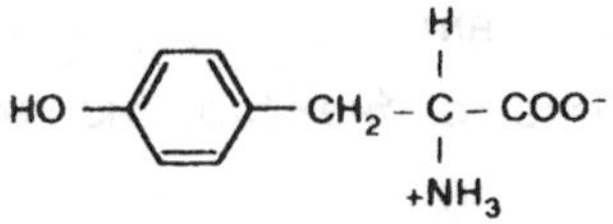

continues

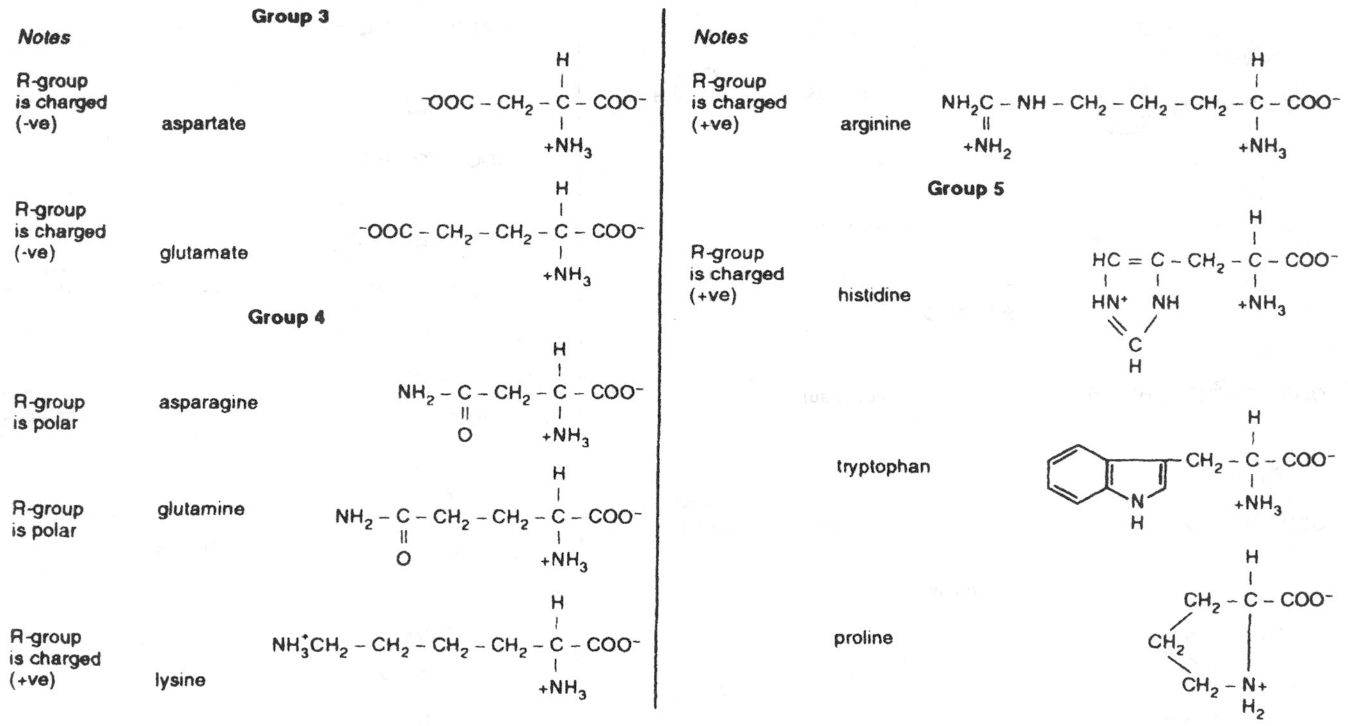

Notes

Group 3

R-group
is charged
(-ve)

aspartate

$$^-OOC - CH_2 - C - COO^-$$
$$| \quad |$$
$$H \quad +NH_3$$

R-group
is charged
(-ve)

glutamate

$$^-OOC - CH_2 - CH_2 - C - COO^-$$
$$| \quad |$$
$$H \quad +NH_3$$

Group 4

R-group
is polar

asparagine

$$NH_2 - C - CH_2 - C - COO^-$$
$$\| \qquad | \quad |$$
$$O \qquad H \quad +NH_3$$

R-group
is polar

glutamine

$$NH_2 - C - CH_2 - CH_2 - C - COO^-$$
$$\| \qquad\qquad | \quad |$$
$$O \qquad\qquad H \quad +NH_3$$

R-group
is charged
(+ve)

lysine

$$NH_3^+CH_2 - CH_2 - CH_2 - CH_2 - C - COO^-$$
$$| \quad |$$
$$H \quad +NH_3$$

Notes

R-group
is charged
(+ve)

arginine

$$NH_2C - NH - CH_2 - CH_2 - CH_2 - C - COO^-$$
$$\| \qquad\qquad\qquad\qquad | \quad |$$
$$+NH_2 \qquad\qquad\qquad H \quad +NH_3$$

Group 5

R-group
is charged
(+ve)

histidine

$$HC = C - CH_2 - C - COO^-$$
$$| \qquad | \quad |$$
$$HN^+ \quad NH \quad H \quad +NH_3$$
$$\backslash\ /$$
$$C$$
$$|$$
$$H$$

tryptophan

$$CH_2 - C - COO^-$$
$$| \quad |$$
$$H \quad +NH_3$$

proline

$$CH_2 - C - COO^-$$
$$| \quad |$$
$$H$$
$$CH_2 \quad CH_2 - N^+$$
$$\backslash CH_2 / \quad H_2$$

Figure 2-6 The classification of amino acid structures. Whether the R-groups are polar, charged positively or negatively or non-polar affects the structure and properties of the protein in which a particular spectrum and sequence of amino acids occur.

letters have nothing to do with the direction of rotation of polarized light. Only L-amino acids occur in proteins. Second, the chemical properties of amino acids do not support the electrically uncharged molecule represented above, because for example, amino acids do not behave chemically as intermediates between amines and acids.

If acid or base be added to a neutral solution of an amino acid, the pH of the solution changes in the same way as described for buffers (Figure 2–7). An amino acid is a buffer, but because it acts both as a weak acid and

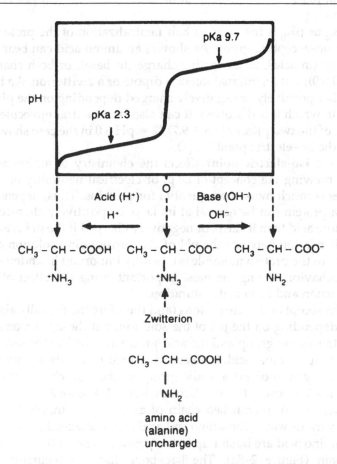

Figure 2–7 The structure of the amino acids in solution is best described as the electrically neutral but charged (+ −) zwitterion. The curve demonstrates the titration of the amino (basic) and carboxylic (acidic) group present in the typical amino acid alanine. The horizontal parts of the curve represent buffering capacity.

as a weak base, there are two plateaus in the titration curve. Buffering is one important function of amino acids in, for example, mashing. Buffers take up or release H^+ ions, depending on whether conditions dictate that they associate or dissociate. For amino acids this is as follows:

$$\underset{\substack{\text{Net charge }+1\\\text{at pH 1}}}{R-CH-NH_3{}^+-COOH} \underset{H^+}{\rightleftharpoons} \underset{\substack{\text{Net charge 0}\\\text{at pH 6}}}{\overset{pK_a\ 2.3}{R-CH-NH_3{}^+-COO^-}} \underset{H^+}{\rightleftharpoons} \underset{\substack{\text{Net charge }-1\\\text{at pH 11}}}{\overset{pK_b\ 9.7}{R-CH-NH_2-COO^-}}$$

The pK_a or pK_b is the point of half neutralization of the proton (H^+)-donating or -accepting group. As shown, an amino acid can bear a positive charge (in acid), or a negative charge (in base), or both charges (at about pH 6.0); it is an internal salt or a dipole or a **zwitterion**. An internal salt is either positively or negatively charged depending on the pH of the solution in which it is dissolved. It can also be a neutral molecule at half the value of the two pKs, or $(2.3 + 9.7)/2 = $ pH 6.0 in the case shown. This is called the iso-electric point (i.e.p.).

While the **iso-electric point** affects the chemistry of amino acids, in practical brewing the concept of i.e.p. or electrical neutrality of charged molecules is much more consequential for proteins. Thus, depending on the pH, a protein can be neutral at its i.e.p. or positively charged at pH values more acid than its i.e.p. or negatively charged if the pH is above its i.e.p. This relation between the pH of the surrounding solution and the properties of the protein molecule is centrally important in understanding protein behavior, among the most important being the effect of pH on enzyme action and protein denaturation.

Proteins accept or donate protons (and therefore incidentally also act as buffers) depending on the pH of the solution, but these reactions do not concern the amino group and the acid group attached to the α-carbon of each individual amino acid. It is caused instead by the dissociation of the extra amino group or extra acidic group in the side chain (R) of some amino acids, especially those in Groups 3 and 4 (Figure 2–6).

A protein is an unbranched chain of amino acids that is folded and coiled in various ways. The amino and acid group attached to the α-carbon of each amino acid are bound up in the peptide bond of the protein backbone chain (Figure 2–8a). The backbone has the recurring pattern NCC–NCC–NCC–NCC–NCC–NCC–NCC, etc., in a very long unbranched chain, where each NCC represents an amino acid comprising, in order, the amino nitrogen (N), the α-carbon (C) and the carboxyl carbon (C); the link (–) between the carboxyl carbon of the first amino acid and the amino

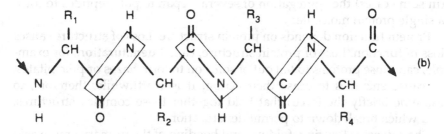

Figure 2–8 (a) Linkage of two amino acids joined by a peptide bond (in the box) to form a dipeptide. (b) Long unbranched chains of amino acids linked through peptide bonds make up polypeptide chains. (The relatively inflexible peptide bond is in the box.)

nitrogen of the second is the **peptide bond** (Figure 2–8b). The polypeptide chain has a direction, which by convention starts at the free amino end and progresses to the free carboxyl end. Because all the amino acids are L-amino acids, the individual R groups all point in the same direction.

The order or sequence of amino acids in this unbranched chain is called the **primary structure** of the protein and is rigidly dictated by the genetic code (the DNA or deoxyribonucleic acid of the cell); this is the only connection between the information contained in the genetic code and the expression of that information. The primary structure of a protein determines its secondary structure. The protein chain emerges from the ribosome within the cell (where it is assembled in a very complex process) rather like a tape leaving an old-fashioned ticker-tape machine, and, like the tape, the amino acid chain can bend and fold in some directions better than others. Particularly, the partial double-bond character of the peptide bond strictly limits rotation at this link, but the other bonds of the backbone rotate easily being single bonds. How far they rotate depends on many factors including the size of the R group. But there is only one form, called the **secondary structure**, in which the twisted chain is most stable. This secondary structure begins to be established at the earliest stage in protein assembly i.e., as it leaves the ribosome. Common secondary structures include the α-helix (Figure 2–9) and the β-pleated sheet (Figure 2–10). Proteins may have regions which lack regular structures such as helices and sheets, but no region has random structure.

The **tertiary structure** extends the secondary structure and is not always easily distinguished from it. Tertiary structure involves the folding of a series of α-helices, for example, on each other, rather as a stranded rope may be coiled on itself or knotted. The **quaternary structure** involves (in some cases) the aggregation of several separate polypeptides to form a single protein molecule.

Protein **function** depends on protein **structure**. Loss of structure causes loss of function. Loss of protein structure, called **denaturation**, for example, can cause proteins to fall out of solution to form hazes or precipitates, or cause enzymes to cease their action. It is worthwhile, therefore, to examine briefly the forces that hold together these complex structures, and which break down to permit denaturation.

The extensive twisting, folding, and bending of the primary amino acid chain, brings together parts of the molecule that were some distance apart in the original chain. In the α-helix, for example, a single turn brings every fourth amino acid in the chain into close proximity. The resulting juxtaposition of the O– of each carboxyl (carbonyl) and the H– of each amino (imido) group of the next third amino acid along the chain, favors the formation of a hydrogen bond at that point. Thus the α-helix is made (conceptually) tubular and held rigid by the H-bonds along its length (Figure 2–9). Eleven amino acids (Figure 2–6) can also form hydrogen bonds through their polar or polar/charged side chains (R). Hydrogen bonds are individually weak but the formation of many of them makes them collectively strong and able to stabilize protein structures. Other bonds include van der Waals' and hydrophobic and ionic interactions and the –S–S– disulfide bridge formed between two molecules of cysteine (Figure 2–11).

Hydrogen bonds are destroyed by heat or by acid or base or other chemicals and even by powerful agitation. Some parts of the protein molecule are more susceptible to these treatments than others and begin to lose form or "unfold." This is denaturation. In an enzyme protein, for example, the place where catalysis occurs (the active site) might lose the form required for binding the specific substrate of its action, and so enzyme catalysis

(a)

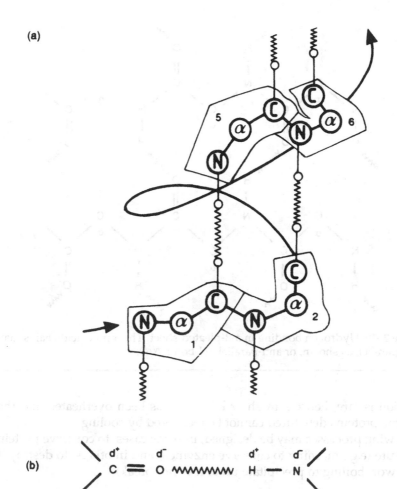

(b)

Figure 2–9 (a) In an α-helix or coiled amino acid chain, amino acid *n* and *n* + 4 become close enough that an H-bond (〰️) forms between the carbonyl dipole of amino acid 1 (or 2, etc.) and the imino dipole of amino acid 5 (or 6, etc.). This is shown in more detail in (b).

slows and then ceases. Because of unfolding, parts of the molecule that were originally bound in the native protein, are revealed for new interactions. As a result, proteins can aggregate with each other to form visible haze particles or precipitates, for example, as hot break (trub) during the kettle boil. Once denaturation happens, renaturation to recover original

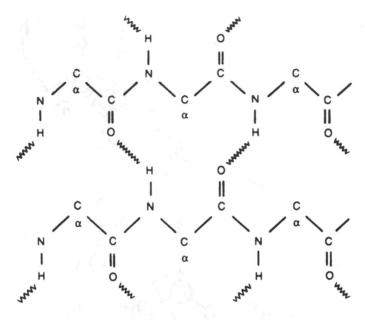

Figure 2–10 Hydrogen bonding in a β-pleated sheet. The amino acid chains may be in parallel, as shown, or anti-parallel. (H-bond, ᨳᨳᨳ).

function is rare; hence a mash or kiln that has been overheated and the enzyme proteins denatured cannot be recovered by cooling.

Brewing processes may be designed, in some cases, to conserve protein structure (e.g., mashing to conserve enzymes) and in others, to destroy it (e.g., wort boiling to precipitate protein as hot break).

2.5 ENZYMES

Enzymes are remarkable protein molecules. Their particular alchemy is that they accelerate chemical reactions under mild conditions of concentration, temperature, pH, and pressure and with exquisite precision. For every chemical reaction of life, there is an enzyme or group of enzymes available.

Enzymes are **catalysts** (a material that participates in a chemical reaction without itself being changed) which, in the presence of suitable substrates and due to the power of specific activation, accelerate thermodynamically possible reactions. For all practical purposes, enzymes can be considered to "cause" (not merely accelerate) chemical changes because

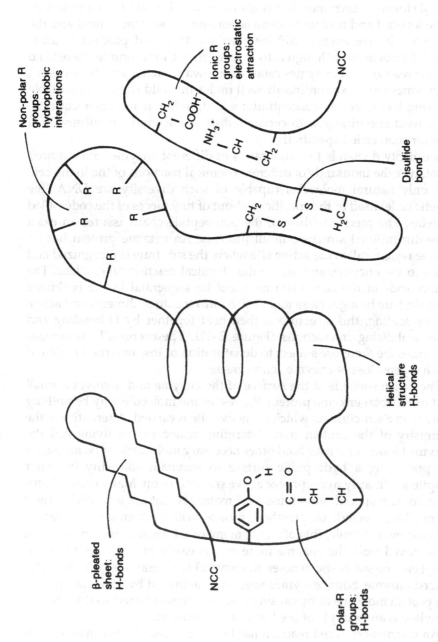

Non-polar R groups: hydrophobic interactions

Ionic R groups: electrostatic attraction

Disulfide bond

Helical structure H-bonds

β-pleated sheet: H-bonds

Polar-R groups: H-bonds

NCC

Figure 2–11 Possible stabilizing forces in proteins.

without amylases, for example, starch degradation would proceed with glacial slowness. Enzymes do not alter the equilibrium of a reaction (that is the forward and reverse reaction accelerate to the same extent) and the difference in free energy (ΔF) between substrate and product is unaffected. In reactions with high activation energies, for example, there is no reverse reaction and enzymes catalyze one-way reactions in those cases.

Enzymes work extraordinarily well under the mild conditions suitable to living tissue, at low concentrations, and, unlike many chemical catalysts, react specifically with certain substrates or kinds of substrates (a phenomenon called **specificity**).

Specificity demands that thousands of different enzymes are required to catalyze the thousands of different chemical reactions of the living cell. The only natural molecules capable of such diversity are DNA (the genetic code) itself or the specific read-out of tiny pieces of the code called proteins. The precise folding of the polypeptide chain assures an exact three-dimensional structure in all proteins. An enzyme protein has an unique region, called the **active site** where the substrate is recognized and binds to the enzyme, and where the chemical reaction takes place. The amino acids of the active site need not be sequential in the backbone chain, but are brought close together to form the three-dimensional active site by folding; this structure is then held together by H-bonding and other stabilizing interactions (Figure 2–11). These crucially important structures are therefore subject to denaturation or loss of structure and, if this happens, loss of enzyme action ensues.

The active site(s) is at the surface of the enzyme and involves a small part of the total enzyme protein; the rest of the molecule may be nothing more than a structure on which the active site is carried. Alternatively, the chemistry of the protein may determine where in the living cell the enzyme locates, or it may bind other necessary molecules, or be necessary for presenting a hydrophilic surface to maintain solubility in water despite a primarily hydrophobic active site, and so on. Many enzyme proteins contain special non-amino acid materials that are essential to their action. These include (a) prosthetic groups, which are an intrinsic part of the enzyme molecule, (b) cofactors, many of which are vitamins and can be removed from the enzyme more or less easily, and (c) metal ions. α-Amylase (the suffix -ase denotes an enzyme) is an example of a Ca^{++}-stabilized enzyme. Some enzymes need to be activated by removing part of the protein molecule or by causing –S–S– bridges to be reduced to –SH. β-Amylase is an example of the latter type of enzyme.

An enzyme-catalyzed reaction has three components that must always be considered: the enzyme itself; the substance(s) it acts on, called the **sub-**

strate; and the conditions of the reaction (especially temperature and pH). Enzymes bind with their substrates briefly at the active site to effect their action. This binding involves hydrogen bonds and other non-covalent bonds, but the unique binding properties of the site are such that some specific molecules are much more tightly and easily bound than related ones. As a result, the vast array of potential substrates are excluded from the site. This defines specificity. Two models illustrate this selective binding phenomenon:

1. The **lock and key** mechanism implies that a substrate is like a key that fits into a lock on the enzyme surface; this model accounts for various levels of specificity because a lock can be made more or less complex and more or less exclusive.
2. The second model is called **induced fit**. Binding of the substrate to the enzyme causes a change in shape which brings the active site into position to catalyze the reaction. Many molecules may bind to the enzyme, but only those that are specific substrates of the enzyme cause the necessary conformational (shape) change. In some cases, this binding is sufficient to cause reaction; in other cases, covalent bonds between the enzyme and substrate temporarily form.

In either case, binding of a substrate to an enzyme creates a new environment around the substrate that effectively lowers the amount of energy needed to drive the reaction (the activation energy of the reaction). An enzyme-catalyzed reaction can derive sufficient energy from the surroundings and does not need high heat or pressure that an uncatalyzed reaction might require. Put another way, the reaction that proceeds via binding of the substrate to the enzyme permits substrate molecules to reach a lower level of excitement (energy content) before they react, than in the uncatalyzed reaction (Figure 2–12).

The **rate** or speed (change over time) of an enzyme reaction depends on many factors, including the amount of enzyme relative to the quantity of substrate. If there is a vast amount of substrate present, as, for example, starch at the beginning of mashing, addition of more enzyme will increase the amount of starch broken down in a given time. Under these conditions, the enzyme is said to be **saturated**. This means that every active site is working as fast as it can. This is the maximum velocity (V_{max}) of the enzyme-catalyzed reaction when the substrate is in large excess (Figure 2–13a and right-hand side of Figure 2–13b). In the reverse case, however, when the enzyme is not saturated (i.e., much enzyme is present but not much substrate), the rate of reaction does not depend on enzyme concentration and adding more enzyme has no effect. In this con-

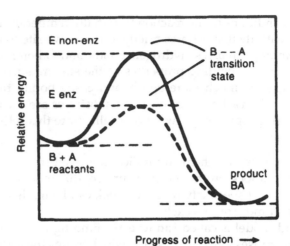

Figure 2–12 To react to form a product, the reactants must reach a sufficient level of energy called the energy of activation (E), which brings them to a transition state. Catalyzed, especially enzyme-catalyzed reactions, require a much lower energy of activation (E-enz) than the same reaction carried out by non-catalyzed methods (E-non-enz). Sufficient energy could come from the ambient surroundings or from mild warming in the case of E-enz. E-non-enz might require the energy input of boiling and pressure, for example.

dition, the reaction rate depends on the concentration of the substrate (Figure 2–13b). The substrate concentration at which the velocity of the reaction is half the maximum velocity ($V_{max}/2$) is called the K_m or Michaelis constant. This value is a fixed property of the enzyme when measured under specified conditions, and it approximates the **affinity** of an enzyme for its substrate. This means that if an enzyme has a low K_m (is half-saturated at a low substrate concentration), it aggressively attaches itself to that substrate or is said to have a high affinity for it. A high K_m implies the opposite.

The curves shown in Figure 2–13 a,b and the K_m have brewing consequences. For example, when a mash begins, a vast excess of starch saturates the amylases and the reaction proceeds at maximum velocity. In this case, of course, a malt with a high diastatic power/dextrinizing units (DP/DU) would be preferred to a low DP/DU malt because more enzyme implies a faster rate of conversion of starch to sugar (Figure 2–13a). Eventually, the starch is substantially hydrolyzed (used up) and the amylolytic reaction slows down because of this decrease in substrate

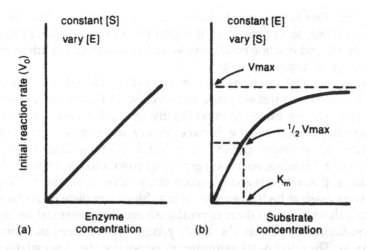

Figure 2–13 Drawing (a): In a large excess of substrate the enzyme is said to be saturated (or fully occupied) and each unit of enzyme is operating at its maximum rate (V_{max}). The overall reaction rate (V_0) depends only on the concentration of enzyme. Drawing (b): The enzyme is saturated (as in drawing a) on the right hand side of drawing b so each unit of enzyme present is operating at V_{max}. When the enzyme is not saturated the reaction rate depends on substrate concentration. The substrate concentration at $\frac{1}{2}V_{max}$ is called the Michaelis constant (K_m): the lower this value the greater is the affinity (attraction) between substrate and enzyme.

concentration (Figure 2–13b). The reaction also slows down for another reason: as α-amylase acts, it cleaves amylose and amylopectin for which the amylases have a high affinity (low K_m), replacing these huge polymers with smaller oligopolysaccharides (dextrins) for which α- and β-amylase have lower affinity (higher K_m). Towards the end of the mash, the relatively high K_m for dextrins causes these to be degraded progressively more slowly. Of course, another factor enters at this stage that also contributes to the progressive slowing of conversion as the mash proceeds: the inactivation of enzymes, especially β-amylase, as a result of denaturation at high mash temperature.

Consequently, it is important to remember that the most significant enzyme-catalyzed reactions in mashing happen in the early stages of the process, because that is where the highest concentration of desirable substrates occurs and where enzymes operate at their maximum concentration and efficiency.

Because enzymes are proteins, their chemistry is influenced by pH and temperature, and most enzymes are said to have an **optimum** pH and temperature. This is true only in the sense that for a given objective and in a given setting, an optimum set of conditions can be defined. However if the objective and reaction conditions were to change, then a different set of optima might apply.

As most chemical reactions do, an enzyme-catalyzed reaction goes faster when warmer. Most enzymes, however, work best at moderate temperatures; above this range, thermal inactivation arises due to denaturation or loss of protein structure. Some enzymes are more sensitive to heat than others: for example, β-amylase is less heat-stable than α-amylase and, in brewers' mashes, which are generally conducted at relatively high temperature, β-amylase is always more at risk than α-amylase (Chapter 13). Brewers mash at high temperature to achieve complete conversion of starch in a short time and there is profligate enzyme denaturation under these conditions and so, as the mash progresses, conversion proceeds more slowly. The effect of temperature on enzyme systems is usually represented as passing through a maximum (Figure 2–14) because heat can accelerate a reaction but also inactivate the enzyme.

The concentration of H^+ ions (pH) dramatically affects the ionization, or electrical status, of a protein and so affects enzyme action in two ways: (a) within a relatively narrow range of pH, enzymes may act faster or slower because the exact ionic status of the protein molecule, especially at the active site, affects binding with the substrate (Figure 2–14), and (b) with large pH change, caused by exposure to significantly acidic or basic conditions, enzyme proteins denature irreversibly.

Enzyme action may be slow as a result of **inhibition**. Some inhibitors, called competitive inhibitors, ape the "true" substrate of the enzyme and compete for the active site and block it. This form of inhibition can be reversed by increasing the level of "true" substrate to dilute and displace the inhibitor. A special form of this inhibition is product inhibition; in such a case, the product of an enzyme reaction, when it reaches high concentration, inhibits the enzyme by binding to the active site. For example, β-amylase is inhibited by high levels of maltose. This is one of the reasons it is difficult to produce highly fermentable worts in dense mashes. In contrast, non-competitive inhibitors react with a site on the enzyme from which it cannot be removed by adding extra "true" substrate.

2.6 CARBOHYDRATES

Carbohydrates are an important source of energy for living organisms and a means by which energy may be stored. The starch of barley endo-

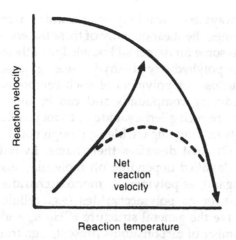

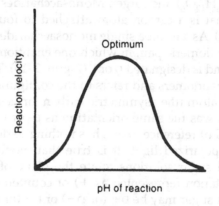

Figure 2–14 Effect of temperature (above) and pH on the reaction rate of an enzyme-catalyzed reaction. Heat accelerates a reaction but also inactivates the enzyme (through protein denaturation). The net reaction velocity therefore has a maximum at quite a low temperature (dashed line). The pH response of reaction velocity usually approximates a bell-shaped curve. This reflects the influence of pH on the ionization states of amino acid side chains in the enzyme protein molecule. Optimal enzyme activity is a function of the charged state of the protein.

---

sperm is an example of stored energy. Carbohydrates also serve as structural components of cells. Examples of structural carbohydrates include the hemi-celluloses of husk and cell walls of barley. Carbohydrates are quantitatively the most important group of compounds that brewers encounter and form the vast bulk of extract in wort. Though an understanding of their structure is in some ways less rewarding for practical brewing than a study of proteins and enzymes, the shear quantity of these materials manipulated by brewers demands some fundamental knowledge of their chemistry.

Carbohydrates are polyhydroxy **aldehydes** (e.g., glucose, Figure 2–15) or **ketones** (e.g., fructose) or polymers of such compounds. Aldehydes and ketones are reducing compounds and can be measured by their reducing power (e.g., reducing ferricyanide to ferrocyanide). Such sugars are called beguilingly *reducing* sugars. Many, though not all, obey the general formula $C_x(H_2O)_n$ that describes their name: hydrates of carbon. Carbohydrates are classified depending on molecular size as monosaccharides (simple sugars), as polymers of monosaccharides (oligosaccharides, e.g., dextrins), or as polysaccharides (e.g. cellulose or starch). Monosaccharides have the general structure $(CH_2O)_n$ and are classified according to the number of carbon atoms present, e.g., trioses (n = three carbons $C_3H_6O_3$, the smallest carbohydrates) to the important hexoses (n = 6, e.g., glucose $C_6H_{12}O_6$) and larger. Monosaccharides contain at least one chiral center, that is a carbon atom attached to four different substituents (Figure 2–2). As a result a single monosaccharide can exist in two forms called an enantiomeric pair, in which one enantiomer is the mirror image of the other and is designated D or L (Figure 2–15). This designation identifies the two enantiomers and refers to the configuration around the penultimate carbon atom (the asymmetric carbon farthest from the aldehyde group), which was the same orientation as D- or L-glyceraldehyde used as the standard of reference. D or L has nothing to do with the direction of rotation of polarized light. It is true that monosaccharides are optically active and sugar solutions rotate the plane of polarized light clockwise (dextrorotation represented as +) or counterclockwise (laevorotation, −). Thus, a sugar may be D+ (or D−) or L− (or L+). The D form of a sugar is the mirror image of the L form: they rotate polarized light in opposite directions. A triose sugar contains one chiral center (Figure 2–16) and there are two forms of that sugar, i.e., one enantiomeric pair. Hexoses contain four chiral centers and there are $2n$ or sixteen aldohexoses (eight enantiomeric pairs; Figure 2–16. Note: this figure omits L-glyceraldehyde and the L-aldoses).

D-Glucose (the suffix *-ose* denotes a sugar) has the formula $C_6H_{12}O_6$ and is an aldose sugar. Fructose has the same molecular formula as glucose, but has a different structure (it is a structural isomer of glucose). Fructose is a ketose sugar.

Although the projection formulae (e.g., Figure 2–16) represent sugars quite well, some aldoses and ketoses form cyclic structures. A ring structure occurs in an aldohexose, like glucose for example, because the alde-

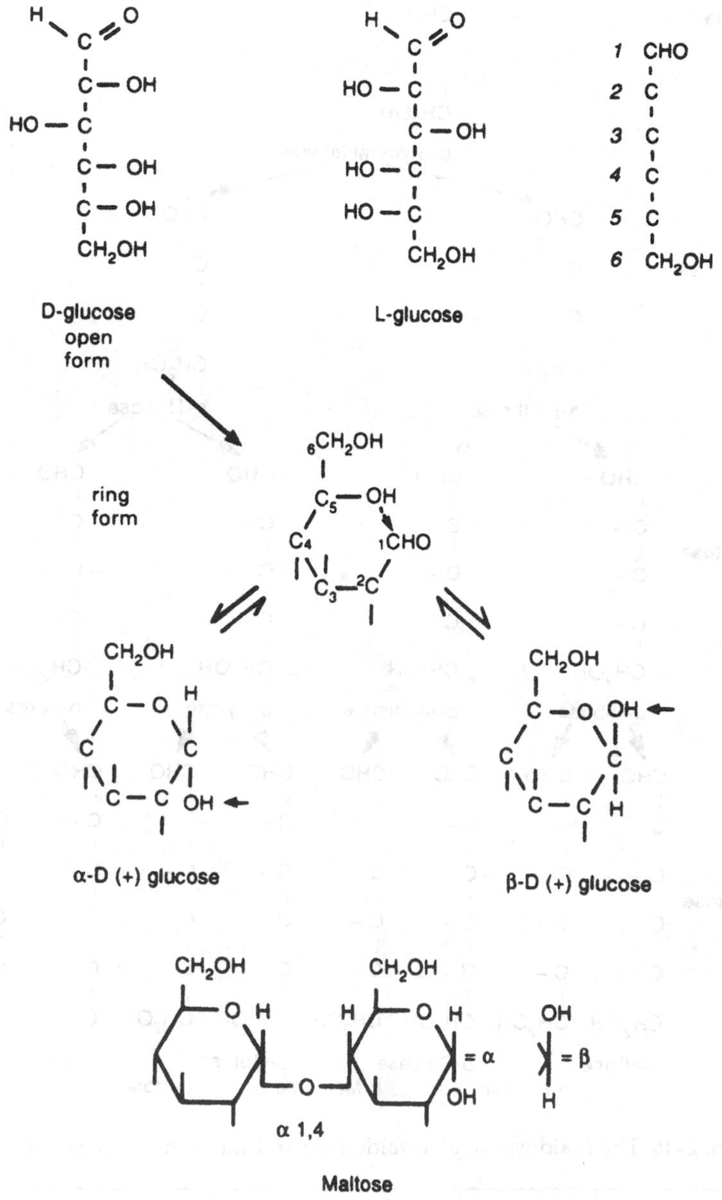

Figure 2–15 Chain or open (Fischer) projection and ring (Haworth) structure of sugars D- and L-glucose and, when the ring forms, the α and β forms of glucose. Maltose is an α-1-4 linked disaccharide. In Fischer and Haworth projections, H atoms on C_2 to C_5 not shown. In Haworth structure OH groups omitted for clarity.

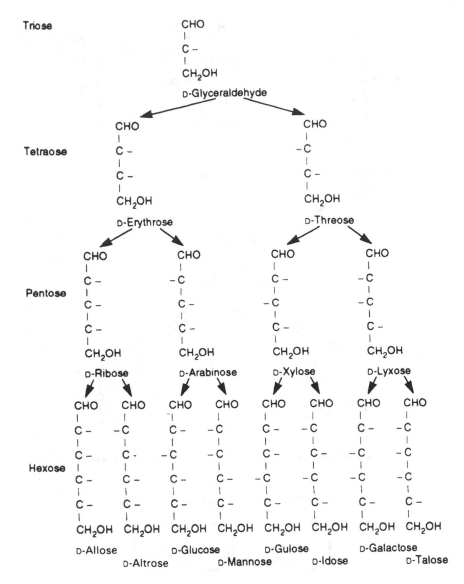

Figure 2–16 The D-aldoses (L-glyceraldehyde and the eight L-aldoses are omitted).

hyde group of C−1 readily forms a hemiacetal with the alcohol group of
either C−4 or C−5 to give a ring with five (furan or furanose) or six
(pyran or pyranose) atoms (Figure 2–15). The ring forms of the sugars
dominate in nature and the pyranose form is more stable and occurs most

commonly. Formation of the ring structure provides a further opportunity for stereoisomerism, because a new chiral center is formed at $C-1$ (Figure 2–15). These isomers are called α or β. Although these are isomers, they are anomers, not enantiomers (i.e., not mirror images). The α form has the same configuration at the anomeric carbon $(C-1)$ *and* at the reference or penultimate carbon atom. The α and β forms of glucose are interconvertible via the straight chain form in a phenomenon called **mutarotation**. Taking all these opportunities for isomerism into account, glucose may be more exactly named α-D(+)-glucopyranose, for example.

The α and β forms become important when monosaccharides link together to form disaccharides, oligosaccharides and polysaccharides, because this configuration determines whether the glucosidic link is in the α or β configuration. The difference is important. Cellulose is a β-linked polymer of glucose whereas starch is an α-linked polymer of glucose. These polymers have vastly different chemical properties and biological functions. Oddly, the α- and β-isomerism also shows up in every-day brewer's parlance: β-amylase was so named because it was thought to leave the product of its action in the β-configuration (β-maltose) through a reaction called the Walden inversion. α-Amylase, on the other hand, leaves its products in the α-configuration. β-maltose was deduced to be the product because of a change in optical rotation that occurs as maltose is formed. However, it appears that this results from a change in configuration not involving the $C-1$ carbon atom and its hydroxyl group. Nonetheless, although it was almost certainly derived from an incorrect interpretation of the chemistry, no one would seriously consider changing the name of β-amylase.

Maltose (glucose–glucose, Figure 2–15) is the primary sugar in wort derived from the breakdown of starch by amylase enzymes. Maltose is an oligosaccharide, specifically a disaccharide, and on hydrolysis, it yields two molecules of the monosaccharide sugar glucose. In maltose, the $C-1$ hydroxyl group of one glucose molecule (in the α configuration) is bonded to the hydroxyl of the $C-4$ of the next glucose molecule in what is called an α-1–4 bond. This same structure is repeated many times in starch (a polysaccharide). When the glucose anomer is in the alternative β-configuration, forming a β-1–4 bond, the disaccharide is called cellobiose and the corresponding polysaccharide is cellulose (Figure 2–17). Glucose molecules can also bond α- or β-1–6 (or β-1–3 as in some β-glucans or α-1–2 as in sucrose, or more rarely α-1–1). The α-1–6 bond is important at the branch point of amylopectin.

Starch is crucially important to brewers. It comprises two glucose polymers, amylose (20–25% of the total starch), an α-1–4-linked straight chain polymer of glucose with a very intense blue-black color with iodine; and amylopectin, a branched structure with a much weaker iodine color. In

(a)

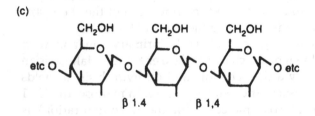

(b)

(c)

Figure 2–17 D-glucose molecules linked together to form chains: (a) In a straight chain of α-1–4 links (e.g., amylose); (b) in α-1–4 linked chains with occasional α-1–6 branch points (e.g., amylopectin); (c) in unbranched β-1–4 linked chains (e.g., cellulose).

barley, starch is contained in crystalline or at least organized granules, which when heated in the presence of moisture, lose their crystallinity and hydrate in a phenomenon called **gelatinization**. Starch is more susceptible to the action of amylases in the gelatinized condition. This topic is explored elsewhere much more fully.

CHAPTER 3

Heat transfer and refrigeration

During brewing, large volumes of liquid are heated and cooled. Generally, the brewhouse processes, addressed in Chapter 15, are hot processes and the cellar operations (Chapter 18) are cold ones. These processes work best when heat is transferred efficiently. Brewers, therefore, need some appreciation of the factors involved in heat transfer and refrigeration. The words "heat transfer" describe the transfer of energy from one place to another as a result of temperature difference.

3.1 HEAT TRANSFER

3.1.1 Heat capacity and phase changes

Heat affects the internal energy of the heated substance and as the vibration of its atoms and molecules increases, so its temperature rises. The temperature of different substances is, therefore, not equally affected by heat. The heat capacity of a substance is expressed as its specific heat (C_p). This is the amount of heat energy (in joules [Note: 1 Btu = 1.055 kJ]) required to raise (or lower) the temperature of one kilogram of a substance one degree Celsius. Water (and so wort and beer) has the highest specific heat of any common substance, at about 4200 J/kg.°C (or 4.2 kJ/kg.°C; the value varies slightly with temperature). A kilogram of water is one liter (not quite a quart). Malt, on the other hand, has a specific heat about 0.4 that of water, and metals (copper and iron and their alloys, for example) generally have about 0.1 that of water.

The heat input (Q) required to raise (or lower) the temperature of any material is therefore a function of (1) its heat capacity or specific heat (C_p), (2) the amount of material to be heated (its mass, M), and (3) the temperature change required (T_1 to T_2 or T°C), i.e.

$$Q = M.C_p.(T) \text{ (kJ)}$$

This is most obviously useful for calculating such things as mashing-in temperatures. For example, what should be the temperature of mashing-in water for a specified mash temperature and mash thickness with malt of known temperature (see section 12.6.2). The equation is equally useful for calculating, for example, refrigeration loads (see below).

Water can be in the form of ice, water, or steam. It requires a great deal more energy to cause a phase change from ice to water or from water to steam than to heat up water. When water undergoes a phase change from solid ice at 0°C to liquid water at 0°C (i.e., there is no temperature change) it requires an energy input of 3.33×10^5 J/kg. This is called the latent heat of fusion (L_f) and is enough energy to raise the temperature of a kilo of water by 80°C (or put another way, to raise 80 kilos of water by 1°C). The phase change of liquid water at 100°C to vapor (steam) at 100°C, called the latent heat of vaporization (L_v), also requires an extra energy input, which is nearly seven times more energy than L_f or 2.26×10^6 J/kg. Herein lies the challenge of raising a kettle to a full rolling boil and achieving a sufficient evaporation rate.

At a phase change of water, therefore, Q, the heat input required (or the amount of heat released), is a function of the latent heat of fusion (L_f for ice) or latent heat of vaporization (L_v for steam) and the amount (mass M) of water changing phase, i.e.,

$$Q = M.L_f \text{ or } Q = M.L_v \text{ (kJ)}$$

These ideas are illustrated in Figure 3–1 in which the total energy for converting ice to steam is apportioned. Melting of ice absorbs 11% of the

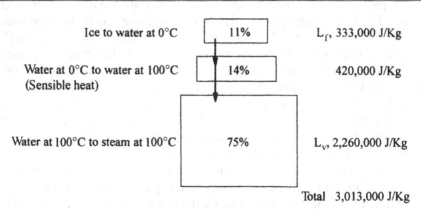

Figure 3–1 The relative amounts of input required to convert ice at 0°C to steam at 100°C. Most of the energy is associated with the latent heat of vaporization L_v forming steam. This heat energy is recovered when steam condenses.

total energy required, heating the water from 0°C to 100°C requires 14% of the energy, and making steam from water at 100°C requires the remaining 75% of the energy. The value L_f is the reason why an ice-bank is a useful means of cooling in a brewery: the melting of ice absorbs much more energy than the same amount of water at the same temperature. Similarly, L_v is the reason why steam is the most common and efficient way of moving heat around a brewery. When steam condenses (phase change), it gives up its enormous content of thermal energy (L_v) and the cool material upon which (or in which) it condenses is thereby heated up.

The high energy content of steam also accounts for the deadly danger of steam scalds to operators.

3.1.2 Steam

Steam tables are used to calculate the amount of steam required for a specific heating task. In such tables, of which five lines are reproduced here as examples, the term enthalpy or heat content is used (Table 3–1). Enthalpy is the total heat content, and is the sensible heat (specific heat at 0°C multiplied by temperature) plus the added latent heat (L_v). Therefore, at atmospheric pressure (0.1 MPa, 100 kPa) and 100°C, enthalpy is 4.2 kJ/kg.°C × (100 − 0)°C + 2260 kJ/kg = 2680 kJ/kg (shown more precisely as 2676 kJ/kg in the table below). (Above atmospheric pressure, L_v is lower; this accounts for deviation from this formula in the steam table.) If water is converted to steam and then further heated, it is called "superheated" steam, and commonly used in breweries.

These ideas can be expressed in a graphical form (Figure 3–2), which relates heat content (enthalpy), pressure, and temperature. Only a narrow range of the domain shown in Figure 3–2 is of ordinary practical interest. Inside the domed curve are all the possible combinations of mixtures of

Table 3–1 Steam Table Example

Temperature	Vapor	Specific volume (m^3/kg) liquid	saturated vapor	Enthalpy (kJ/kg) saturated liquid (H_c)	vapor (H_v)
°C	pressure kPA				
80	47.4	0.0010291	3.4	336	2644
100	100	0.0010435	1.67	420	2676
120	198	0.0010603	0.9	504	2706
140	316	0.00108	0.51	589	2734
200	1554	0.0011565	0.13	852	2793

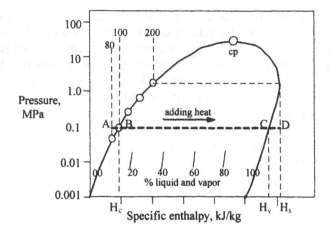

Figure 3–2 Relating heat content (enthalpy), pressure, and temperature as described in the text. The line A,B,C,D represents a course similar to that shown in Figure 3–1.

liquid water and steam at various temperatures and pressures. Outside this curve, the liquid is subcooled on the left (having very little heating power), and on the right (above the critical point, CP) is superheated steam, which is a very common and useful commodity. The line of the dome itself links all pressure-enthalpy points where the steam (or liquid) is exactly saturated (that is 100% water or 100% steam). Inside the dome, the (more-or-less) vertical dotted lines represent the percent (%) of steam in the steam-water mixture. This evaluates steam "quality"; inside the dome the steam is "wet steam." As the graph implies, wet steam contains less energy (enthalpy) than dry (superheated) steam.

If we draw the line ABCD to represent heating water at atmospheric pressure from 80 to 200°C, we can read off the graph (or from steam tables, which are more accurate) the increase in enthalpy that arises. This is the amount of heat that can be recovered when the steam is used for heating.

At A, enthalpy is 335 kJ/kg (i.e., 80°C × 4.2 kJ/kg°C) but as the **liquid** reaches saturation at 100°C (point B), the enthalpy rises to 420 kJ/kg; this is the value for H_c (heat content of condensate at 100°C). Continued input of heat raises the percentage of steam content (quality) of the steam-water mixture; its enthalpy (but not its temperature) increases until point C is reached. Here, the system is pure 100% (saturated) steam for the first time. The enthalpy is H_v (heat content of the vapor) or 2676 kJ/kg, and $H_v - H_c$

is the latent heat of vaporization (L_v). If heating is continued up to 200°C (point D), the enthalpy rises a little more to 2793 kJ/kg (H_s, the heat content of steam) and the steam is superheated.

As pointed out above, the heat requirement for a heating task (kJ) can be calculated from the formula

$$Q = M.C_p.(T_2-T_1) \text{ plus } Q = M.L_v,$$

(if some part of the material is being boiled off, e.g., evaporation in the kettle).

Then, assuming a supply of steam of known quality and temperature, the amount of heat available per kilogram of that steam can be derived from the steam tables. The total weight of steam needed can then be calculated and converted to volume of steam through the data for specific volume (m³/kg), which is also given in the steam tables. Thus, engineers can size the steam boiler suitable to the heating task at hand.

Figure 3–2 makes it clear that the valuable energy for heating is in the condensation of steam (H_s to H_c), and relatively little heating value, e.g., for wort boiling, resides in the super-heat or remains in the condensate. Therefore, the steam condensate must be efficiently removed from a steam-heated device, because it effectively reduces the heating area available for heat transfer. However, there is still valuable heat in the condensate (H_c), which gives ample reason to recover the condensate and return it to the boiler for regenerating steam. Steam traps are responsible for this and must always be working efficiently.

3.1.3 Conduction

The discussion above best applies to steam condensing *directly in* the material being heated. This is often the case, but more usually heat moves through barriers (the walls of a calandria in a kettle, for example) into mixed liquids (e.g., moving wort). Heat moves by conduction, convection, and radiation. Radiation is not considered here because it is usually not a major factor in heat transfer in breweries.

Conductive heat transfer takes place at the molecular level and there is no movement of the materials involved. An example is heat transfer through a metal plate in a heat exchanger or through a wall of a calandria or refrigerator. Heat flows from hot places to cold ones until (eventually) both are at the same temperature.

It is perhaps intuitively obvious that the amount and rate of heat transfer depend on four factors (Figure 3–3). These are: (1) the temperature dif-

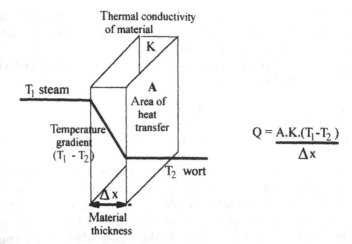

Figure 3–3 The factors involved in conductive heat transfer and the equation for heat transfer of this kind. Note T_1-T_2 is the driving force of heat transfer; all the other factors are fixed by the properaties of the system.

ference or temperature gradient between, say, the wort and the steam, (2) the area of the heat transfer surface, (3) the thickness of the material, and (4) the nature of the material itself, i.e., whether it is a good heat conductor (such as most metals) or a poor one (such as wood or cork).

This last factor is called the **thermal conductivity** (k) of the material, which is the amount of heat transferred (joules) under defined conditions of material thickness (one meter), temperature gradient (1°C), and time (one second), or

Thermal conductivity, $k = J/s.m.°C$ or

k = joules per second per meter per degree Celsius, or

$k = W/m.°C$ (because J/s = Watts)

The thermal conductivity of many materials is known and available in published tables. The thermal conductivity of stainless steel, for example, is about 17 W/m.°C (depending on the composition of the alloy); pure copper k = 386 W/m.°C; and for some common insulating materials k = 0.04 W/m.°C.

Taking these ideas together, the heat transfer by conduction, Q, can be calculated from a relatively simple formula:

$Q = A.k.(T_1 - T_2)/x$ (W)

In which (Figure 3–3):

Q is the rate of heat flow (W = J/s)
A is the area through which the heat flows
k is the thermal conductivity of the conducting material
$T_1 - T_2$ is the temperature of the surfaces involved
x is the thickness of the conducting material, and
$T_1 - T_2/x$ is the temperature gradient or the driving force of conductive heat transfer.

While a brewer may never make a calculation of this sort, this formula and Figure 3–3 usefully highlight and interrelate the issues involved in heating and cooling, i.e., can help explain why a system works well (or poorly!). For example, fouling of a heat transfer surface introduces additional thickness (x) and a different (and likely much smaller) thermal conductivity, k. Similarly, too few plates in a heat exchanger cause insufficient heating or cooling (small area, A). It explains why too many plates are inefficient, why heat transfer through copper (higher k) is more efficient than through stainless steel, and why high pressure steam is effective, and so on. High pressure steam affects the driving force of the heat exchange $T_1 - T_2$. A special case of interest, especially for nucleate boiling at stainless steel heat transfer surfaces in kettles, concerns $T_1 - T_2$. This term suggests that the greater the temperature gradient (e.g., the higher the temperature/pressure of the steam) the faster the heating will be. This is true up to a point. However, stainless steel is a rather poorly wetted surface and so steam bubbles that form on the heat exchange surface (nucleate boiling) do not quickly let go, but (depending on temperature) coalesce to form a layer of steam right at the heat exchange surface. Such a film is an effective barrier to heat transfer, because it increases x and much reduces k (in the formula above), and so substantially reduces the flow of heat. For this reason, $T_1 - T_2$ for boiling wort should not exceed about 25°C above boiling or 4 bar of steam pressure in a stainless steel vessel. This is much less of a problem with copper, which wets much more easily; the steam layer, therefore, does not form so readily.

Some breweries use hot cleaning solutions for vessels in cold cellars. Although there is loss of heat to the mass of the vessel as it warms up and the practice seems intuitively wrong, the air in the cellar is stationary and acts as an insulator (see free convection, below). The heating of the cellar is, therefore, much less than might be expected.

Calculating the size of a calandria for *maintaining* vigorously boiling wort can usually be considered a simple problem of conduction because heat transfer is so rapid under these conditions; T_1 is the temperature of the steam and T_2 is the temperature of the boiling wort. However, heating up wort and cooling it down are more complex processes because the rate of heat transfer into or from the liquids is much slower and much affected by the flowing nature (or turbulence) of the fluids involved. Such systems require applications of the concepts of convective heat transfer.

3.1.4 Convection

While heat may be **conducted**, for example, through the copper wall of a calandria from steam into boiling wort or through the plate of a heat exchanger from hot wort into the flowing cooling water, the heat is **convected within** the steam or wort or coolant, because they are *moving*. Natural or "free" convection is much less efficient than "forced" convection in which the fluid is moved or stirred by a fan or pump, for example. Thus, when a batch of wort is heating up in a kettle, or a pocket of hops is cooling down in a refrigerator, heat transfer is much improved by stirring (moving) the wort in the kettle or the air in the refrigerator; this form of heat transfer is called **convection** (Figure 3–4 a,b). Physical motion is always involved in convective heat transfer. Engineers have tables that estimate the values of the convective (or surface) heat-transfer coefficient, h, expressed as $W/m^2 \cdot {}^\circ C$, which can be used in such calculations. The rate of heat transfer Q (W) by convection can be calculated from

$$Q = h.A.(T_1 - T_2)(W),$$

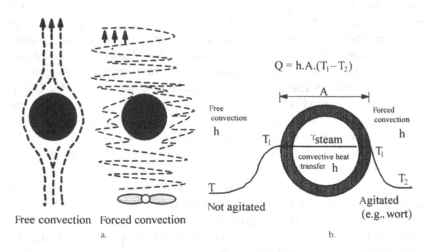

Figure 3–4 (a) Representation of heat transfer by free or forced convection and (b) the factors involved in such heat transfer. Convection (as opposed to conduction) always involves movement, and so heat transfer is much improved by agitation or stirring. Note that heat transfer through the pipe walls is by conduction.

which is the same form as for conductive heat transfer (above). Indeed, in this equation, h can be conveniently thought of as a combination of k and x in the conductive heat transfer equation. That is, heat must be transferred through the mass of the liquid (thermal conductivity important) and across a "stagnant" layer of liquid close to the heating (or cooling) surface, of x thickness (Figure 3–4 b). The value of x (and so of h) is much smaller in vigorously stirred systems where the "stagnant" layer is thin.

3.1.5 Insulation

The effectiveness of insulation material might be thought of as a combination of two ideas: (1) the low thermal conductivity of *dry* air (k = 0.026 W/m.°C) and (2) the low convective heat transfer coefficient of *stationary* air. Although insulation material conducts and convects heat poorly (that is why it insulates), it *does* transfer heat according to the principles outlined above. The formula for conduction and convection above shows that heat loss is greatest where there is potential for a high temperature difference ($T_1 - T_2$, e.g., a steam pipe), where there is high potential for convection over a large area (hA, e.g., a long pipe in drafty location), and where insulation is thin (low x). To be effective, insulation must remain dry (water has a much higher thermal conductivity than air) and, therefore, condensation inside the insulating material must be avoided. For this reason, cold tanks (those below ambient temperature) are insulated with water-proof insulators such as foams, which have a closed pore structure, but hot tanks (those above ambient temperature) are insulated with fibrous materials with an open pore structure, such as fiberglass. Also, insulation materials should be free of chloride ions to avoid corrosion of stainless steel.

3.1.6 Overall heat transfer coefficient

It is possible to consider situations in which several heat transfer conditions exist in relatively complex cases such as Figure 3–5. This brings into play the overall heat transfer coefficient $U_{overall}$, which can be calculated for any situation from the several coefficients involved. In the case of Figure 3–5, these are the convective heat transfer inside the pipe, h_{inside}, the conductive heat transfer through the metal of the pipe, k, and the metal thickness, x, (also k and x for the insulation), and the convective heat transfer outside the pipe $h_{outside}$. Then (Figure 3–5):

$$1/U_{overall} = 1/(hA)_{inside} + (x/kA)_{pipe} + (x/kA)_{insulation} + 1/(hA)_{outside}$$

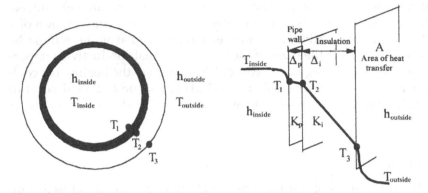

Figure 3–5 Practical systems are often quite complex heat transfer situations such as this representation of heat transfer from an insulated pipe. The figure represents the factors involved in calculating the overall heat transfer coefficient (see text).

With $U_{overall}$ so quantified, the usual expression can be used again for quantifying the overall rate of heat transfer, thus

$$Q = U_o.A.(T_1 - T_2) \text{ (W)}$$

Note that calculations with pipes are more complex than with plates because the inside area (A_{inside}) and outside area ($A_{outside}$) are different and this must be taken into account.

3.2 REFRIGERATION

By reputation, Milwaukee became the capital of American brewing because of a large German population craving lager beer and an ample supply of ice from Lake Michigan with which to manufacture it (by prolonged cold storage in refrigerated cellars). The rise of Milwaukee as a brewing center is doubtless much more complex than that. However, brewers became intrigued with mechanical refrigeration when it first became a practical reality in the nineteenth century, and it has remained a fact of life in breweries ever since. Refrigeration was quickly adopted by brewers, first for making ice and later for air-conditioning and cooling secondary coolants such as brine or glycol. The origins of refrigeration remain with us even today: "one-ton of refrigeration" is equal to the cooling capacity of one ton of ice (2000 lb. $\times$ 144 Btu/lb.)/24 hr = 288,000 Btu/24 hr = 304,000 kJ/24 hr = 3.52 kW).

3.2.1 Phase change of ice

Raising the temperature of one kilogram of ice from $-1°C$ to $+1°C$ "should" require about 6.3 kJ of heat (C_p of ice at 0°C is about half that of water; C_p of water = 4.2 kJ/kg.°C). However, the remarkable H-bonding capacity among water molecules in ice and the breaking of them as the ice melts to water requires an *additional* absorption of 333 kJ of heat energy per kilogram associated with the phase change from ice to water. This is the extra cooling power of ice expressed as the latent heat of fusion, L_f (333 kJ/kg, see above). A kilo of ice at 0°C and a kilo of water at 0°C, therefore, have different cooling capacities. This is the principle of an "icebank" sometimes used for cooling of wort or beer.

3.2.2 Phase change and ammonia

Ammonia has parallel qualities to water and for the same reason—the formation of H-bonds (Chapter 2)—and similarly undergoes a phase change that can be harnessed for cooling. However, the phase change of ammonia, from a liquid to a gas, takes place at minus 33.3°C at atmospheric pressure. The latent heat of vaporization of ammonia (at minus 15°C) is 1314 kJ/kg. Ammonia can also be condensed back to a liquid at reasonable pressure at ordinary temperatures, freezes at a low temperature (minus 77°C), and is reasonably manageable from the point of view of availability, cost, heat-carrying capacity, and safety. Thus, ammonia (and other materials meeting similar criteria, such as Freon, where legal) can usefully be employed in refrigeration.

When a suitable material has the opportunity to undergo a phase change, e.g., from liquid to gas, it must absorb energy (the latent heat of vaporization, L_v) from within itself and/or from the surroundings in order to gasify. It thereby cools. This absorption of energy can be harnessed to provide a stream of cold coolant for use about the brewery.

3.2.3 The refrigeration cycle

A refrigerator has four components (Figure 3–6): expansion valve, evaporator, compressor, and condenser. The refrigeration cycle can be considered to have four stages. Ammonia (or other primary refrigerant such as Freon) under high pressure at ambient temperature is a saturated liquid, A. Stage 1: If that pressure is released through an expansion valve B (called "flashing"), a colder liquid/gas mixture ("flash-gas") is created at lower pressure because the heat for gasification (phase change) is

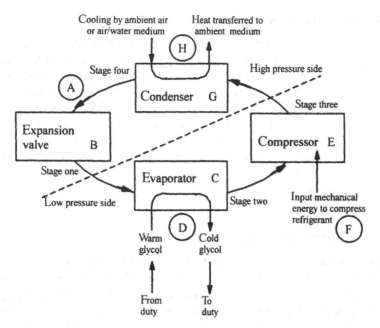

Figure 3–6 The refrigeration cycle (see text).

---

absorbed from within the refrigerant, which is thereby cooled. Stage 2: The cold liquid portion then gasifies by absorbing energy provided to the evaporator C (which is a heat exchanger) in the form of a stream of secondary coolant. When gasified, the ammonia has no further cooling capacity. Stage 3: The ammonia gas must now be recovered and reconditioned for re-use, by vigorously compressing the gas in the compressor, E, which pressurizes and heats up the gas. Stage 4: Under high pressure, the hot gas is now cooled by a stream of ambient air (or water or a water-air mixture) in the condenser G (also a heat exchanger), which, *at this pressure*, serves to liquefy the ammonia back to the starting position of the cycle, A. The ammonia can now be reused, as required, by repeating the cycle.

The schematic in Figure 3–6 envisions a reservoir of glycol or brine (D) providing a source of energy for L_v (for ammonia gasification) to the evaporator/heat exchanger, and so creating a reservoir of cooled glycol (secondary refrigerant) for use about the brewery. The volume of the cold secondary refrigerant is chosen so that sudden cooling loads do not overwhelm the refrigeration capacity of the plant. Depending on demand and duty, breweries may have several refrigeration plants at different locations for flexibility and economy. A secondary refrigerant would most likely be used for cooling wort or beer in a heat exchanger, cooling fermenters, maturation vessels and yeast storage tanks, supplying the green beer chiller, the beer pasteurizer, and likely providing air conditioning. However, direct expansion of the primary refrigerant could be used for space cooling in the fermenting and lager cellars and hop storage rooms, for example.

The cost of power for compression of gasified ammonia (or other primary refrigerant) in the compressor E (e.g., by an electric motor or internal combustion engine, F) is the primary energy cost of refrigeration and a determinant of efficiency of refrigeration (see below).

The expansion valve, B, is the point of control in the refrigeration cycle because the amount of liquid primary refrigerant entering the evaporator determines the amount of cooling available. The opening and closing of the expansion valve is therefore controlled by information from the low-pressure side of the system, e.g., the temperature of the secondary refrigerant. This also controls the compressor because its capacity must match the amount of ammonia gas produced.

Calculations about refrigeration, as far as brewers are concerned, center on estimating the total refrigeration capacity required. This is decided from an estimate of the total heat energy to be removed. This relates to the mass (weight) of material to be cooled, its temperature and specific heat, and the rate of cooling required. In fermentation, heat is released by the action of yeast and its growth, which is an additional load. This has been estimated to be about 3.6 kJ/kg.hr (1W/kg of fermentable sugar). Heat might also be gained through windows, open doors and from many other sources.

The total refrigeration load or total heat to be removed can be represented as:

$$Q = M.\ C_p.(T_1 - T_2)(kJ)/303{,}825\ (kJ/24hr) = \text{tons of refrigeration needed}$$

where C_p is the specific heat of the material to be cooled,
M is mass or weight of material to be cooled (kg/24hr) and
T_2 and T_1 are the temperatures (°C) before and after cooling.
One ton of refrigeration is 303,825 kJ/24hr.

3.2.4 Refrigeration and pressure-enthalpy diagrams

As for steam, the characteristics of refrigerants can be described in pressure-enthalpy diagrams. The refrigeration cycle can be drawn onto such a diagram yielding necessary data for calculations. In the schematic form of such a diagram (Figure 3–7), the line of the dome represents, on the left, saturated (pure) liquid and (beyond the critical point, CP) saturated vapor (gas) at all the temperatures shown on the curve. Under the dome are all

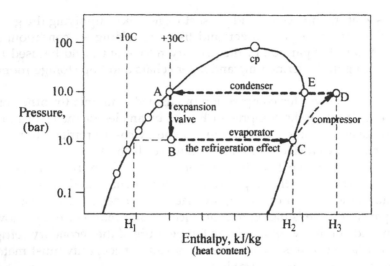

Figure 3–7 The four stages of the refrigeration cycle (Figure 3–6) are represented on this pressure-enthalpy diagram, which is specific for each refrigerant. The concepts contained in the diagram are described in the text. Such data are used for calcuations about refrigeration.

the possible combinations of gas-liquid mixtures for the refrigerant at these temperatures. From any positions on the curve or under the dome, values for temperature, pressure and H (enthalpy, kJ/kg) are interrelated. Thus, if we assume the refrigerant is held at, say, 30°C (point A on the graph, and in the refrigeration cycle, Figure 3–6), the pressure on the saturated liquid refrigerant is P_1 or, say, 10 bar and enthalpy is H_1. When the refrigerant enters the expansion valve, there is a pressure drop to P_2 or about 1 bar; there is no change in enthalpy (still H_1). However, L_v is abstracted from the refrigerant by partial gasification and the temperature drops to, say, minus 5°C or point B. The refrigerant is now a mixture of gas and liquid, however, and so point B is within the dome. The liquid portion now absorbs heat from the evaporator, C, and, as the liquid gasifies at minus 5°C toward the saturated (pure) gas at point C, the enthalpy of the refrigerant increases to H_2. $H_2 - H_1$ is the energy absorbed to create cooling in the evaporator and is the **refrigeration effect** or the latent heat of vaporization, L_v, of the liquid portion of the "flash-gas." Dividing this value into the cooling load gives the reprinted refrigerant flow rate. At point C, the refrigeration capacity of the refrigerant is exhausted.

To recover the refrigerant for reuse, it is compressed in the compressor. Its pressure rises to P_1 and it becomes superheated (point D). Its enthalpy increases somewhat to H_3. The refrigerant now enters the condenser where the superheat (to point E) and heat of vaporization (or condensation, to point A) is removed by transfer of heat to a stream of ambient air (or other coolant, e.g., again at 30°C). The refrigerant returns to being a saturated liquid at high pressure (point A), where the cycle began.

Engineers use the values of H_1, H_2, and H_3 for many calculations surrounding refrigeration, e.g., to size the compressor and evaporator and calculate the coefficient of performance. Thus, by means of mechanical refrigeration, heat energy is transferred from the low-temperature of the evaporator to the higher temperature of the ambient air in the condenser. The energy input of the compressor permits this. The efficiency of refrigeration is reported as the ratio of these two effects called the coefficient of performance (COP):

$$COP = H_2 - H_1 / H_3 - H_2$$

which is:

$$\frac{\text{heat absorbed by the refrigerant in the evaporator}}{\text{heat equivalent of the energy supplied to the compressor.}}$$

The refrigeration effect $(H_2 - H_1)$ is much greater than the heat equivalence of the work required to produce this effect $(H_3 - H_2)$.

CHAPTER 4

Water for brewing

W ater is the lifeblood of many industrial processes and brewing is no exception. Brewers use water in three main ways: (1) as **product water**, which is the major (92%+) component of beer and includes water for mashing, sparging, dilution of beer brewed at high gravity, and for making additions such as hop extract or diatomaceous earth in filtration, etc., (2) as **process water** used for cleaning and sanitizing the plant and packing beer transfer lines and so on, and (3) as **service water** for raising steam for transport of heat energy around the brewery. Each use requires somewhat different water quality. Altogether, depending on their efficiency, brewers use between four and perhaps even twelve volumes of water for each volume of beer produced; process water and service water represent the primary demand for water in a brewery for which quite soft water (see below) is preferred. Large breweries tend to be more efficient in water usage than small ones.

4.1 SOURCES AND COMPOSITION OF WATER

Brewers take for granted the extraordinary properties of the H_2O molecules that render water a liquid at room temperature (Chapter 2.2). They are much more interested in the small quantity of dissolved material that makes water suitable (or not) for brewing, that is, as product water.

Natural waters contain dissolved CO_2, which forms carbonic acid, H_2CO_3, in solution. Much of this CO_2 arises from the action of bacteria in the strata through which ground water percolates. Although silica, aluminum, and iron are the most common components of the lithosphere, these are generally present as very insoluble oxides. Calcium, sodium, potassium, and magnesium are the next most prevalent elements in that order. Carbonic acid helps to dissolve these elements as their bicarbonate salts from the rock strata from which the water is abstracted before reaching the brewery. Other salts such as $CaSO_4$ dissolve too, if present in the underground strata. Water from lakes and rivers is similarly affected by the local geology, though generally lower in dissolved substances. However, such supply is prone to surface contamination from agriculture and industry.

Brewers prefer an independent supply of borehole or well water because of its relatively purity, constant composition, and temperature, though municipal or city mains water may be the primary supply or a useful back-up. However, the city is usually free to mix and switch several sources of water so that the salt composition of the water supplied to the brewery can vary from time to time and require adjustment.

Water for brewing must be potable (drinkable) as judged by appropriate national or international standards. It must be free of contaminants that might be harmful to humans from natural sources or from agricultural, industrial, or domestic pollution. The bacterium *Escherichia coli*, which is an indicator of fecal pollution and hence the possible presence of pathogens, must be absent. The water must also be free of taints, because these are especially easy to detect in carbonated products, and iron. Water that meets these fundamental standards is then treated by brewers to meet the particular requirements of their brewery and products.

After necessary pretreatments, e.g. and filtration almost all brewers chlorinate the incoming water to kill any possible contaminating bacteria and to oxidize any possible organic material present. This water is then passed through carbon filters to remove the chlorine and other adsorbable materials that might give the water odor, flavor, or color. The resulting water is a dilute solution of dissolved salts, the composition of which is then adjusted as necessary for its particular use. The carbon filters must be assiduously maintained to prevent them from becoming part of the problem, especially as a source of microbial contamination.

4.1.1 Water hardness

The amount and composition of dissolved salts in water govern **hardness** and **alkalinity,** which are of central interest to brewers because they influence the efficiency with which brewing processes operate and the way the water extracts desirable and/or undesirable substances (including flavor compounds) from brewing raw materials.

Water is either hard or soft. Hard water contains calcium and/or magnesium ions (up to hundred(s) of milligrams per liter) and soft water contains very little or even none (tens of milligrams per liter or less). The total hardness of water is its calcium plus magnesium content. Hard water forms a scum and no foam when lathered with soap (that is traditional soaps that are the sodium salts of long-chain fatty acids, such as sodium palmitate, and indeed, this is the origin of the somewhat antiquated term, "hard"). The scum is the insoluble calcium salt of the fatty acids. Soft water lathers easily. Soft water contains little or no calcium and magnesium, but may commonly contain sodium and potassium ions.

Hard, calcium-containing or calcareous, waters may be "temporarily" hard or "permanently" hard. The hardness of *permanently* hard water is not affected by boiling because the calcium is present as the salt of strong acids such as calcium sulfate or calcium chloride. The hardness of *temporarily* hard water is affect by boiling because the calcium is present as the salt of a weak acid (carbonic acid) as calcium bicarbonate. This salt breaks down with heating to form calcium carbonate, which is very insoluble in water, and so much of the calcium (the hardness) *and* alkalinity is precipitated and removed by boiling (see below). This is the "fur" of calcium carbonate that accumulates in tea kettles and hot-water pipes of many hard-water areas and forms sludges and scales in boilers.

4.1.2 Water alkalinity

Bicarbonate ions (plus carbonate and hydroxyl ions, if present) represent the total alkalinity of water. These ions are usually present as the calcium, magnesium, sodium, and potassium salts. Total alkalinity can be measured by titrating water with dilute acid of known strength to the end point equivalent of **Methyl orange** (about pH 4.3, the so-called M alkalinity). If a water changes to pink on the addition of **Phenolphthalein**, it contains P alkalinity caused by carbonates and OH- ions, and is unlikely to be a good product water for brewing (though it might be satisfactory for other applications). P alkalinity is measured by titration with acid to the end-point of phenolphthalein (which changes from pink to colorless, at about pH 8.4). Though waters rarely contain P alkalinity, M−2P = bicarbonate alkalinity.

The pH of water is easily measured, but is not a very useful estimate of brewing potential or the total alkalinity of the water. The pH of an alkaline water depends on the **ratio** of ions present in the water that represent the dissociation of carbonic acid:

$$H_2CO_3 \Leftrightarrow H^+ + HCO_3^- \Leftrightarrow 2H^+ + CO_3^-$$

At pH 8.4, the system is almost 100% bicarbonate ion (Figure 4–1). However, **total alkalinity**, a most important brewers' measure, depends on the **concentration** of these ions in water, which pH does not reveal. Bicarbonate ions plus carbonic acid, (and carbonates at high pH) are a buffer system, and in most waters, bicarbonates dominate. Waters commonly, therefore, have a pH of, say, 7.4 to 8.2. Further, calcium and magnesium ions (if pres-

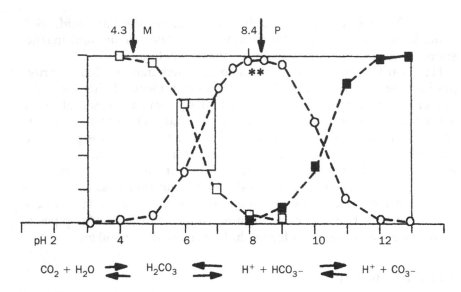

Figure 4–1 The effect of pH on the distribution of the different forms of CO_2 in solution. The total alkalinity is titrated with dilute acid to the end point of methyl orange at pH 4.3 and will include carbonate as well as hydroxyl ions. Bicarbonate alkalinity is titrated between pH 8.4 (Phenolphthalein) and pH 4.3, the end-point of Methyl orange (see text). Most waters have a pH at about 8.2 (**) where the dominant ion is bicarbonate (-o-). At low pH, CO_2 gas easily escapes the system. Where the concentration of carbonic acid and bicarbonate is about equal (e.g., box) the system is a buffer, the strength of which depends on the concentration of the ions present.

ent) affect the pH of brewing *processes*, but have no effect on *water* pH. The pH of water, therefore, is only an interesting piece of information if the value is unusual in some way, e.g., indicating contamination with acid or alkali. Alkalinity is generally undesirable in brewing, having the opposite effect from hardness, and brewers control both factors with care.

4.1.3 pH effects of ions

Hardness in brewing water has the effect of reducing the pH (raising the acidity) of the mash and wort compared to distilled or alkaline water. This generally favors mash enzymes and there is a some improvement in extract yield; slightly increased free amino nitrogen (FAN); lower wort color; better and more rapid mash/wort separation; superior formation of breaks during boiling; brighter worts; and possibly more rapid onset of fermentation, yeast growth, and (later) flocculation. Calcium precipitates oxalates, which might otherwise crystallize in beer. Lower pH also serves to extract less astringent material from malt and (though somewhat less *total* bitterness) less *harsh* bitterness from hops, especially in highly hopped beers. Wort pH itself, however, is not necessarily a harbinger of beer pH—this depends on the concentration of the buffer systems present in wort and their removal during fermentation. In contrast, alkalinity, extracts astringent, harsh, bitter, and colored substances from raw materials, and is especially to be avoided in sparge water where the buffering action of wort is minimal or absent. It is not surprising, therefore, that brewers wish to select and control the balance between water hardness and alkalinity for consistent and efficient brewing commensurate with their brewing objectives.

$CaSO_4$ (gypsum, sometimes called Burton salts and named for Burton-upon-Trent in England, which is famous for its gypseous brewing waters) is commonly added to mash water to raise the Ca^{++} content to that required, especially in the brewing of pale ales. Calcium and, to a lesser extent, magnesium (the hardness ions) affect pH by reacting with the phosphate of the malt, which dissolves during mashing. The following ions represent the dissociation of phosphoric acid

$$H_3PO_4 \Leftrightarrow H^+ + H_2PO_4^- \Leftrightarrow 2H^+ + HPO_4^{--} \Leftrightarrow 3H^+ + PO_4^{---}$$

Ca^{++} reacts with the PO_4^{---} ion to form insoluble $Ca_3(PO_4)_2$ and release acid (H^+), therefore:

$$3Ca^{++} + 2H_3PO_4 \rightarrow [Ca_3(PO_4)_2] + 6H^+$$
$$3Ca^{++} + 2H_2PO_4^- \rightarrow [Ca_3(PO_4)_2] + 4H^+$$
$$3Ca^{++} + 2HPO_4^- \rightarrow [Ca_3(PO_4)_2] + 2H^+$$

The molecule in square parentheses leaves the system as a precipitate, and so as a result of the reaction the equilibrium is drawn to the right with the release of H^+ (acidity).

The primary acid salt NaH_2PO_4 dominates at wort pH in equilibrium with the alkaline secondary salt Na_2HPO_4, and it is quite usual to represent the effect of acidifying and alkalizing as the balance between these two salts:

$$Na_2HPO_4 \text{ (alkalizing)} \Leftrightarrow NaH_2PO_4 \text{ (acidifying)}$$

Magnesium is less effective at changing mash/wort pH than calcium because magnesium phosphate is much more soluble than calcium phosphate in these reactions.

The chemistry of the pH effect of alkalinity can be shown in a variety of ways:

$$H^+ + HCO_3^- \xrightarrow{heat} [CO_2] + H_2O$$

$$HCO_3^- \xrightarrow{heat} [CO_2] + OH^-$$

$$NaHCO_3^- \xrightarrow{heat} [CO_2] + NaOH$$

$$Ca(HCO_3)_2 \xrightarrow{heat} [CO_2] + [CaCO_3] + H_2O$$

The molecules in square parentheses leave the system as gas or precipitate. Note that the last reaction is not alkalizing but demonstrates the removal of alkalinity by boiling in the presence of Ca^{++} ions. The full alkalizing effect of bicarbonates (i.e., formation and removal of CO_2) is revealed in hot systems in the brewhouse especially during wort boiling (hence inclusion of heat above), and this is why gypsum (a source of Ca^{++} as $CaSO_4$ and therefore acidifying) is frequently added to the boil as well as the mash water to counteract rising pH at this stage. A similar rise in pH during the decoction stage (mash-boil) of lager brewing might account for the very low alkalinity of water used for lager brewing.

Alkalinity is usually removed from brewing water by adding acid. One equivalent of alkalinity is much more effective in raising the mash pH than one equivalent of hardness is in lowering it.

4.2 EVALUATION OF WATER

Brewers are interested in judging whether a product water is suitable for brewing a particular kind of beer, and (if not) how it should be treated for brewing use. This usually involves comparing water to some ideal analysis, e.g., that used at the company's flagship brewery or used in a center famed for brewing a particular kind of beer. The standard waters usually quoted are the water of Burton-upon-Trent for dry, hoppy pale ales; the water of Pilsen for delicately late-hopped pale dry lagers; the water of Dortmund for medium ales or lagers; the water of Dublin for stouts; and that of Munich for dark lagers (Table 4–1). The last two are similar waters.

4.2.1 Water analysis and evaluation

Water analyses are expressed in a variety of ways, most of which are not conducive to ready interpretation. The most useful information is the con-

Table 4–1 Analyses of different waters

	Burton-on-Trent		Pilsen (soft)		Dortmund		Dublin/Munich		London MWB		London deep well	
	ppm	millivals	ppm	millivals	ppm	millivals	ppm	millivals	ppm	millivals	ppm	millivals
Magnesium (Mg^{++})	62	5.2	—	—	23	1.9	19	1.6	4.0	0.3	19	1.6
Calcium (Ca^{++})	268	13.4	7.0	0.35	260	13.0	80	4.0	90	4.5	2.6	52
Sodium (Na^+)	30	1.3	3.2	0.14	69	3.0	—	—	24	1.0	4.3	99
Carbonate (CO_3^{--})	141	4.7	9.0	0.3	270	9.0	168	5.6	123	4.1	156	5.2
Sulphate (SO_3^{--})	638	13.7	5.8	0.12	283	5.9	—	—	58	1.2	77	1.6
Chloride (Cl^-)	36	1.0	5.0	0.14	106	3.0	—	—	18	0.5	60	1.7
Nitrate (NO_3^-)	31	0.5	—	—	—	—	—	—	—	—	—	—
Ca^{++} plus Mg^{++} to CO_3^{--} ratio	100:25		100:71		100:60		100:97		100:86		100:124	

centration of ions present, expressed as ppm (mg/liter). The most rational interpretation of these data are then by computing these values as normality through their equivalent weights. The equivalent weight of an element is its atomic weight divided by its ion charge. Thus, sodium (Na^+) has an atomic weight of 23 and an ion charge of one ($^+$). Its equivalent weight is then $23/1 = 23$. Calcium (Ca^{++}) has an atomic weight of 40 and an ion charge of two ($^{++}$) and, therefore, an equivalent weight of 20. The concentration of an ion in ppm or milli-grams per liter is then easily converted to milli-equivalents (millivals or mv) by division. For example, a solution containing 322 ppm sodium contains $322/23=14$ mv of Na^+, but one with 322 ppm of calcium contains $322/20=16.1$ mv of Ca^{++}. An ion such as sulfate (SO_4^{--}) is treated similarly using the sum of the atomic weights of its components divided by the ion charge. For sulfate $[32 + (4 \times 16)]/2 = 48$ equivalent weight and for bicarbonate (HCO_3^-), the equivalent weight is $[1 + 12 + (3 \times 16)]/1 = 61$.

The first great advantage of this method is that cations (positively charged ions such as Ca^{++}) and anions (negatively charged ions such as HCO_3^-) must have the same total number of millivals because one millival of any anion reacts with one millival of any cation (hence "equivalence" or "equivalent weight"). If this is not the case, the data are incomplete or wrong.

The first comparison can then be made to famous waters by comparing the hardness-to-alkalinity ratio, that is the total millivals of Ca^{++} plus Mg^{++} to millivals of carbonate (bicarbonate) or the {$Ca^{++} + Mg^{++}$}: CO_3^{--} ratio (Table 4–1). Notice that Burton water has a very high hardness ratio of 4:1. Dortmund water is still dominated by hardness (nearly 2:1) and is fully suitable for brewing pale ales and lagers, though not such dry and pale beers as at Burton or Pilsen, and more moderately hopped. In contrast, water from Dublin and Munich have a hardness-to-alkalinity ratio of 1:1, implying that all the calcium present could be there as the bicarbonate salt and none as gypsum ($CaSO_4$). Such water is ideal for making full, dark, and possibly sweet beers, whether lagers or stouts. The reason for this is that roasted materials are rather acidic in nature and tend to lower the mash pH (the bicarbonate content ameliorates that). The alkalinity assures excellent extraction of color and flavor from the malt and hops to an extent that is necessary in dark beers, though probably undesirable in pale ones. Water that has an intermediate value for the hardness-to-alkalinity ratio between Dortmund and Dublin might be most suitable for porters, mild ales, some lighter stouts, and rich-colored lagers. Water in which the hardness-to-alkalinity ratio is less than 1:1 (e.g., 1:1.5), which implies the presence, e.g., of sodium (bi)carbonate, is probably not satisfactory for brewing without proper treatment.

The overall effect of water on mash pH can be estimated through its residual alkalinity. By subtracting the acidifying effect of Ca^{++} (as mv/3.5) plus Mg^{++} mv/7.0) from the alkalizing effect of total alkalinity (bicarbonates and carbonates as mv/1) the residual alkalinity can be roughly projected, and is useful for comparing waters. The Ca^{++} and Mg^{++} are discounted because they are less effective than carbonates in changing mash pH. For each mv of residual alkalinity so calculated the mash pH should rise (positive value) or fall (negative value) about 0.1 pH unit. Thus for untreated Munich water (Table 4–1): Residual alkalinity (mv) = 5.6 − (4.0/3.5 + 1.6/7.0) = +4.2 mv which implies a pH rise of about 0.4 units compared to a mash with distilled water.

4.2.2 Other ions

Other ions must be taken into account when considering water quality for brewing. Generally, total dissolved solids (total salts) should decrease in step with the hardness-to-alkalinity ratio; the salinity (sodium salts) should always be low, no more than indicated in the tables, especially in alkaline waters. Although sodium chloride is sometimes added to brewing water or wort at the kettle for fullness of flavor, sodium plus sulfate ions are usually counter-indicated for pale ales and lagers, because of a sour taste, though progressively more acceptable for brewing darker and sweeter beers. Chloride ions tend to contribute fullness and also, reputedly, mellow bitterness and $CaCl_2$ is sometimes preferred to $CaSO_4$ for this reason, as well as for its ready solubility. Mg^{++} is reputed to give an astringent bitterness and SO_4^- a dry and bitter palate. Ferrous or ferric ions (along with, e.g., copper ions, Cu^{++}) that survive into beer are likely to participate in oxidative reactions leading to stale flavors and haze formation (where they concentrate), as well as metallic tastes. Ammonium ions, NO_2^- and NO_3^- are likely indicators of water pollution.

4.3 WATER TREATMENT

Water treatment comprises two actions: reduction of alkalinity and adjustment of hardness. Reverse osmosis is an effective modern means of water treatment in which water is pressed under high pressure (some 20–50 bar) against a thin membrane. Pure water penetrates the membrane, called the permeate, and is used for brewing. The retentate retains the dissolved materials and is discarded though it could be treated by other methods for use, e.g., in cleaning, depending on its composition.

De-mineralization or ion exchange removes ions by swapping them, e.g., H^+ and OH^-, on suitable exchange resins. These techniques remove sodium and potassium ions (as well as hardness), which are not removed by traditional treatments for hardness and alkalinity.

4.3.1 Treatments for hardness and alkalinity

Boiling is an useful traditional way to reduce the hardness and alkalinity of water, especially when the $\{Ca^{++} + Mg^{++}\}$ to CO_3^{--} ratio significantly favors the cations, or gypsum is added before boiling. The carbonate does not precipitate completely, even so, leaving perhaps 0.5 to 0.7 mv in solution.

By heating/boiling the bicarbonate decomposes:

$$Ca(HCO_3)_2 \xrightarrow{heat} [CaCO_3] + [CO_2] + H_2O$$

Generally stated, the alkalinity is driven off as CO_2 (and CO_3^-) and the Ca^{++} hardness is precipitated as calcium carbonate, especially when the water is hot. Water could also be aerated to drive off CO_2 efficiently and promote the reaction. The Burton water shown in Table 4–1 was traditionally boiled before use, increasing its hardness-to-alkalinity ratio considerably.

If there is a significant content of sodium salts present as the bicarbonate (i.e., the hardness-to-alkalinity ratio is less than 1:1), there can be a rise in pH associated with boiling because of the formation of sodium hydroxide, thus

$$NaHCO_3 \xrightarrow{heat} NaOH + [CO_2]$$

which is prevented by adding gypsum before boiling to provide an acceptable hardness to carbonate ratio (1:1 or better). This then applies:

$$2NaHCO_3 + CaSO_4 \xrightarrow{heat} [CaCO_3] + Na_2SO_4 + [CO_2] + H_2O$$

In such case $NaSO_4$ could be an undesirable (sour) component of the beer. On the other hand, adjusting alkalinity with acid does not affect the hardness of a water.

$$Ca(HCO_3)_2 + H_2SO_4 \rightarrow CaSO_4 + [CO_2] + 2H_2O$$

$$2NaHCO_3 + H_2SO_4 \rightarrow Na_2SO_4 + [CO_2] + 2H_2O$$

Also phosphoric, lactic, or hydrochloric acid could be used, with reputedly different effects on beer flavor. Note that bicarbonates are buffers and so, when adjusting with acid, the pH will initially resist change, then decrease greatly. It is necessary to add a calculated amount of acid based on the measurement of the total alkalinity.

Traditional methods also include treatment with lime water, $Ca(OH)_2$, to improve water quality by removing alkalinity and hardness (temporary hardness), and with lime and soda-ash (sodium carbonate) to remove permanent as well as temporary hardness:

$$Ca(HCO_3)_2 + Ca(OH)_2 \rightarrow [2CaCO_3] + 2H_2O$$

$$Ca(HCO_3)_2 + Ca(OH)_2 + CaSO_4 + Na_2CO_3 \rightarrow [3CaCO_3] + Na_2SO_4 + 2H_2O$$

4.3.2 Other uses of water

Removal of all hardness is desirable for process water (e.g., for cleaning) and boiler feed (service) water. Calcium reduces the efficiency of alkaline cleaners and deposits mineral "stones" on surfaces during cleaning. In boilers, calcium salts deposit as sludges or scales, and can foul other heat exchange surfaces. Some hardness can be counteracted in such waters by adding a sequesterant such as Calgon (a poly-phosphate) or EDTA (Versene: ethylene-diamine-tetra-acetate), which chelates calcium ions and can serve to remove, as well as prevent, stones and scales under some circumstances. Corrosion inhibitors in boiler feed water and antifungals in pasteurizer water, for example, are necessary to maintain the efficiency of these units. The presence of most sodium salts in process or service water, because they are soluble, and high pH, is not usually a problem.

4.4 EFFLUENTS

Waste streams from the brewery include wastes such as: (a) broken glass and damaged cans, cardboard, pallets and paper bags, and so on, which can be partially reprocessed (recycled) or disposed of to the landfill, (b) spent grain and other process wastes such as trub and yeast mass, which are usually recovered as animal feed, (c) gaseous effluents such as steam and CO_2, for example, which often carries organic matter and aromas that are subject to control (CO_2 is a "greenhouse" gas, but can be usefully recovered), and (d) waste water or effluent. This last item is a major challenge for breweries because water is expensive to acquire and expensive to dispose of. Modern brewing practice therefore demands wise use of water by (a) minimizing waste creation, (b) controlling discharge of organic matter to the drain, and (c) by extensive re-use of water and cleaning solutions and rinses.

4.4.1 Sources of effluents

The brewhouse is a source of low volumes of waste water that are high in dissolved and suspended organic matter. Discharge of these wastes to drain is most undesirable and brewers have found ways to recycle last runnings ("sweetwater") from the lauter, spent grain pressings and drainings from trub, and so on, as mash foundation water, sometimes called "weak wort recycling." Water used for cooling wort can also be used for mashing. Tank bottoms from fermenters are potentially highly polluting. They can be centrifuged or pressed (cross flow filtration has also been tried) to recover beer and the yeast can be disposed of as distillers yeast or, after inactivation, in spent grain (at suitable levels for fodder) or for other downstream uses. Yeast glycogen could be fermented to alcohol.

The cellars also produce significant flows of moderate pollution potential, significantly associated with cleaning and sanitizing these vessels. By using built in CIP (cleaning-in-place) systems, which can recover, recondition, and reuse cleaning and rinsing streams, discharge of waste water can be much reduced. Intermittent rinsing, rather than continuous flow rinsing, of cleaned tanks is very effective and final rinse water can be recycled several times, e.g., as cleaning post-rinse and cleaning pre-rinse water before discharge. On-site processing of separable waste streams can significantly reduce their volume and pollution potential and might even yield valuable by-products such as grain alcohol distilled from waste beer. The causticity of alkaline cleaners can be somewhat reduced before disposal by injecting CO_2 from fermenters that is unsuitable for recovery, e.g., because of its air content. Solid wastes should always be separated out for disposal where possible. Dilution is no solution.

The packaging plant, in contrast to the brewery, is a source of relatively high flow volumes that (except for wastes from washing kegs and returnable bottles) is not especially polluting. These volumes need to be minimized, for example, by minimizing carry-over and intensive recycling of pasteurizer cooling water.

Where necessary wastes should be "bulked up" for disposal and discharged at a steady rate to avoid shock loads at the waste treatment plant. Many breweries operate on-site waste treatment plants of varying complexity, depending on the analytical limits imposed by the governing jurisdiction, to reduce the pollution potential of the waste before discharge to the sewer for final treatment.

4.4.2 Treatment of wastes

Effluent is characterized by its volume, its BOD (or COD), TSS (total suspended solids), pH, and temperature. Upper limits may be set for these parameters because waste streams that are too strong will incapacitate the bacteria at the waste water treatment plant. Charges for processing the waste will be based on volume, BOD, and TSS and so all three parameters need to be kept to a minimum. Nevertheless, even with stringent waste control, the cost of disposal can be intimidating and many breweries find it advantageous to operate their own waste water treatment facility.

BOD stands for biological oxygen demand and COD for chemical oxygen demand. They both purport to estimate the pollution index of the waste in the water, that is the amount of oxygen required ("demand") for the complete oxidation of the organic waste to CO_2 and water. The methods usually yield different numbers, and the problem of acquiring a reasonable sample for analysis make the values a rough guide at best. Domestic (household) waste might have a BOD of, say, 300 mg/liter or so. Brewery values are much higher, say, between 5,000 and 15,000 mg/l. Typical materials in the brewery that can find their way into the effluent, such as beer or wort (e.g., associated with trub or yeast or tank bottoms), have a huge BOD of up to 100,000 mg/l or more, as well as dense values for suspended solids. Such values not only represent a loss of potentially salable substance, therefore, but also raise disposal costs.

The purification of brewery waste streams by bacterial oxidation in aerated systems such as biological filters, oxidation ponds or activated sludge systems can be represented thus:

(a) Assimilation:
Organic matter (BOD) + oxygen + NH_3 + Bacteria
$$\longrightarrow \text{New cells} + CO_2 + H_2O \text{ (Growth)}$$

(b) Auto-oxidation
Bacterial cells + oxygen $\longrightarrow CO_2 + H_2O + NH_3$ (Self-digestion)

The term "bacterial action" represents the activity of bacteria at the waste water treatment plant. This mixed population of bacteria must grow to form a mucilaginous floc and respire to achieve waste purification. Therefore, the quality of the waste streams must be within reasonable bounds of concentration, composition, pH, temperature and so on, and sufficient time must be allowed for bacterial growth. This is the function of the waste water treatment plant.

Anaerobic treatment is a popular alternative to aerobic treatment for breweries because such plants are cheaper to build, can handle stronger waste streams and shock loads, and yield a flow of methane that can be used as a source of energy in the plant. About 90% of the carbon in the waste is transformed into gas (methane CH_4, 70% and carbon dioxide CO_2, 30%) and so there is relatively little sludge formed. The process has two parts (a) transformation of complex material into simple organic acids (acetification) and (b) generation of methane from organic acids.

Most municipal treatment plants comprise three parts: (a) pre-treatment, e.g., by settlement, (b) oxidation with bacteria, and (c) final settling followed by discharge. The pre-treatment process settles the waste in large shallow vessels and often achieves remarkable improvement in the waste quality. The settled waste is then aerated with a high concentration of a mixed bacterial floc either floating free (e.g., activated sludge or oxidation pond methods) or growing on the surfaces of solid supports (biological filter methods). The adsorption, growth and respiration of these bacteria, in the presence of high aeration to provide the "oxygen demand," purifies the waste. The treated waste is then again settled, perhaps filtered, aerated, usually chlorinated, and discharged to a natural body of water. The settled sludges are partially returned to the process for digestion or dried and buried.

CHAPTER 5

Microbiology and microbial contaminants of brewing

5.1 INTRODUCTION

The science of microbiology (micro = very small, bios = life) is concerned with the study of organisms less than 1 mm in size. Such organisms may be acellular (for example, viruses) or cellular. All cellular life forms are considered to have evolved from Progenotes (unknown forms) with branches leading to the Archaea (representatives of this group of microorganisms are still around today occupying places of high temperature and pressure, e.g., thermal springs); the Eubacteria (Bacteria) and the Eukarya. The Archaea have characteristics in common with both the Bacteria and the Eukarya. All Archaea and Eubacteria are microbes. The Archaea and Eubacteria collectively are termed prokaryotes (pro- before, karyon- nucleus). All beer spoilage bacteria belong to the kingdom of the Eubacteria. Members of the Eukarya (= true nucleus) are characterized by possessing a true nucleus. The nucleus is an organelle, surrounded by a double membrane, that contains the chromosomes of the organism. The Eukarya may be subdivided into two groups, lower and higher. Lower Eukarya of interest to the brewer are the brewing yeast, wild yeasts, and other fungi (molds).

5.2 BASIC PROPERTIES OF MICROBES

Microorganisms are ubiquitous. They are found anywhere where the environment will support their growth. They require a growth medium of water containing dissolved nutrients and trace elements. The medium must usually be at moderate temperatures and pH. Different organisms show different tolerances to temperature and pH but ranges of 2 to 40°C

71

and pH 2 to 9 would encompass most. It is the individual characteristics of a microbe that fit it to specific environments. For example, acid-tolerant bacteria, which grow in the absence of oxygen, would, given sufficient nutrients, survive and grow well in an acidic, anaerobic environment (for example, beer). Such surroundings would provide an ecological niche for these organisms and prove hostile to those without the required properties.

Bacteria grow by division; cells increase in size and divide in two (binary fission). Most yeast reproduce by **budding**: a small protrusion (bud), which is formed on the side of a cell, grows and eventually separates. Both types of reproduction result in a doubling of the population at each division (generation). When cells double at regular intervals of time they are said to grow **exponentially** (Figure 5–1). Under ideal conditions, doubling times vary considerably from the fastest growing bacteria (about 10 minutes) to slower-growing yeasts (1 to 2 hours). A single bacterium, capable of doubling every 30 minutes, alighting on a growth medium would be capable of producing 16.8 million cells in 12 hours. A clear liquid would become very cloudy! About 100 000 bacterial cells/ml give just-visible turbidity.

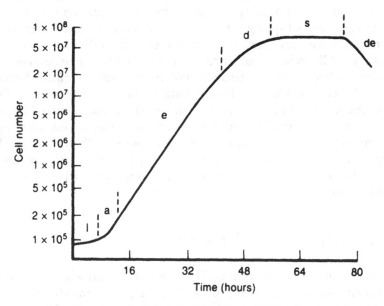

Figure 5–1 Growth curve of yeast in batch culture. 1, lag phase (period of adaptation); a, phase of accelerating growth rate; e, phase of exponential (logarithmic) growth; d, phase of decelerating growth rate; s, stationary phase of growth; de, death (autolytic) phase.

Some microbes have life cycles which include a sporulation phase. The single cells of bacteria (vegetative cells) produce **spores** containing their genetic information. The spore is resistant to adverse conditions, especially lack of nutrients, and remains alive (viable), but non-growing. In this form, the organism may resist relatively high temperatures (100°C) and extremes of pH. When the environmental conditions become favorable, spores germinate into a vegetative cell that then reproduces. The spores of many fungi have similar properties. Some yeasts can sporulate, but the spores are not usually greatly more tolerant of environmental conditions than are the vegetative cells.

Some bacteria possess flagellae which are used to propel the organism through its environment. Yeasts do not possess this property.

Vegetative microbes survive in their environment by maintaining within the cell the relatively constant pH and concentrations of ions necessary to support enzyme activity. This process is referred to as **homeostasis**. The concentration of intracellular components results in a high internal osmotic pressure and a resulting turgor. The cell wall (mainly composed of cross-linked carbohydrate polymers) counteracts this force. To a large extent, maintaining homeostasis is a function of the properties of the cell membrane and enzymes associated with it and requires the expenditure of energy. Damage to the cell membrane results in a breakdown in homeostasis and death of the cell. This damage may be induced by heat and pressure, by extremes of pH, and by action of detergents and other active chemicals that either denature proteins or disrupt the composition of the membrane. Typical of such substances are alcohols, phenols found in antiseptic agents, and quaternary ammonium compounds found in biocides (Figure 5–2). Some of these agents are used to sanitize surfaces and equipment in the brewery. In addition, cells may be killed by chemicals that block essential metabolic processes. These substances, often antibiotics, find use in the formulation of selective media for detecting particular microbes in the brewing process.

5.3 MICROBIOLOGICAL PRACTICE RELEVANT TO BREWING

The requirements of the brewer are threefold: (1) to maintain and store yeast; (2) to monitor yeast quantity and vitality; and (3) to monitor the equipment, process, and product for the presence of unwanted and undesirable microbes. This chapter discusses the third requirement. The first and second are discussed in Chapter 16.

Monitoring the plant and process requires samples for microbiological analysis. These are usually taken manually, but may be taken automati-

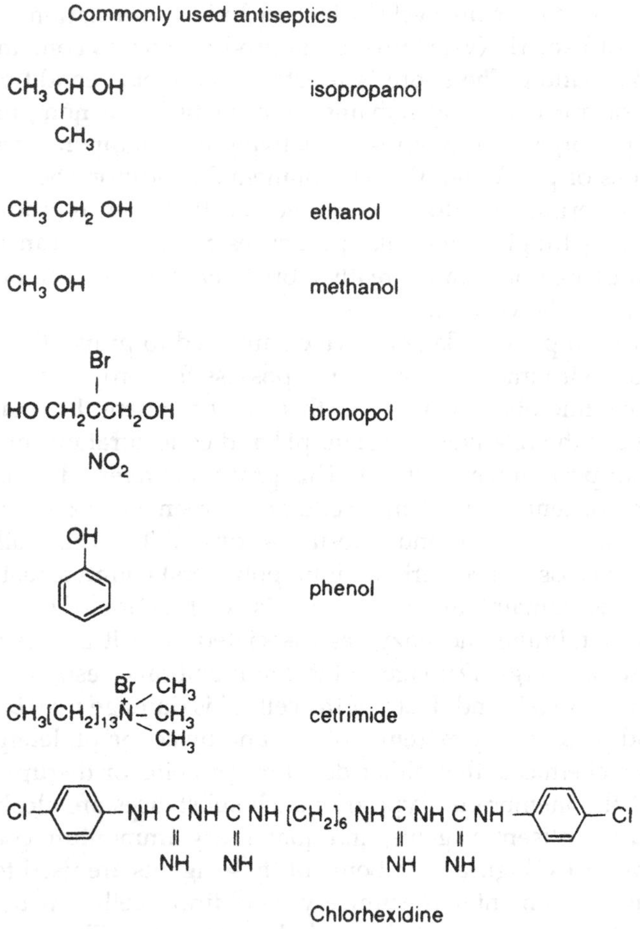

Commonly used antiseptics

CH₃ CH OH   isopropanol
|
CH₃

$$CH_3\ CH\ OH,\ CH_3$$

CH₃ CH₂ OH   ethanol

CH₃ OH   methanol

Br
|
HO CH₂CCH₂OH   bronopol
|
NO₂

OH
⬡   phenol

Br   CH₃
CH₃[CH₂]₁₃N–CH₃   cetrimide
CH₃

Cl–⬡–NH C NH C NH [CH₂]₆ NH C NH C NH–⬡–Cl
‖   ‖   ‖   ‖
NH NH   NH NH

Chlorhexidine

Figure 5–2 Typical antiseptic agents.

cally at intervals or even continuously, especially from pipes carrying wort and beer (see below). The sampling points must be capable of being sterilized by heat or swabbing with an anti-microbial agent. Sample points are typically of two designs, simple cocks (taps) or dairy type. The latter incorporate a replaceable septum that can be swabbed to sterilize before being pierced with a sterile needle through which a sample is withdrawn. The vessel into which the sample is taken must be pre-sterilized and the whole process done rapidly, taking precautions to minimize the possible ingress of microbes from the air. The container is then sealed and transferred to a laboratory.

Samples may be processed by a variety of techniques depending on the nature of the sample, facilities available, and the time required to obtain a result.

Direct microscopic examination may be used with yeast samples to detect contaminating bacteria. Often the sample will be stained using the Gram stain technique. This process separates bacteria into two main groups, Gram-positive and Gram-negative. The procedure detects differences in the composition of the cell wall. Gram-positive cells absorb the dye crystal violet and it is fixed in the cells by reaction with iodine. The fixed dye is not easily removed by solvents such as alcohol so that the cells, stained with crystal violet, appear blue-black. Gram-negative cells also absorb the dye, but iodine does not act as a fixative. Accordingly, the dye is easily removed in alcohol and the then colorless cells can be stained with another dye, usually safranine, in which case they appear pink. Yeasts are Gram-positive.

As discussed in Chapter 16, microscopic analysis is an insensitive process subject to interference from particulate matter. Furthermore, it does not distinguish between living and dead cells. Samples may therefore be inoculated into suitable media and incubated. Living cells grow, producing a turbid culture. In practice, these tests are not favored because they do not give a reliable indication of the numbers of organisms present. They also depend on the presence of at least one living cell per sample and, where low levels of microbes are expected (wort, finished beer), very large volumes have to be processed to detect contaminating microbes.

Accordingly, various **plating techniques** are commonly used. Here, medium is solidified by the inclusion of agar (melts at 100°C, solidifies at 47°C). Molten sterile medium is poured into sterile Petri dishes and allowed to set. Samples may be spread on the surface of the medium or rapidly suspended in thin agar (0.5%) molten at 49°C and poured onto the surface of agar medium in a Petri dish. Alternatively, larger volumes of sample may be mixed directly with thin molten medium and poured directly into the dish (pour plate). The spread plate technique is typically used to analyze up to 1 ml sample per plate, whereas the pour plate procedure can accommodate a 10–15 ml sample.

Viable microbes grow by division (or budding) and produce visible colonies of cells. Each colony represents one original cell or dividing cell (or sometimes a clump of cells), and thus the numbers of living cells (strictly the number of colony forming units) in the original sample may be estimated. Once again, this process is not sensitive enough where low

levels of microorganisms are encountered. It is therefore routine procedure to membrane filter samples before processing further. The membrane is manufactured to contain pores of average diameter (0.22–0.45
μm), smaller than that of the typical bacterium or yeast. A sample is filtered through a sterile membrane (Figure 5–3), the filtration process being
carried out at a microbiologically clean workstation. A vacuum is usually
used to draw the sample through the membrane, which is then removed
and placed face up on solidified medium. Nutrients in the medium diffuse through the membrane to the cells on the surface, which then grow
to produce colonies.

Semi-automated processing of samples by this method has been
described. One system uses a ribbon of membrane in a cassette. A continuous bleed of a very small proportion of the process liquid is drawn
through the membrane, which eventually is cut into sections and placed
on suitable medium. An alternative process uses a circular plate of individual membranes and the process stream is automatically sampled at
pre-determined intervals. After each sample, the plate rotates to bring a
fresh membrane into the sampling position. Membranes are eventually
removed and placed on solidified media. As an alternative to agar media,
absorbent pads impregnated with dehydrated media can be used. Each
pre-sterilized pad, in a sterile Petri dish, is reconstituted by the addition
of sterile water. A membrane filter may then be placed on top of the pad.

Two key factors in all the techniques that allow microbes to grow are
the composition of the medium and the method of incubation. These
affect the recovery of microorganisms, the selectivity towards them
(which microbes grow), and the time taken for growth and hence to
obtain a result.

5.3.1 Design of microbiological media

Two different philosophies influence the design of media. The first is
that "any microbial contamination is undesirable and warrants action."
The second is that "action is only necessary if the microbes detected are
likely to cause spoilage."

The first philosophy leads to the development of general purpose,
catch-all, media and the second to selective media. In practice, both types
of media are necessary. Medium design and technology are continually
developing. Many commercial suppliers provide excellent literature and
advice about media for specific purposes. When consulting these sources,
however, it is well to remember that a perfect medium for all microbes
under all circumstances does not exist.

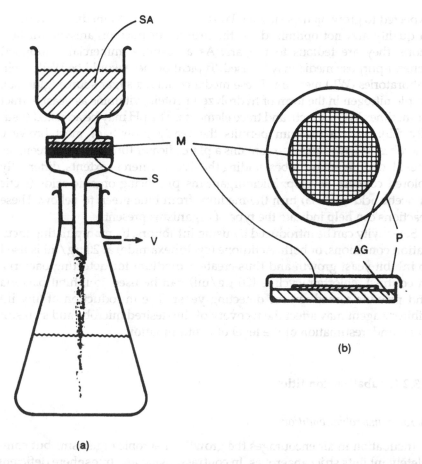

Figure 5–3 Membrane filtration. (a) Filtration unit comprising a Buchner flask connected to a vacuum line (V) and a glass filtration unit containing a sinter (S) with a small-pore (0.46 mm or less) membrane on top. The upper part of this unit contains the sample (SA, hatched area). The upper and lower parts are clamped together (clamp not shown). (b) After filtration, the membrane is placed on the surface of medium (AG) solidified with agar in a Petri dish (P). After suitable incubation, microbes present in the sample grow to form colonies on the surface of the membrane.

The basic requirement of all media for culturing a given microbe is to provide nutrients for the organism at an appropriate pH. The simplest media may be based on brewer's wort at pH 5.5 or beer (with alcohol added post-sterilization) at pH 4.3. Clearly, organisms found in breweries would be expected to grow on these media. Wort- and beer-based media are variable in quality and not optimized for the growth of microorganisms; furthermore, they are tedious to prepare. As a result, commercially supplied general-purpose media may be used. Typical of these would be Wallerstein Laboratories (WL) medium. These media contain a sugar source (glucose), ample nitrogen in the form of hydrolyzed protein, vitamins as yeast extract (sometimes liver extract), and trace elements. The pH may be adjusted (usually 5.5). Such a medium permits the ready growth of most brewery microbes. WL medium also contains a pH indicator (bromocresol green): as a result of different microbes binding the dye to different extents, differently colored colonies develop. Microorganisms producing organic acids (lactic or acetic acid bacteria) turn the medium from blue-green to yellow. These reactions can help indicate the type of organisms present.

Selectivity can be introduced by using inhibitors, by manipulating incubation conditions, or both. Actidione (cyclohexamide) at 20 μg/ml is used to inhibit yeast growth and thus create a medium for detecting bacteria. In contrast, chlortetracycline (20 μg/ml) can be used to inhibit bacteria and provide a medium for detecting yeast. The introduction of any inhibitory agent may affect the recovery of the desired microbe and so result in an underestimation of the level of contamination.

5.3.2 Incubation conditions

Aerobic/anaerobic conditions

Incubation in air encourages the growth of aerobic organisms but completely inhibits strict anaerobes. In contrast, using an atmosphere deficient in oxygen encourages the growth of anaerobes and suppresses aerobes. Culture in the absence of air is achieved in liquid media by almost completely filling containers with medium and sealing, using screw caps. With plated samples, incubation may be in a specialized incubator from which air is evacuated and carbon dioxide (CO_2) introduced. The simplest technique is to place plates in a metal can containing a candle, light the candle, and carefully fit an air-tight lid; as the candle burns, the oxygen is used. Another procedure is to place an absorbent pad containing an oxygen-absorbing agent (pyrogallol) into the lid of a Petri dish. The dish is sealed with tape or plastic film. The most convenient technique is to use gas generating kits in anaerobic jars (Figure 5–4). Petri plates are placed in a jar and a generating kit opened and water added. The jar is sealed. The hydrogen and CO_2 are generated by the action of water on the contents of the kit. A catalyst included with the kit accelerates the reaction between

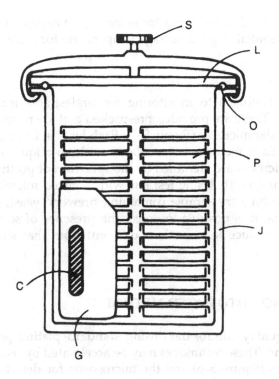

Figure 5–4 Establishing anaerobic conditions for incubation. The jar (J) contains plated samples of Petri dishes (P). A gas generator (G) is filled with water before the jar is sealed by clamping the lid (L) using the screw (S). The screw causes the attached clamp to pull down onto the seal (O). The catalytic strip (C) on the outside of the generator causes evolved hydrogen to react with oxygen in the jar to produce water. The generator also produces CO_2.

hydrogen and the oxygen in the jar to give water. The end result is that the atmosphere of the jar is enriched with CO_2 and contains very little oxygen. An indicator solution is usually included in the jar to show that anaerobiosis has been achieved. Creating anaerobic conditions is important in culturing several brewery bacteria. Where absolute anaerobiosis is needed, reagents such as thioglycollate are incorporated into media to counteract the effects of the traces of oxygen left in an anaerobic jar system.

Temperature

Temperature of incubation is also important. Where possible, the optimum for the organism is desirable to minimize the incubation time. For many bacteria, 30°C is good, but for some (e.g., *Pediococcus*), no higher than 25°C is essential. A good average temperature for yeasts is 28°C.

Media test kits

A different approach to monitoring for undesirable microbes uses media test kits. These are provided pre-packed and are normally used to identify particular microbes (Figure 5–5). Each kit is in fact a range of different media. Each medium is inoculated with an aliquot of the same sample and microbes are identified by the spectrum of positive and negative results for growth. Using test kits with a mixed microbial population, such as a brewery sample containing brewers' yeast, wild yeasts and/or bacteria, it is possible to detect the presence of some contaminants if they produce test reactions different from that shown by the brewing yeast.

5.4 RAPID AND AUTOMATED METHODS

Obtaining quality control data using standard plating procedures is time consuming. These techniques may be accelerated by using slide culture techniques (Figure 5–6) and the microscope for detecting growing colonies. The technique is used to detect wild yeasts by incorporating optical brighteners in the medium. These chemicals (the "blue whiteners" of the soap powder industry) adsorb to yeast cell walls. The cells then fluoresce in the visible spectrum. Using this type of approach shortens the time from 2–5 days to perhaps 8–16 hours. Other rapid processes using microscopy are the immunofluorescent procedure (IMF), the fluorescein diacetate (FDA), and the direct epifluorescence filter techniques. Although some fluorochromes can be excited by light from a halogen bulb, best results are obtained using fluorescence microscopes equipped with a mercury vapor light source and epifluorescence condenser. Such equipment is very expensive, so these procedures are for large sophisticated breweries.

5.4.1 Immunofluorescent techniques

IMF uses the specific interaction between antibodies raised (usually in rabbits) to microbial cell walls as the detection mechanism. The antibodies react on a microscope slide with any microbes present, after which a second reaction is conducted to detect bound antibodies. The second reac-

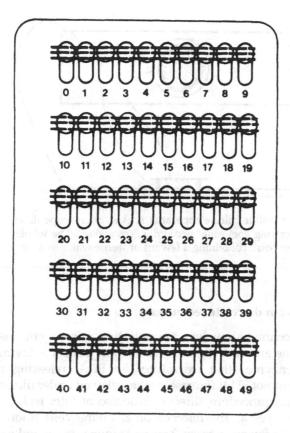

Figure 5-5 Kit produced by API Laboratory Products for identifying yeasts. Each plastic bubble (0 to 49) contains a different dehydrated medium. Each bubble is inoculated with the same suspension of microorganisms. Growth (turbidity, measured by the ability to obscure the black lines passing behind each bubble) or acidity shown by a color change of an acid indicator in the media is recorded. Different organisms give different patterns of growth. Contaminants in a fermenter may be detected because they produce patterns differing in one or more respects from the primary fermenting strain.

tion uses a fluorescent tag that, when the slide is viewed under ultraviolet light, fluoresces such that the cell walls of the microbes are clearly seen. The technique was developed primarily for detecting wild yeasts but because it does not distinguish between living and dead cells (and many components of the brewing process contain dead wild yeasts, for example, brewing sugars), it has fallen out of favor.

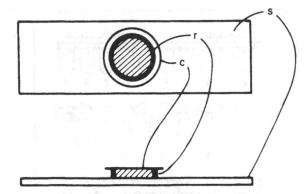

Figure 5–6 Simple slide culture apparatus. s, glass microscope slider; r, 1 mm deep, 15 mm diameter ring enclosing medium; c, coverglass. The whole is placed on a support in a Petri dish containing a few ml of sterile water to maintain humidity.

5.4.2 Fluoroscein diacetate techniques

The FDA technique involves concentrating microorganisms on a suitable membrane and reacting with fluorescein diacetate. Living cells contain active enzymes that hydrolyze the FDA, releasing fluorescein. Fluorescein, but not FDA, fluoresces when viewed under ultra-violet light. A more general procedure, direct epifluorescent filter technique (DEFT), uses acridine dyes as the fluorochrome. Living cells fluoresce orange, while dead cells fluoresce green. This technique is rapid and much used in the food and dairy industries. Although it requires expensive equipment, the technique can be partially automated using, for example, continuous flow through a tube irradiated with low-wave length light to excite fluorescence and scanned by a fluorescence microscope (flow-cytometry).

5.4.3 Bioluminescence techniques

Microscope-based fluorescence techniques suffer from background noise due to the presence of particular matter, and this is particularly a problem in samples taken from the brewing process. An alternative approach is to detect the presence of adenosine triphosphate (ATP) in samples. ATP, the energy currency of living cells, is hydrolyzed when cells die so its detection in a sample is indicative of the presence of living cells at the time of analysis. It is possible to detect ATP using the luciferin–luciferase system, which fireflies employ to emit light. In the presence of ATP and molecular oxygen, the enzyme luciferase produces oxyluciferin and light. The energy for the process is obtained from the hydrolysis of ATP. This biolu-minescent technique can be used to detect picogram amounts of ATP. Up to 1,000 microbes per sample can be detected in a few minutes. This tech-nique is used routinely to check for the cleanliness of vessels by analyzing post-cleaning samples. Coupling the analysis to membrane filtered sam-ples increases the sensitivity of detection.

In most cases, the ATP present is likely to be that extracted from brew-er's yeast, since (it is hoped) this is the microorganism most likely to be present in the process. It is likely however if the cleaning regime has removed all the yeast, it will also have removed much lesser numbers of other organisms. Although expensive, the technique is very rapid and rel-atively simple to apply. Consequently, it can be conducted by plant oper-atives and it provides information that can be used to direct immediate action (e.g., to re-clean a vessel). Thus, the microbiological status of the sample can be used to pass/fail the cleaning regime. The bioluminescence procedure may be used to detect organisms on the surface of a membrane filter. Computer controlled scanning with a fluorescence microscope is used. This procedure provides the brewer with rapid information on the microbiological status of any sample.

5.4.4 Other techniques

Automated methods for use in large laboratories employ **biophysical** or **bioelectrochemical** methods. In the former, organisms growing in selec-tive media are detected by the change in impedance (or alternatively con-ductance) produced as medium constituents are used and end-products excreted. In the latter, a biosensor couples the metabolism of the growing microbes to an amperometric detector. The impedance and conductance techniques are at present relatively insensitive and the time taken to ob-tain a result depends on the amount of contamination in the sample. In use, they typically show responses in less than a day. They are limited in choice of medium by the requirements of the sensing system; neverthe-less, they do offer some degree of selectivity.

Detection of microbial growth by heat output (microcalorimetry) is also a feasible procedure.

5.5 SOURCES OF MICROBES IN BREWING

Microbes abound in air. They grow on aqueous solutions of nutrients. All the raw materials used in brewing, especially water, malt and cereals, sugars, and hops are potential sources of microbes. Once in the process, microbes grow when in contact with wort, beer, or both, so brewing-plant surfaces (internal and external) harbor them. Pitching yeast that is re-used may be a major source of accumulated contaminating microorganisms. Brewing water should be potable (fit for human consumption) and therefore will contain very low levels of microbes and none harmful to humans. In most operations, water must be analyzed for the presence of lactose-fermenting bacteria that grow at 43°C. Specialized media (e.g., Mac-Conkey agar) incorporating the sodium salts of taurocholic and glyco-cholic acids are used. These acids are the constituents of mammalian bile and are inhibitory to most bacteria. Bacteria associated with the intestines of animals and humans are resistant. A water supply containing such microbes is judged to have been fouled with sewage and is therefore unfit for consumption.

Raw materials, which will be extracted at high temperature and boiled (malt, cereals, sugars, hops), represent a hazard only in store. They should therefore be stored away from those parts of the process occurring after the kettle boil, and especially from fermentation and yeast-handling areas. Any material added to any part of the process after wort has been cooled (diluting water, sugar, etc.) must be free of microbes and "sterilized." Holding sugar solutions at 70°C or boiling are effective. Ultra-violet radiation may be used to sterilize water. Routine cleaning-in-place (CIP) of brewing plant with detergents and sterilants is essential to maintain a high standard of hygienic operation.

Yeast needs special consideration and the microbiological quality control is rigorous. Contaminating bacteria (but not yeasts) can be attacked by treatment with acid. Alternatively, new pure culture may be obtained (Chapter 16).

It is possible, where the highest production standards operate, to store beer cold (-1°C) at the point of sale. This practice restricts the growth of any microorganisms in the product. It is more usual, however, to treat finished beer by either sterile filtration or pasteurization to remove or kill, respectively, any microbes present and so extend the shelf-life of the product.

5.6 MICROBES OF THE BREWING PROCESS

Excluding malt, which when contaminated with molds can cause reduced gas stability (gushing) in beer, there are two distinct environments for microbes in the brewing operation: (1) wort; and (2) beer. The former is nutritionally rich, pH 5.5, contains oxygen, and many microbes can grow in it. Those that grow fastest and are unaffected by hop constituents will be the ones most commonly encountered. Beer is typically nutritionally poor (yeast has used most of the wort nutrients), at pH 4.5 or less, contains 3–5% alcohol, hop components, and is anaerobic and cold. Environments between the two extremes occur in the fermentation process albeit for short periods of time.

The range of microbes found in the brewing process is small, with three broad groups occurring: (1) Gram-positive bacteria; (2) Gram-negative bacteria; and (3) wild yeast. Table 5–1 lists the stages in the process where each of the most often encountered members of each group is found. Schematic representations of the typical appearance of the major groups are shown in Figure 5–7, while the main characteristics of each group are listed in Table 5–2.

5.6.1 Gram-positive bacteria

The main members of this group include lactic acid bacteria, *Leuconostoc, Streptococcus, Microccocus* and *Bacillus.*

Lactic acid bacteria

These bacteria undoubtedly present the greatest problems to the brewer. Species of two genera (*Lactobacillus* and *Pediococcus*) are routinely encountered as potent beer-spoilage microbes. They are Gram-positive, non-spore-forming, non-motile bacteria which react negatively in the catalase test (10% hydrogen peroxide fails to effervesce in the presence of the culture).

Table 5–1 Occurrence of spoilage organisms at different stages in the brewing process

Stage	Organism(s) mainly found
Mashing	Heat-tolerant lactic acid bacteria; Enterobacteria*
Cooled wort	Enterobacteria; *Hafnia protea*
Early fermentation	Enterobacteria; *Hafnia protea*; acetic acid bacteria; *Lactobacillus*; *Pediococcus*
Late fermentation	*Lactobacillus*; *Pediococcus*; (acetic acid bacteria)
Post-fermentation, maturation, storage,	*Lactobacillus*; *Pediococcus*; acetic acid bacteria; *Zymomonas*; *Pectinatus*; *Selenomonas*, *Zymophilus*, dispense *Megasphaera*

*under mash/lauter tun plates.

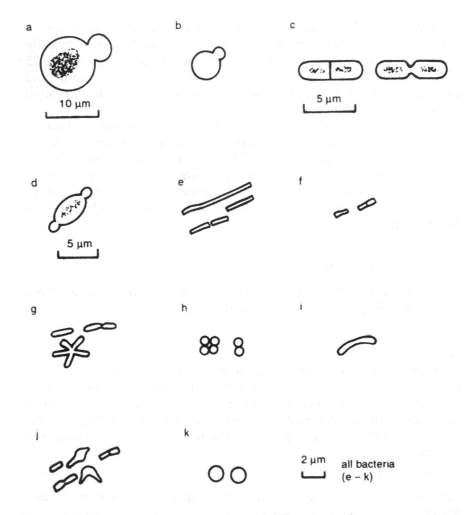

Figure 5–7 Diagrammatic representations of different microbes encountered in brewing. a, Brewer's yeast; b, wild yeast (*Saccharomyces*); c, wild yeast (*Schizosaccharomyces*; d, wild yeast (*Kloeckera*); e, Lactobacillus; f, Enterobacter; g, *Zymomonas*; h, *Pediococcus*; i, *Pectinatus*; j, *Hafnia* (*Obesumbacterium*); k, *Megasphaera*.

Lactobacillus

These are rods with dimensions $1.0 \times 2{-}120$ μm. Long rods are particularly associated with organisms grown in beer. The bacteria are acid-tolerant, inhibited by molecular oxygen and some strains fix CO_2. The

Table 5–2 Main characteristics of microbes found as spoilage organisms in brewing

Organism	Shape	Dimensions (µm)	Gram Reaction	Catalase Reaction x	Motility	Anaerobic Growth	Spoilage Characteristics
Enterobacteria	Rod	0.3 to 1 x 1 to 6	Negative	NA	Variable	Yes	Sulphur off flavours (including dimethyl sulphide) in wort pass into beer.
Hafnia protea	Rod *	1 x 2	Negative	NA	Variable	Yes	Usually none occasionally diacetyl in wort.
Acetic acid bacteria	Rod	0.6 x 1 to 4	Negative	NA	Yes	No	Acetic (vinegary) beer
Zymomonas	Rod	1 to 1.5 x 2.5 to 6.5	Negative	NA	Young cells	Yes xx	Acetaldehyde (apple) and in some cases hydrogen sulphide (bad eggs) in beer.
Lactobacillus	Rod	1 x 2 to 120	Positive**	Negative	No	Yes	Lactic acid (sour) and occasionally diacetyl in beer.
Pediococcus	Coccus	0.8 to 1	Positive**	Negative	No	Yes	Lactic acid and often diacetyl in beer.
Pectinatus	Curved Rod	0.7 x 2 to 32	Negative	NA	Yes	Yes (obligate anaerobe)	Acetic and propionic acids and hydrogen sulphide
Selenomonas	Curved Rod (crescent shaped)	0.6 x 5-15	Negative	NA	Yes	Yes (obligate anaerobe)	Lactic, acetic, propionic acids
Megasphaera	Coccus	1.2	Negative	NA	No	Yes (obligate anaerobe)	Butyric and caproic acids

continues

Table 5–2 continued

Organism	Shape	Dimensions (µm)	Gram Reaction	Catalase Reaction x	Motility	Anaerobic Growth	Spoilage Characteristics
Zymophilus	Rod	0.9 x 5 to 30	Negative	NA	Yes	Yes (obligate anaerobe)	Acetic and propionic acids
Wild Yeasts	Various (Spherical, Oval,) Lemon	5 to 10	Positive	NA	No	Usually	Phenolic, specific esters (pineapple, banana, black currant etc)

*   In many situations this organism may show various bizarre shapes (see Figure 5–7) and is said to be Pleomorphic.
** When stained these organisms may appear mixed (both Gram positive and Gram negative cells) and are said to be Gram variable.
x   NR, test not relevant. In general, facultative anaerobes and microaerophilic bacteria are catalase positive whereas obligate anaerobes are catalase negative.
xx Grows in the presence of traces of oxygen said to be microaerophilic

growth is enhanced in the presence of an atmosphere of this gas. Optimal growth is achieved at 30°C. The metabolism of the organisms may be homofermentative; that is, predominantly lactic acid is produced as the end-product. Such strains use the Embden–Meyerhof–Parnas pathway from hexose sugar to pyruvate and reduce the latter to lactic acid. Alternatively, heterofermentative strains utilize the phosphoketolase pathway (via 6-phosphogluconic acid) and produce a mixture of mainly lactic and acetic acids. Both hetero- and homofermentative strains are found in breweries. Strains of *Lactobacillus* may use glucose, fructose, maltose, or ribose as carbon sources. Most strains require amino acids (or peptides) and vitamins. In most beers there is usually sufficient residual carbohydrate, nitrogen, and vitamins to support their growth. Not all of these organisms grow well in beer. Those that do grow, spoil beer by souring (lactic taste), producing turbidity (characteristically "silky" strands), and the production of diacetyl. Diacetyl is generated by a different mechanism to that produced by brewer's yeast and most probably involves the condensation of "active acetaldehyde" and acetyl coenzyme A. The production of diacetyl (butane 2,3 dione) is the most serious problem, especially in lighter-flavored beers where the characteristic aroma and taste is considered highly undesirable. *Lactobacillus* strains are isolated on media containing several sugars (including maltose), amino acids, and vitamins (meat digests and yeast extracts). Many surveys of media have been conducted; most conclude that no medium is perfect and many brewing companies develop their own specialized media. However, commercially available Racka–Ray and MRS (de Man Rogosa Sharpe) media are popular. The antibiotic polymixin may be included to inhibit the growth of acid-tolerant Gram-negative bacteria. Anaerobic conditions, in the presence of CO_2, at 30°C are preferred for incubation.

Pediococcus

These are cocci 0.8–1.0 μm in diameter, which often appear in groups of four cells (tetrads). Species of *Pediococcus* are homofermentative. *P. damnosus* is especially common in breweries and well suited to growth in beer. Strains require vitamins for growth or stimulation of growth. Optimum temperature for growth is about 25°C. *Pediococcus* is a more common contaminant than *Lactobacillus*, world-wide, in beer at the end of fermentation and during storage. It is particularly prevalent as a spoilage organism in beers fermented at low temperatures. Spoilage results mainly from the production of diacetyl. *Pediococcus* may be isolated on the same media used for *Lactobacillus*. The inclusion of 2-phenylethanol (20 μm/ml) to suppress the growth of *Lactobacillus* is used in some media. Incubation is at 25°C under anaerobic conditions.

Other Gram-positive bacteria

Other Gram-positive bacteria not belonging to the lactic acid group are often associated with the process and may be isolated on occasion from samples. Their presence may give false-positive results in plating tests or on microscopic examination because they are mistaken for lactic acid bacteria. They include the homofermentative *Lactococcus* and the aerobic spore forming *Bacillus* and aerobic *Micrococcus* and *Staphylococcus*. Detailed microbiological analysis is needed to identify these organisms. If brewers detect gram positive rods or cocci in beer and/or yeast, they would be wise to assume that beer spoiling organisms are present and take appropriate action.

5.6.2 Gram-negative bacteria

The main members of this group include Enterobacteriaceae, acetic acid bacteria, *Zymomonas, Pectinatus, Selenomonas, Zymophilus,* and *Megasphaera.*

Enterobacteriaceae

Members of this family are Gram-negative rods, 0.3–1.0 μm × 1.0–6.0 μm. They are resistant to bile salts and many strains are motile. They can grow in the presence or absence of oxygen (facultative anaerobes). They are sensitive to pH; growth does not occur below pH 4.3, but organisms may survive. They can be potent spoilage organisms in wort where they grow very rapidly and some can produce dimethyl sulfide and diacetyl. Early in fermentation, they produce off-flavors including those described as "herbal phenolic" (e.g., guiacols) and can inhibit yeast growth. Some strains can metabolize nitrate to nitrite, which may increase nitrosamine levels in beer. This serious problem (there are legal limits on beer nitrosamine levels) can be exacerbated by growth of these microbes under the plates of mash and lauter tuns in the brewhouse. This almost certainly results from inadequate cleaning beneath the plates and growth of bacteria in between brews.

Enterobacter agglomerans, Citrobacter freundii and *Klebsiella* species are the most commonly encountered. Also found is *Obesumbacterium proteus* (Hafnia protea). Strains of this organism can spoil beer by producing both diacetyl and dimethyl sulfide. It is a noted contaminant of pitching yeast in ale breweries where it has often been found at a level of 1 cell per 100 yeasts. It is slower growing than the other Enterobacteria; however, it appears to bind to yeast and is concentrated in pitching yeast. *E. agglomerans* may also be concentrated in this way. In bottom fermentations, many, if not all, bacteria will settle with yeast and their concentrations will be maintained. All of these organisms are killed by acid washing or pasteurization. They may be selectively isolated on media containing bile salts to inhibit other microbes, actidione to inhibit yeast and amino acids as source of carbon and nitrogen (MacConkey's medium plus actidione). On this medium, *H. protea* takes 24–48 hours to produce small, colorless colonies. Other strains produce colonies overnight at 30°C; some produce red colonies. There is some concern that many members of the Enterobacteriaceae may not be detected on isolation media and although failing to grow in pitching yeast, remain viable.

Acetic acid bacteria

These Gram-negative bacteria are rod-shaped, 0.6 μm × 1.0–4.0 μm; there are both motile and non-motile strains. Two genera are recognized, *Acetobacter* and *Gluconobacter*. Both are acid- and alcohol-tolerant and unaffected by hop components. They spoil beer by acetifying it (turning it to vinegar), converting alcohol to acetic acid. They are typically contaminants of live beer where air replaces beer drawn from a cask. Both organisms are obligate aerobes and do not usually spoil beer packaged and dispensed under carbon dioxide. Strains isolated from beers seem to require very little oxygen for growth. The highest concentrations are found not in beer, but around leaking areas of kegs and in bar taps.

These bacteria may be isolated on media containing alcohol as carbon source, protein hydrolysate and yeast extract as nutritional supplements at pH 4. Incubation is in air at 30°C.

Zymomonas

This organisms is a short, fat rod 1.0–1.5 μm × 2.5–6.5 μm. It is a contaminant of primed beers and restricted to breweries using the priming process (mainly ale producers). It is acid- and alcohol-tolerant, and resistant to hop constituents. Members of the genus may or may not be motile. Its metabolism (Entner–Doudoroff pathway) enables it to convert sucrose, fructose, or glucose to ethanol and CO_2. Beer spoilage results because most strains also produce high levels of acetaldehyde (ethanal) and many produce hydrogen sulfide. Members of the genus are used to make fermented beverages in some countries (notably Mexico and Central Africa). The organism cannot use maltose as a carbon source. Its occurrence in ale breweries is thought to arise from the practice of using invert sugar or glucose syrup as primings in beer. In the U.K., *Zymomonas* contamination has been particularly associated with failures in keg and cask cleaning, since large numbers of the organism may grow in residual beer, overloading the sanitation processes.

Pectinatus

The organism is a slightly curved Gram-negative motile rod with rounded ends, which occurs singly or in pairs. Young cells are actively motile and appear "X"-shaped due to their motion. Older cells move in a snake-like manner. The dimensions of the rod are 0.7–0.8 μm × 2.0–32 μm. The flagellae of the organism protrude from one side of the cell only. *Pectinatus* is an obligate anaerobe identified in American, Finnish, and German lager beers. Beer spoilage results from the production of acetic, propionic acids, and hydrogen sulfide. Isolation is under strictly anaerobic culture on thioglycollate agar.

Selenomonas and Zymophilus

These obligately anerobic organisms are implicated in the spoilage of finished beers. They are motile rods and produce acetic and propionic acids from glucose.

Megasphaera

Megasphaera is a Gram-negative coccus, non-motile, and an obligate anaerobe. Beer spoilage is by the production of considerable amounts of butyric acid and lesser amounts of caproic acids. *Megasphaera* may be grown under anaerobic conditions on beer medium enriched with glucose and peptone (a commercially available peptic digest of meat).

5.7 WILD YEASTS

By definition, a wild yeast is any yeast other than the brewing strain(s) that is found in the brewing process. Wild yeasts can, in principle, represent a major hazard to the brewing operation for two main reasons: firstly, they are often difficult to detect; and secondly, they, unlike bacteria, are not susceptible to acid washing and are therefore impossible to eradicate by acid treatment of pitching yeast. In fact, the only practical way of ensuring pitching yeast is free of wild yeast is to operate a regular regime of introducing pure culture yeast into the brewery (Chapter 16).

It is usual to consider two general types of wild yeast: those belonging to a different genus to brewing yeasts (wild non-*Saccharomyces*) and those belonging to the genus *Saccharomyces*. The former include members of the genera *Pichia, Hansenula, Debaryomyces, Kluyveromyces, Schizosaccharomyces* and many others. Most wild yeast neither flocculate well nor interact with finings so they readily pass into beer in storage. They are, of course, acid-tolerant organisms, well suited in many cases to survival under anaerobic conditions. Their growth eventually produces turbid beer, but off-flavors are generated before this stage. Many wild non-*Saccharomyces* produce estery off-flavors (banana, pineapple). Such flavor components are, of course, normally present in the product, but the distortion of the overall profile and accentuation of one or more specific aroma notes taint the product. Wild *Saccharomyces* strains may also spoil beer in this way, but are also especially noted for the production of smoky, peaty, or phenolic flavors.

Wild non-*Saccharomyces* strains are often (but not always) different in shape from brewing yeast. For example, they may be elongated or lemon-shaped. They may reproduce differently, for example, budding only at the ends of the cell or in the case of *Schizosaccharomyces* species by binary fission. Preparations may be Gram stained (yeasts are Gram-positive). Careful microscopic examination of samples may therefore detect some wild yeasts but the technique is very insensitive and unreliable. Wild yeasts often sporulate, producing spores of characteristic shapes (e.g., hat, kidney). Inoculating pitching yeast on medium designed to induce sporulation rarely produces many yeast spores (often detected by staining), but the presence of wild yeast spores is much more obvious. In general, however, brewers use various types of agar media to detect the presence of wild yeasts. Because of the close similarity between wild Saccharomyces and brewing strains, these tend to be more difficult to detect and can produce the greatest problems in beer spoilage. Problems have been experienced, particularly with strains of *S. cerevisiae*, formerly known as *S. ellipsoideus* (fails to fine) and *S. diastaticus* (ferments dextrins not used by brewer's yeast).

Many non-*Saccharomyces* strains can use the amino acid lysine as the sole source of nitrogen for growth (brewing yeasts and the majority of wild *Saccharomyces* cannot). Lysine medium is therefore a general-purpose medium for detecting some wild non-*Saccharomyces*; inclusion of chlortetracycline inhibits the growth of any bacteria in samples. Because of its availability and ease of use, this medium is popular but, of course, gives limited information. Media for wild *Saccharomyces* often incorporate inhibitors. Thus the dye crystal villet (20 μg/ml), Cu^{++} ions at 10 μg/ml all inhibit growth of resistant wild yeast strains, both non-*Saccharomyces* and *Saccharomyces*. An alternative to the selective approach, especially when examining pitching yeast, is the use of **differential media**. The principle is that all yeasts grow, but different colony sizes, morphologies (rough or smooth) or, in the presence of a suitable indicator, different colors distinguish wild yeasts from the uniform appearance of brewing cells. A combination of media is used in practice and as with media for bacteria, no one medium is perfect.

CHAPTER 6

Beer quality and flavor

6.1 INTRODUCTION

There are two distinct aspects to beer quality. The first is its relative nature, kind, or character and the second, its degree of excellence. In the former case, we are, for example, talking of beer types in terms of the method of production, place of origin, or style. Examples would be top fermented, double decoction, Pilsner, and Burton ale. This aspect would also include many quality parameters that are perceived even before the beer is consumed. For example, color, clarity, degree of carbonation, and presence (or not) of foam. The brewer might also refer to beer types as lager, ale, or dark quality. In recent years, there has been an increased market share for low and no alcohol beer types. Clearly the quality referred to in these ways is largely determined by the brewing practice. The same applies to other beer parameters such as pH and alcohol content. Traditionally, beers arose as characteristic styles depending on the water supply, available raw materials, method of temperature control in brewhouse, fermentation, and storage. Today, it is technologically possible for a single brewer to produce a whole variety of beer qualities from the same equipment. Quality assurance procedures are used to set overall specifications and often techniques are available for fine tuning them. Table 6–1 gives some examples.

These readily measured and controlled parameters also contribute to the degree of excellence of the product. This aspect of quality is, of course, largely subjective and additionally takes into account the flavor of beer. With the exception of control of bitterness, control of overall flavor is the most difficult aspect of quality assurance. It still largely relies on the selection of raw materials and yeast, consistent processing, and hygienic operations. Nevertheless, the majority of brewers would hold, and sensory science shows, that their beers have unique and characteristic flavors and every effort is taken to maintain this aspect of quality. Flavor evaluation is used to monitor uniformity, changes in raw materials, or process and development of new products.

Table 6–1 Methods of measurement and adjustment of typical beer quality assurance parameters

Parameter	Method of measurement	Method of adjustment
Color	Spectrophotometer or color comparitor	Grist composition (colored malts to darken; non-malt starch to lighten); kettle adjuncts (caramel to darken, non-malt syrups and reduced boiling to lighten)
Clarity	Spectrophotometer or haze meter	Filtration following prolonged storage at 0 to −1°C, fining in tank or cask
		Removal of protein precursors with adsorbents (silica hydrogels, nylon filters) or proteases (papain), adding tannic acid
		Removal of polyphenol precursors with adsorbents (polyvinyl pyrrolidone, AT) adding antioxidants (ascorbic acid)
Foam	Generate foam with gas bubbles and time its collapse	Higher malt in grist; use wheat flour adjunct, alginate extracts
		Adjustment to dispense gases and taps
Hop bitterness	Spectrophotometric, chromotographic	Use pre-isomerized hop extracts after fermentation
Hop aroma	Sensory evaluation	Late hop addition to kettle; dry hop; add aroma extracts to beer
Carbon dioxide content	Measure volume of gas absorbed by alkali	Lower by venting fermenter; increase by sealing fermenter In line carbonation

6.2 FLAVOR

Perception of flavor involves the individual senses of touch, taste and smell. The sense of touch is used to perceive the so-called mouth-feel characteristics, which include smoothness, astringency (drying), temperature, and the tingling sensation given by effervescent CO_2. Taste is perceived by the taste buds of the tongue and four primary tastes are recognized (Figure 6–1). These are sweet, salt, sour, and bitter. Standard solutions of sucrose (cane sugar), sodium chloride (table salt), dilute hydrochloric acid, and quinine define these tastes. More recently, a fifth taste standard has been recognized and named umami, and the standard for this is glutamic acid. It is commonly depicted (as in Figure 6–1) that sweetness is perceived predominantly at the front of the tongue and bitterness at the back, with salty and sour tastes at the front and rear edges of the tongue, respectively. However, it is likely that this is not the case and the taste receptor cells may not have such a localized distribution. Taste buds (small onion-shaped structures) contain 50 to 100 taste receptor cells and there are only a few thousand buds on a tongue. It is also clear that taste receptors operate in different ways. Salt and sour receptors respond to ionic stimuli (influx of sodium and hydrogen ions respectively), whereas sweet and bitter receptors respond to complex organic molecules. From genetic experiments in mice, it is clear that there is a large number of receptors for identifying different bitter substances. The

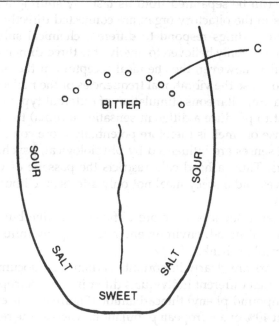

Figure 6–1 Simplified diagram of the taste sensory areas of the human tongue. C, circumvallate papilli at the back of the tongue.

same situation almost certainly will hold in humans. So, individually, we may have up to 90 different receptors for bitterness, each perhaps recognizing a specific compound. However, we cannot distinguish the individual compounds and all of them are perceived simply as bitter. Perhaps the taste receptor cells produce all of the receptors, thus providing a broad range and high sensitivity. The taste response is most sensitive to bitter compounds. This is claimed by some to have a selective advantage since many plant poisons (alkaloids) are bitter and their presence may be detected before a lethal dose is reached! Sensitivity to bitterness is clearly important as far as beer flavor is concerned since one of the most characteristic flavor notes is hop bitterness. The receptors responsible for detecting sweetness remain to be elucidated.

It is quite common for individuals to confuse the senses of taste and smell. The flavor of any beverage taken into the mouth is virtually simultaneously discerned by both senses. This happens as volatile vapors pass up into the olfactory organ at the back of the nasal cavity (Figure 6–2). It is possible to test the validity of this by taking a mouthful of beer (or other suitable beverage) and pinching the nose to stop the circulation of the vapors (taking due care not to choke in the process!). Similarly, the aroma of a beverage can be separated from its taste by sniffing a sample. The nerve endings in the olfactory organ are connected directly to the brain. Different nerve endings respond to different chemical substances. The interaction is commonly believed to involve the three-dimensional shapes of the molecules; however, the chemical receptors in the nerve endings may be able to sense the vibrational frequencies of the molecules and this defines aroma. Simultaneous stimulation of different types of nerve ending is thought to produce a different sensation (aroma) to single interactions. The sense of smell is therefore potentially more complex than that of taste. Both senses are influenced by physiological, psychological, and genetic factors. Thus, a head cold restricts the passage of vapors to the olfactory nerves and a spicy meal not only affects smell but desensitizes the taste buds.

Psychological influences are more difficult to define but most people would accept that mood, environment, and company influence the perception of a meal or drink.

Genetic factors are clearly important. Certain well-documented examples illustrate that different individuals differ in their perceptive abilities. Thus, the compound phenyl thiocarbamide (Figure 6–3) is excruciatingly bitter to about 40% of a European population, whereas the remaining 60% cannot taste it at all. Some individuals are born without a sense of smell (anosmics) and many individuals experience specific anosmias (unable to

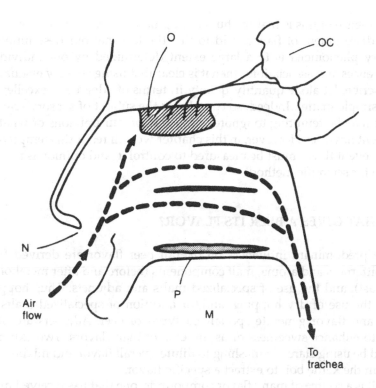

Figure 6–2 Diagrammatic representation of the olfactory organ. Arrows (broken lines) show the air flows. N, nostril; P, upper palate (roof of mouth); M, mouth; O, olfactory lobe (sensory nerve endings); OC, olfactory center in the brain.

perceive certain aromas). Given instances like these, it is quite plausible to propose that some individuals will have different sensitivities to the same substances (indeed this can be readily demonstrated by experiment) or even, for example, lack the ability to perceive some particular aromas. The

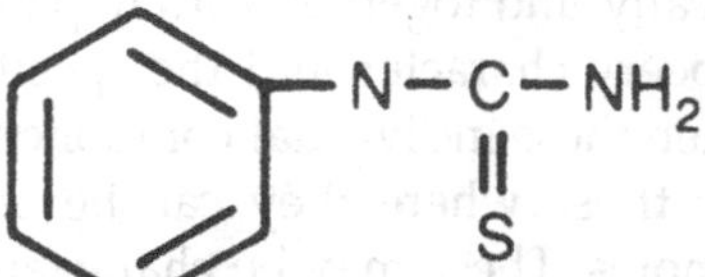

Figure 6–3 Phenyl thiocarbamide. Some 40% of Europeans find this compound excruciatingly bitter; the remainder cannot taste it

consequence of this is that the human population is not uniform in terms of its discernment of flavor. Add to this the fact that our description of sensory phenomena is to a large extent determined by our individual experiences and associations, then it is clear that using sensory evaluation to describe, let alone quantify, quality in terms of "degree of excellence" is no simple matter. Indeed, so complex is the subject of sensory science, that it may be tempting to ignore it and rely on our opinions of what an excellent flavor is. However, in this chapter we will resist this temptation. In any event flavor must be measured to control it, and opinion is no substitute for scientific method!

6.3 WHAT GIVES A BEER ITS FLAVOR?

The predominant influences on overall beer flavor are derived from hop bitterness and aroma, malt components (before and after metabolism by yeast), and the use of specialized malts and adjuncts. Thus hopping rates, the use of dry hopping, and/or addition of specialized malts for color and flavor generate specialized types of beer. Adjuncts are often used to enhance sweetness or as diluents of malt flavors. Two extremes would be using starch in mashing to dilute overall flavor and mixing fruit pulp in the kettle boil to extract a specific flavor.

Malt is a source of many flavor compounds; one that has received much attention is dimethyl sulfide. Its cabbagy-vegetable, sweetcorn aroma is a predominant flavor note in many European lager beers. The characteristic flavors of all beers also derive both directly and indirectly from the products of yeast metabolism. Alcohol, by virtue of its high concentration, makes a major contribution, but it is the minor products of yeast metabolism that give a beer its characteristic flavor. Table 6–2 gives an indication of the range and concentrations of materials encountered; some specific examples are referred to in Chapter 18. There are certainly in excess of 700 such constituents of beer. Most of these are present at levels just below those at which they are easily perceived. However, acting both synergistically and antagonistically, and together with hop and malt constituents, they give the overall beery character and the specific flavors associated with the beverage. When these individual components are present at levels of double or more those where they can be easily discerned, they exhibit specific flavor notes. These may be characteristic (therefore desirable, flavors) or uncharacteristic of the beer (and therefore undesirable, taints). It follows therefore that one brewer's desirable flavor may well be another's taint!

Table 6–2 Some constituents of a typical beer

Component[a]	Typical amount (g/l)
Ethanol (alcohol)	35
Carbon dioxide	5 (30 produced)
Organic acids (e.g. pyruvate)	<0.1
Aldehydes (e.g. acetaldehyde)	<0.1
Esters (e.g. isoamyl acetate)	<0.01
Higher alcohols (e.g. isobutanol)	<0.01
Diketones (e.g. diacetyl)	<0.0002
Sulfur compounds (e.g. dimethyl sulfide)	<0.00005

[a]The other main product of fermentation is of course yeast, about 1.5 g wet weight/l is produced

6.4 ANALYSIS OF FLAVOR

An experienced individual may smell and taste a beer and declare that it is within specification. It is as well to remember, however, that she/he is subject to physiological and psychological influences like anyone else!

Two basic types of objective analytical test may be used: these are the **difference test** and the **profile analysis**. The value of difference tests is greatly enhanced if tasters are trained to identify different characteristic beer flavors. Profile analysis must be conducted by a trained panel.

Training involves firstly identifying individuals who are capable of discriminating different beer characters. For such a process, reference materials are used (some examples are given in Table 6–3). These may be presented as aroma samples in small vials or added to a beer with no strong flavor characters of its own. In this case, the aroma and taste of the compound may be evaluated. As well as using pure chemical standards, the enterprising sensory analyst may choose to use raw materials (e.g., ground malt) or other foodstuffs (e.g., liquid from canned sweetcorn) as standards. Such materials are often more readily obtained and have the distinct advantage that they themselves are foodstuffs.

The taster is also trained to identify flavors using a common set of descriptive terms. In brewing, the terminology of the Flavor Wheel (Figure 6–4) has gained general acceptance. However, it is still not unusual to find different terms being used in specific instances. Proposals have been made to extend the flavor wheel to include the importance of mouth-feel characteristics (Figure 6–5). Training in identification of particular taints may also be given. The diverse origins of many beer taints are indicated in Table 6–4.

Table 6–3 Some reference standards for evaluation of flavor and aroma

	Standard[a]
Taste	
Sweet	Cane sugar (sucrose) 0.75% in water
Salt	Table salt (sodium chloride) 0.18% in water
Sour	Sour milk (or citric acid 0.05% in water)
Bitter	Isohumulone (or quinine sulfate 0.001% in water)
Aroma	
Diacetyl (butterscotch, toffee, buttery)	Diacetyl (2,3 butane dione)
DMS	Dimethyl sulfide
Cooked cabbage	Over-cooked green vegetables
Phenolic, hospital-like (medicinal, antiseptic)	Trichlorophenol
Smoky	Guaiacol
Floral (rose)	2-Phenylethanol
Estery (apple)	Ethyl hexenoate
Estery (peardrop, banana)	Isoamyl acetate
Fatty acid (cheesy, old hop)	Isovaleric acid (old hop oil)
Sulfury (yeasty)	Commercial or home-made yeast autolysate
Sulfury (sufitic, sulfur dioxide)	Potassium metabisulfite

[a]For aroma standards the concentrations used depend on whether sniff samples or beers spiked with the standards are evaluated. Sniff samples are better for initial training. Samples are dispersed in odor-free solvent in sealed vials

Objective tasting requires specialized facilities and a clear code of practice. Randomized procedures are used to code samples and the preparation and presentation are done by different individuals. Each taster conducts the analysis in isolation, in dim light in a room of constant humidity with filtered and odorless air. Sessions are held at times away from those used for meals and coffee breaks. Two types of differential tests are often used: the **duo–trio** and the **triangular**. Both use three samples, the former uses a control beer as a reference and a control and test beer; the taster is asked to match the reference. The latter does not use a reference and two glasses contain the control and one the test (or *vice versa*). In the absence of a difference, 50% of answers will be correct in the duo–trio test and 33% in the triangular test. Statistically significant deviation from these scores indicates a difference in the beers under test. These tests are often used to compare the effects of process changes on beer flavor. Typically 15 to 30 individuals, after adequate training, may be used in such tests.

A typical procedure for sensory evaluation of beer as recommended by the American Society of Brewing Chemists is shown in Table 6–5. Clearly

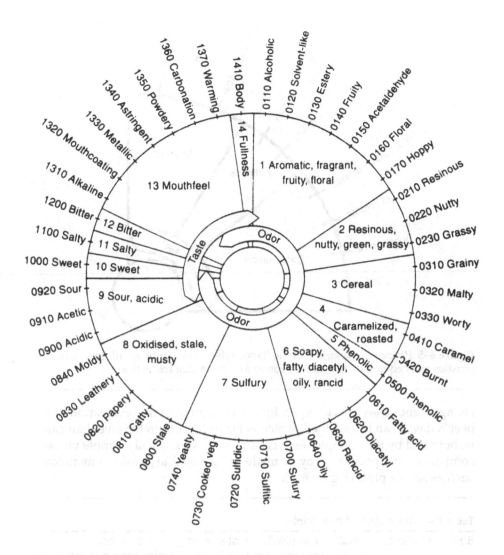

Figure 6–4 Beer flavor wheel. Copyright American Society of Brewing Chemists; with permission.

from the foregoing, the conditions used to evaluate beer flavor are far removed from those in which the product is normally consumed!

Flavor profile analysis uses small panels of three to five trained tasters. They will score a sample for particular attributes. In this way, a profile of attribute scores will be obtained. This analysis is more comprehensive and is used to monitor flavor consistency, look for specific effects of process

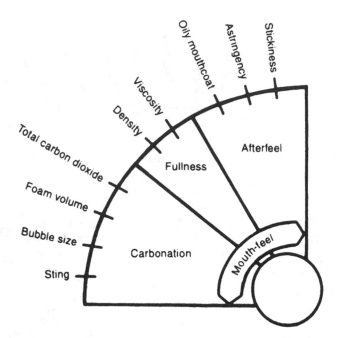

Figure 6–5 Proposed expansion of the flavor wheel to include mouth-feel characteristics as of equal status to odor (aroma) and taste characteristics.

---

changes, and analyze beer types. In this way, an objective evaluation of a beer's flavor can be obtained, which is far more comprehensive than can be obtained by for example chemical analysis. Additionally, simple visual comparison of "profiles" may be made using suitable graphical methods such as spider plots (Figure 6–6).

Table 6–4 The origins of beer taints

Basic ingredients	Water; malt (DMS); roasted malt (phenolic); yeast (contamination; autolysis); poor quality hops and extracts
Process aids	Clarifying agents; finings (sulfur); primings (iron and infection)
Process	Filter aid; trub; fermentation; conditioning; mashing; high gravity
Contamination	Coolant; detergent; air; sterilant; poor hygiene and sanitation
Packaging	Air; leaking containers; poor washing of containers; microbiological contamination
Dispense	Tubing; detergent; sterilant; air; glass rinsing and drying

Table 6-5 Sensory evaluation of beer, as recommended by the American Society of Brewing Chemists (ASBC)

Some basic conditions for flavor evaluation
1. Silence
2. Dim light
3. Participants separated in booths
4. Room ventilated with odorless air of constant humidity
5. No smoking
6. Participants not to use perfumed cosmetics
7. Participants not to have consumed food or drink immediately prior to the session

Rules for triangular tasting (ASBC)
1. Compare odors before and after swirling
2. If an odor difference is detected, taste the one (or two) samples having the least odor
3. Decide on the difference
4. Fill in tasting form

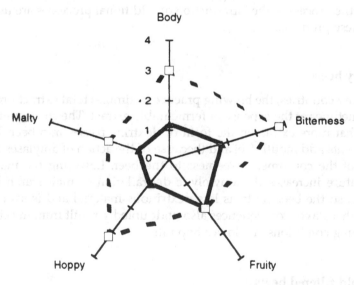

Figure 6-6 Spider plot (radar plot) profiles of two beers. Intensity scores (0 none to 4 very intense) for different attributes (body, hoppy, bitterness, malty and fruity) radiate from the center. Note that 'hoppy' refers to the aromatic contribution of hops. The inner solid line is a U.S. light beer brewed by a major company; the outer dashed line is a U.S. pale ale brewed by a microbrewery. Although the trained panel scored them both equal in fruity character (esters), they are markedly different in the other four characters!

6.5 BEER TYPES

The popular brewing literature abounds with recipes for traditional beer types. These products are produced largely by varying the composition of the malt (and adjunct) grist. This is achieved by including specialized malts, caramels, and sugars. Other variations involve changing the nature of the hop varieties used, varying the times of hop addition to the kettle, and dry hopping. The nature of the conditioning process and use of priming sugars also markedly affect beer flavor.

The strength of the beer is also a great variable. In most markets, there has always been the requirement for beers of lighter flavor and less intoxicating properties (e.g., small beers and mild beers produced from weaker worts). These, typically, were consumed by men involved in heavy manual work. In response to the concerns of the modern consumer about diet (calorific intake) and regulations on blood alcohol and driving, the brewer produces a range of different beers. These include dry beers, low calorie (low carbohydrate, lite beers), low alcohol, and no alcohol beers. Conventional modern brewing plant is used and either simple modifications to the process or the introduction of additional processes are used to make these products.

6.5.1 Dry beers

In many countries, the brewing practice optimizes total extract production sometimes at the expense of fermentable extract. The consequence of this is that more extract in the form of dextrins passes into beer. These components add mouth-feel and (because of the action of amylases in the mouth of the consumer) sweetness to the beer. Lowering the mashing temperature increases the amylolytic degradation of malt starch by β-amylase, so the beer contains less dextrinous material and is said to be drier. Other flavor consequences also undoubtedly result from the change in mashing conditions and lower hop rate.

6.5.2 Cold filtered beers

These products are produced using efficient filtration processes to ensure removal of yeasts and contaminating microorganisms from finished beers. The practice ensures that pasteurisation is not required so any characteristic flavors that may result from the latter are no longer present.

6.5.3 Lite, low calorie, low carbohydrate beers

These terms are applied to a whole range of products. At their simplest, they are beers brewed from wort of lower gravity (containing less fermentable extract) than the "normal beer." Alternatively, extracting malt, containing high soluble protein, at low temperature (35–40°C), leaves starch behind but solubilizes many malt components contributing to beer flavor. These include simple sugars, but small amounts of additional highly fermentable sugars can be added to obtain a fermentable wort. The majority of lite beers are however produced by greatly lowering (in some cases to zero) residual dextrin content. The dextrins in beer arise because the malt amylases are limited in their action on malt starch (see Chapter 13). However, if enzymes capable of degrading these dextrins are added to the process, the polymers are removed and the sugar produced is fermented by yeast. Enzymes capable of hydrolyzing the α-1–6 and α-1–4 glycosidic bonds of the dextrins are found in unkilned or lightly kilned malt. Extracts of this material may be added during mashing or at the fermentation stage. Alternatively, enzymes produced commercially from fungi or bacteria may be employed (often referred to as amyloglucosidases).

As a rough approximation, the calorific content of beer can be expressed as:

Calories in 10 cl = 4 × % (w/v) solids × 7 × % (w/v) alcohol

In typical beers, residual dextrin accounts for the bulk (75%) of the solids. The calorific value of beer is largely due to alcohol, but dieticians do not consider alcohol to be a carbohydrate. Using enzymes therefore removes the carbohydrate; hence, the beers are low carbohydrate. The calorific value of the alcohol produced by the fermentation of the additional sugars is less than that of the sugars themselves; therefore, the beer would be higher in alcohol but lower in calories (but not by much). In practice, less malt would be used in this process to achieve a typical alcohol concentration in beer, thus further lowering the calorific value. How much reduced in calorific value such beers are clearly depends on the calorie content of the standard brew. In general, most low calorie beers are 20–30% lower in calories than the standard product. The term "lite" is also applied to beers with lower alcohol content.

6.5.4 Low alcohol beers

Unfortunately, world-wide, there is no agreed standard terminology for these products. "No alcohol" beer ("alcohol-free" beer) in most instances will contain less than 0.05% alcohol by volume. A category of beers of alcohol content 0.05–0.5% is referred to in the U.K. as "dealcoholized" and beers with up to 1.2% alcohol as "low alcohol." Additionally, there is a category of beers from 1.2–3.0%, often considered to be reduced in alcohol.

Beers with alcohol content of 2–3% are usually produced by normal processing of appropriate grists or by fermenting weak worts and adding back the volatile constituents collected from the vent of fermentations of high-gravity worts. Beers with alcohol content of 1.2% and below are generally produced using additional technologies to remove alcohol. However, this requires considerable investment in specialized equipment. Smaller brewing operations can achieve these objectives by other techniques. Thus restricting fermentation of normal worts by using yeast unable to ferment maltose is a practical proposition (down to 0.5% ethanol); compared to diluting high-gravity wort before or after fermentation (down to 2% alcohol); or compared to mashing at high temperature (80°C) to produce a wort of low fermentability (1% alcohol); and compared to the cold contact process of mixing wort with yeast from a normal fermentation at low temperature (-1°C) for several days. In this latter process, provided that alcohol-free yeast is used, beers of less than 0.05% alcohol can be produced.

6.5.5 Ice beers

In a patented process introduced in North America, green beer is continuously chilled to form tiny ice crystals in a scraped surface heat exchanger (without any significant beer concentration) and then exposed to beer-ice in a reactor. It is claimed that this rapid process stabilizes the beer physically and matures its flavour.

6.5.6 Removal of alcohol

Three technologies are used to remove alcohol from beers brewed to a normal specification. They are: (1) vacuum evaporation (also called vacuum distillation); (2) dialysis; and (3) reverse osmosis.

Vacuum evaporation

Vacuum evaporation (Figure 6–7) uses plate evaporators, which are similar in appearance to plate heat exchangers. Beer pre-heated to 35°C is passed under vacuum through a plate evaporator heated with steam at 50°C. The evaporated stream passes to a separator where the dealcoholized beer (DAB) is withdrawn and the alcohol-rich vapor passes to a condenser

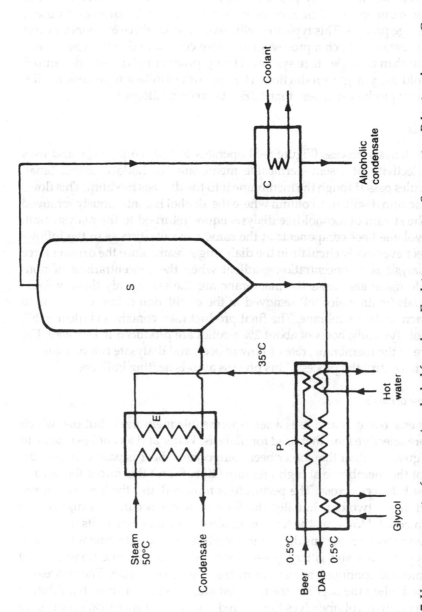

Figure 6–7 Vacuum evaporator for removing alcohol from beer. E, plate evaporator; S, separator; P, heat exchanger; C, condenser; DAB, dealcoholized beer.

to yield an alcohol-rich by-product. The DAB can be re-circulated to obtain further reduction in alcohol content. It is more usual, however, to use a multistage process. This typically will have in series three evaporators and three separators. Such a process gives more control and better energy utilization than a single stage system. The by-product is typically denatured and sold for vinegar production. The vacuum distillation process readily produces product with less than 0.05% by volume alcohol.

Dialysis

The dialysis process (Figure 6–8) operates at low temperature and uses the selectivity of a semi-permeable membrane (or hollow fibers). Small molecules pass through the membrane into the dialysis medium. This flows to a vacuum distillation column where the alcohol is continuously removed and the stream of alcohol-free dialysis liquor returned to the dialysis unit. Non-volatile beer components at the same concentration as in the inflowing beer eventually circulate in the dialyzing stream. Since the driving force for dialysis is a concentration gradient when the concentrations of non-volatile materials across the membrane are the same, only those volatile materials (mainly alcohol) removed in the distillation column continue to pass across the membrane. The final product may contain as little as 0.5% alcohol. Typically, beers of about 2% alcohol are produced in this way. The nature of the membrane, rate of flow of beer, and dialysate can be manipulated to control the process. This process avoids heating bulk beer.

Reverse osmosis

Reverse osmosis also uses a semi-permeable membrane, but one which is more selective than that used for dialysis. Water and alcohol permeate to a far greater extent than other beer components. Beer is passed to the surface of the membrane at high pressure (40 bar) and this forces the permeable substances across. The permeate is removed and the beer is concentrated about twofold. Initially, the flow of water occurs at a higher rate than alcohol. However, the transfer of alcohol increases with its increasing concentration on the inside of the membrane. Pure, demineralized, and deoxygenated water is then passed outside the membrane (diafiltration) and alcohol continues to pass from the beer concentrate. The process is stopped when the beer has the required alcohol content such that dilution to the original volume gives the desired low alcohol specification (usually around 2%). Membrane processes require scrupulous sanitation and regular replacement of membranes which deteriorate and become blocked. Reverse osmosis has the principal advantage that no heat is applied to the system, but it suffers from the disadvantage of operation at high pressure.

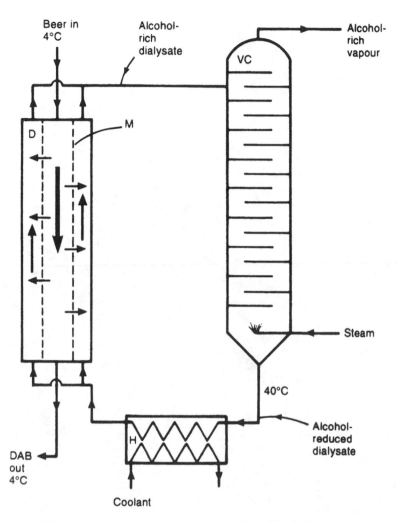

Figure 6–8 Dialyser for removing alcohol from beer. D, dialysis unit; M, membrane; VC, vacuum distillation column; H, heat exchanger; DAB, dealcoholized beer. Arrows indicate directions of flow.

Low alcohol products lack the preservative properties of ethanol and as a result are less stable especially in microbiological terms. Consequently, rigorous CIP and microbiological quality control are needed. The production of low alcohol beers (especially less than 0.05% alcohol) obviously incurs extra cost. However, there is usually a significant saving since tax is not paid at the same rate as for standard beers (less alcohol); moreover, no tax is levied on no-alcohol beers.

CHAPTER 7

Analytical methods and statistical process control

The importance of Quality Assurance has been recognized by the creation of standards under the auspices of the International Organisation for Standardisation (ISO). The ISO 9000 standard operates within the European Community and affects trade within the EC and with the EC and other countries (e.g., U.S.). Implementation of ISO 9000 commits a company to operating a quality management system, to installing the necessary structures of a system, to consider the customers needs, to register in accordance with those needs. and to provide a means of continually improving the system. Registration verifies a commitment to quality and states that a certain level of performance has been successfully implemented. Increasingly. customers expect companies to have their quality systems registered to ISO standards (ISO 9001 being the most demanding and derived from U.K. B55750 version of U.S. MIL-Q-9858).

A fundamental axiom of brewing is that consistent quality of raw materials and consistent reproducible processing lead to consistent product quality.

To assure product quality, therefore, it is necessary to produce a specification for the product and this includes specifications for the raw materials and processes to be used in its production. The quality control process is then designed to monitor the ability to meet the specifications and thus ensure product quality. Total Quality Management embraces far more than the analytical analysis of the product and process; however, only these aspects will be considered here.

7.1 CONTROLLING QUALITY

The specifications set for raw materials define the price the brewer will pay. Paying a high price for inferior quality is clearly bad business and

113

there may well be adverse effects on processing and/or product quality. Malt, hops, and yeast inevitably show variability and must be carefully monitored. Control of process specifications ensures batch to batch consistency and failure in this regard will often result in economic inefficiency and variability in the final product. Routinely meeting the specification of the end product gives a consistency that must impact on sales as well as testifying to the professional competence of the brewer. A well-controlled process yielding consistent quality is rightly the basis of professional pride in achievement.

In developing a quality assurance regime, it is necessary first to define which parameters are important and second to decide on methods to measure and monitor them. The expense of quality control measures can be considerable, so it is important to concentrate effort on those areas having a maximal impact on final product quality where prompt action can be taken to address any problems. A typical example would be rapid microbiological analysis of a cleaned bright beer tank. Identifying key steps and controlling them is the principle behind Hazard Analysis and Critical Control Point (HACCP) strategies.

Although the brewer sets most specifications, some are required by law and must be met. Some examples of where QC methods are applied are given in Table 7–1.

A small selection of important methods is given in Table 7–2. Additional ones can also be found in Table 6–1 Chapter 6. The force of law backs the methods described by the first and last entries in Table 7–2, potability of water and alcohol content of beer respectively.

Many methods of brewing analysis are developed within individual companies and then become established at the national level. So, National or International compendia of methods are published. Table 7–3 lists some of these.

Table 7–1 Some important points for application of quality control

Raw materials	Water, Malt, Hops, Yeast, Cereal adjuncts
Process aids	Primings, Finings, Cleaning agents
The process	Milling (grist composition)
	Mashing (wort specific gravity)
	Fermentation (yeast vitality)
	Conditioning (final specific gravity)
	Packaging (volume)
	Vessels (cleanliness)
The product	Finished Beer (dissolved oxygen)

Table 7–2 Some Important quality control methods

Applied To	Method	Principle
Water	Potability	Microbiological analysis
Malt	Appearance	Experience of Buyer
	Odor	Experience of Buyer
	Hot water Extract	Small scale laboratory mash
Hops	Alpha acid content	Solvent extraction and spectrophotometry
Wort	Specific gravity	Hydrometer or density measurement
	Fermentability	Laboratory fermentation using high yeast concentration
Yeast	Viability	Count stained cells under a microscope
Vessels	Potency of cleaning agents	Measure pH
	Cleanliness	Microbiological analysis
Beer	Alcohol content	Distillation

One very important point to note is that methods used under the different systems to measure the same parameter are extremely difficult (in some cases impossible) to compare. This is partly because different units of measurement are used, but mainly because the methods differ in execution. As a means to eliminating such problems, the various agencies listed in Table 7–3 have agreed and recommended that some methods be designated as International Methods.

Notwithstanding this however, many large companies use specific in-house methods for their quality control. This is justified on the grounds that the method must be tailored to their specific brewing process. Thus, they either wish better to match the production process or there is no standard method for a parameter of importance to them (e.g., dimethyl sulphide concentration is important in some beers but not in others). Clearly when buying raw materials, the brewer must use the same methods to

Table 7–3 Compendia of analytical methods

Origin	Compendium of Methods
U.S.	American Society of Brewing Chemists, Methods of analysis Laboratory methods for craft brewers (ASBC)
Europe	Analytica—European Brewing Convention
Germany	Methodensammlung der Mitteleuropäischen Analysenkommission (MEBAK)
UK	Recommended Methods of the Institute of Brewing

evaluate quality as those used by the supplier even though different specific in-house methods may be used in production.

7.2 ANALYTICAL METHODS

The methods used should be simple to conduct, accurate, reproducible, and inexpensive. The brewer then needs to establish how reliable (accurate and repeatable) the method is in his/her hands. Then the method is used over a period of time and the values obtained recorded. In this way, the variability in actual values can be used to set realistic quality control values to be used in drawing up specifications. Of course, this can only happen when using the method shows that the parameter measured has a low variability. Established methods are then compared between laboratories to establish their wider validity.

7.3 ACCURACY AND PRECISION

How is the precision of a method evaluated? First of all, consider what an analytical method is desired to achieve. We could say that we want it to give

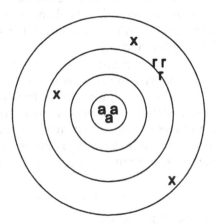

Figure 7–1 Accuracy and precision. Different methods a, p and x are used each to make three replicate measurements. An accurate analytical method is on target (close to the "true" value) and replicates are closely grouped (a). A precise (repeatable) method (p) gives good replicates (closely grouped) but the measurements are offset by a consistent amount from the "true" value. Unreliable methods give scattered vaues both in relation to replicates and the "true" value (x). In practice a method giving results like (a) is the ideal but one giving values like (p) may be acceptable. A method giving values represented by (x) is useless.

an accurate measure of the true value every time we make a measurement. Figure 7–1 illustrates the concepts of accuracy in terms of a target. Any of the shots labeled **a** is accurate because it falls within the central area. If an archer shot three times in succession giving all three **a** hits on the target, we would say that it was excellent shooting. The archer shows accuracy and close grouping. We could add that the close grouping shows the archer's action is repeatable—all arrows hitting the target close to each other. A high quality analytical method has similar characteristics. That is to say, it is both accurate and repeatable. However, like the example, it is not perfect and contains error (variation) displayed by the fact that the three arrows are not in the same place. Error-free archery would see the second arrow sticking the end of the first and the third in the end of the second—an extremely unlikely (therefore improbable) event. Measurements with an error-free analytical method would all be the same (also a highly improbable event).

The shots represented by **r** in Figure 7–1 are a fixed distance from the target, but show close grouping so the archer's action is still repeatable. In this case, it is as if the archer needs to make an adjustment to the sights. In terms of brewing methods, one showing a fixed bias like this can be acceptable. The fact that it is off-target does not matter so long as it is consistently so (an example could be measurement of hop α-acids by solvent extraction, not only the α-acids are extracted and measured). Accuracy is, however, very important in inter-laboratory trials. Here, although different laboratories may precisely determine the value of the parameter, they may be off target in different ways leading to overall discrepancies in analysis. A method is said to be reproducible when different analysts in different laboratories get values closely grouped around the target.

Shots **x** in Figure 7–1 exemplify what cannot be tolerated in a method. Here, the archer is neither accurate nor able to repeat the shots and in terms of an analytical method, such a result is useless.

No method is perfect; all are subject to error (variation). What is needed is a method of evaluating the error or variation in measurement. To do this, we must enter the area of probability and statistics.

7.4 STATISTICAL ANALYSIS OF ERROR

The sources of error in any analytical process can be described in the following way:

Total error = sampling error + systematic error (bias, reproducibility)
      + repeatability (replicative) error.

Total error can be minimized by taking steps to eliminate sampling and systematic errors.

Sampling error often arises from the fact that actual measurements are made on a small sample from a large batch. Examples would be a kilogram of malt from a rail car full or a liter of wort from hundreds of hectoliters. Sampling error can be minimized by adopting standardized sampling procedures, e.g, taking several samples from different positions in the rail car and making them homogeneous (mixing). Compendia of methods (Table 7–3) also include detailed instructions for taking samples and thus minimizing sampling error. In inter-laboratory trials, the batches of the same sample are sent to all the laboratories involved.

Systematic error or bias (compare **r** versus **a** in Figure 7–1) results in values being off-target. Frequently they result from poorly calibrated instruments, failure to control potential variables such as temperature, or may reside in the experimental technique of a particular analyst. Competent analysts learn by training and experience to minimize such errors.

It is standard practice for possible sources of error from bias (systematic or reproducibility error) to be identified and minimized by carefully defining the protocol, equipment, etc., to be used. This is known as establishing the robustness of the method and is a key objective for interlaboratory trials.

Repeatability (replicative) error is the random occurrence of variation in results that cannot be explained by sampling or systematic errors. The lower this error, then the smaller the variation in the results obtained from repeated analysis of the same sample.

Taking the mean or arithmetic average ($\bar{x}$) of a series of measurements gives the best estimate of the "true" value of an analysis.

$\bar{x}$ = Sum of all measured values/number of measurements.

In mathematical notation this is shown for any mean as:

$$\bar{x} = \frac{1}{n} \sum_{i=1}^{i=n} x_i \qquad \text{[Equation 1]}$$

Where $\sum_{i=1}^{i=n} x_i$ is the sum of all measured values from the first, $x_i = x_1$ to the last $x_i = x_n$, and n is the total number of measurements.

EXAMPLE 1

The following 10 measurements of beer bitterness were obtained: 21.7, 22.9, 20.3, 21.5, 21.6, 22.2, 19.5, 20.7, 21.8, 19.6

Average value $\bar{x}$ = (21.7 + 22.9 + 20.3 + 21.5 + 21.6 + 22.2 + 19.5 + 20.7 + 21.8 + 19.6) / 10 = (211.8)/10 = 21.18

The best estimate of the bitterness (value rounded up) is therefore 21.2. Intuitively, we would expect a random variability of the individual measurements about the mean. We would also expect that small differences would be more likely to occur than large ones, and the more measurements we make, the closer to the true value the mean will be (as the number of measurements approaches infinity, so the closer the mean becomes to the "true" value).

In statistics, a distribution of this sort is termed a normal distribution. Such distributions are quite common (e.g., IQ measurements) and apply to continuous variables like results from an analytical method. A normal frequency distribution is shown in Figure 7–2 and has the form of a bell-shaped curve. The frequency of occurrence of a value (number of times it occurs as a fraction of all values) is plotted on the y axis and the value on the x axis. The area of the curve encloses all possible frequencies and indi-

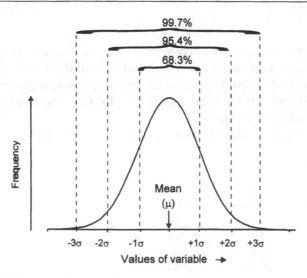

Figure 7–2 The normal curve. See text for discussion.

vidual values extend from −infinity to +infinity. The mathematical relationship which gives rise to this distribution is:

$$y = 1 \backslash \sigma \sqrt{2\pi}\, e^{-(x-\mu)^2/2\sigma^2}$$

where y is the height of the curve at any point x along the scale of the variable (x axis), σ is the standard deviation of the population, μ is the average value of the variable (x) for the distribution and π is the mathematical constant.

The standard deviation is directly related to the spread of the distribution: 68.3% of all values lie within $\pm\,1\sigma$ of the mean, 95.5% of all values within $\pm\,2\sigma$ and 99.7% within $\pm\,3\sigma$ (Figure 7-2). The standard deviation, therefore, gives a number, derived from a series of replicate observations, which defines the distribution of the variation of values. The standard deviation can be used to indicate the variation in an analytical method. The smaller the variation, the smaller the standard deviation, and so the less variation (error) there is in the method. Clearly, it is not possible to make an infinite number of measurements so our statistical analysis will be based on a number of measurements and a best estimate S made of σ.

The equation for calculating the best estimate S of the standard deviation (σ) of a number (n) of measurements is:

$$S = \sqrt{\frac{1}{(n-1)} \sum_{i=1}^{i=n} \left(x_i - \bar{x}\right)^2} \qquad \text{[Equation 2]}$$

Most modern calculators and computer spreadsheet programs contain built in functions that make the calculation of standard deviation very straightforward. Many packages give the option of using n or $(n-1)$ in the calculation, the latter is the most appropriate for the sample sizes (small) usually encountered in brewing.

Returning to the example of beer bitterness to calculate S.

EXAMPLE 2

x	$(x - \bar{x})^2$
21.7	0.2704
22.9	2.9584
20.3	0.7744
21.5	0.1024

EXAMPLE 2 CONTINUED

	x	$(x - \bar{x})^2$
	21.6	0.1764
	22.2	1.0404
	19.5	2.8224
	20.7	0.2304
	21.8	0.3844
	19.6	2.4964
Sum	211.8	11.256
Mean ($\bar{x}$)	21.18	

So $S = \sqrt{\dfrac{1}{9} \times 11.256} = 1.12$

The best estimate of the mean (rounded up) was 21.2 and the best estimate of the standard deviation 1.1 (rounded down).

We can now express our results for bitterness of the beer sample as mean ± 1S or 21.2 ± 1.1.

In practical terms (Figure 7–2) we can say that if we made further measurements we would expect nearly all of them (99.7%) to lie within ± 3S (3 × 1.1 or 3.3) of 21.2 units; 95.4% would lie within ± 2S (2.2). Thus, by chance it would be unlikely that two consecutive values would differ by more than 2.2 units from the mean and would therefore lie in the range 18.0 to 23.4.

By chance alone, we would only expect a single value to deviate more than 2S from the mean one measurement in 21 (100/4.6) and for 2 consecutive measurements once in 441 (21 × 21).

Table 7–4 gives the probability that a measurement made at random from a normally distributed population of measurements will fall outside a given number of standard deviations from the mean.

Table 7–4 Probability values for the normal distribution

Value of S	Probability of obtaining a value outside Mean +S **or** mean −S	Probability of obtaining a value outside Mean ± S
1.96	0.025 (1/40)	0.05 (1 in 10)
2.57	0.005 (1/200)	0.01 (1 in 100)
3.2	0.0005 (1/2000)	0.001 (1 in 1000)

Although S is the best estimate of the error, many analysts prefer to express it as a percentage of the mean and call this the coefficient of variation (CV) of the method.

$$CV = 100 \times \frac{S}{\bar{x}}$$ [Equation 3]

In the example, the CV for bitterness is $(1.1/21.2) \times 100 = 5.2\%$. This value makes it easier to compare the precision of methods because methods with the same CV have the same degree of repeatability (the variation about the mean is the same), even though the absolute values of S and the mean differ.

The statistic gives us a clear indication of the variation in the method used. The smaller the standard deviation (and therefore CV), the more repeatable the method. This is illustrated in Figure 7–3 where two normal distributions of the same mean and different standard deviations are shown. In practical terms, there is no point setting a specification tighter

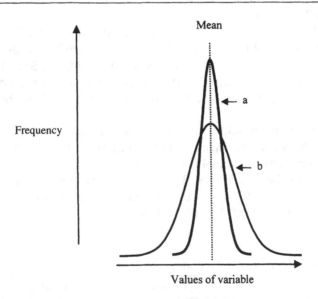

Figure 7–3 Normal distributions for two different populations of values. The two distributions have the same mean but different standard deviations (spread). If they represented the distributions of results from two different methods analysing the same parameter then the method giving curve a is preferred because it has a smaller spread (smaller standard deviation) than b.

than the variation in the method will allow. If this is done, it will be impossible to maintain the specification because chance variation will be greater than that set by the specification.

If the variation of the method is larger than desired, then the only recourse is to refine the method or seek an alternative with a lower amount of variation. Ideally, we would look for a CV value as low as practicable. Many analysts would consider 3% to be a desirable value. In this case 99.7% of observations would fall within ±9% of the mean.

Given that S is known for an analytical method, we know what variation (from random, non-subscribable causes) to expect when we make further analyses (under the same conditions). However, making judgements on the basis of single measurements is not ideal. Any one measurement may lie toward one extreme end of the distribution. If several measurements are made, then it is extremely unlikely that all of them will lie at the extreme. So, if we take a sample of, say, 4 or 5 measurements and take the average of these as our measured value, then we shall have a more reliable indicator. Means of samples of 4 or 5 will vary with each sample taken but the spread of the variation will not be as great as for individual measurements.

It can be shown that the standard error (measure of spread) of means

$$\text{SEM} = \frac{S}{\sqrt{n}}$$ [Equation 4]

Where S is the standard deviation calculated from the values of all individual measurements. So for $n = 4$, the SEM is half of the standard deviation obtained using individual measurements. The take-home message is that when critical judgments are to be made, it is better to use the average values of a small number of samples rather than rely on individual measurements because a more sensitive analysis results. Such an analysis is also more satisfactory because even if the individual values are not normally distributed, it can be shown that the means of samples as small as 4 are (very nearly). This is a central tenet of statistics called the Central Limit Theorem. The application of this knowledge underpins the use of Quality Control Charts (see below).

Computers and spreadsheets make it easy to calculate S. However, a simple calculation based upon the range of the values obtained is an alternative. So when small sample numbers are used, the range of the samples (difference between highest and lowest values) divided by a factor d will estimate S.

$$S = \frac{R}{d}$$ [Equation 5]

Table 7–5 includes values of d for different sample sizes. As an example, for the first 5 samples of bitterness discussed above, the lowest value is 19.5 and the highest 22.2, so the range is the difference between these namely 2.7. Taking d (from Table 7–5) for 5 samples as 2.326, then the estimated standard deviation is $2.7/2.326 = 1.16$ sufficiently close, for practical purposes, to that calculated from Equation 2.

If we take all 10 values the lowest is 19.5 and the highest 22.9 the range is therefore 3.4. The appropriate value for d is 3.078 thus the estimated standard deviation $3.4/3.078 = 1.1$ (the same as calculated using Equation 2).

7.4.1 Interlaboratory collaborative trials

Method development either between different laboratories of one company or different companies at the international level requires evaluation between, as well as within, laboratories. While it is reasonable to expect one trained analyst in one laboratory to show a high degree of accuracy and precision, we would expect that the accuracy and precision will vary between different analysts and laboratories. Collaborative trials are used to check the reliability of methods between different laboratories and identify any problem areas that can be resolved by further development.

The practice currently adopted by the Institute of Brewing and the European Brewery Convention (based on ISO 5725) determines the repeatability and reproducibility for a standard test method by inter-laboratory tests. Careful preparation and distribution of samples is used to effectively eliminate sampling error and, as far as possible, all of those parameters whose proper control is critical to the performance of the method have been identified. In this way, errors due to bias are also minimized.

The precision of a method is defined both in terms of its repeatability and reproducibility. Repeatability corresponds to the within laboratory

Table 7–5 Statistical parameters for establishing action and warning limits on control charts

Sample size (n)	d	A_2	$D_{0.001}$	$D_{0.025}$	$D_{0.975}$	$D_{0.999}$
2	1.128	1.880	0.00	0.04	2.81	4.12
3	1.693	1.023	0.04	0.18	2.17	2.98
4	2.059	0.729	0.10	0.29	1.93	2.57
5	2.326	0.577	0.16	0.37	1.81	2.34
6	2.534	0.483	0.21	0.42	1.72	2.21
7	2.704	0.419	0.26	0.46	1.66	2.11
8	2.847	0.373	0.29	0.50	1.62	2.04

variation and this, when combined with between laboratory variation, gives the reproducibility. The standard deviations for within laboratory (Sr) and between laboratory S_L are calculated. These are then combined to give the reproducibility standard deviation S_R.

Two types of design of collaborative trial are specified, one using replicate analysis of a single sample or two or more identical samples, each analyzed once. This is known as a uniform level experiment. The second, the split level experiment, analyzes two similar but slightly different samples, each analyzed once. The latter is identical with the procedure recommended by the American Society of Brewing Chemists.

Data is first analyzed for outliers (values which are statistically improbable), which are excluded from further analysis. Then the repeatability and reproducibility standard deviations are calculated.

For a uniform level experiment, the repeatability standard deviation Sr is given by:

$$S_r = \sqrt{\frac{\Sigma \, d_i^2}{np}} \qquad \text{[Equation 6a]}$$

Where d_i is the difference in each pair of results, n the number of replicate samples and p the number of laboratories.

For a split level experiment

$$S_r = \sqrt{\frac{\Sigma \, (d_i - \bar{d})^2}{n(p - 1)}} \qquad \text{[Equation 6b]}$$

where $\bar{d}$ is the mean difference.

EXAMPLE 3

Consider 4 laboratories receiving a sample of beer for analysis of diacetyl content. Each measures the diacetyl content of 2 replicate samples (a and b) and reports the results shown below (Table 7–6). This is then a uniform level experiment and Equation 6a is used.

Summing the data in the d_i^2 column we have 0.0055, the number of replicates n is 2 and the number of laboratories p is 4, substituting these values in Equation 6 gives the within laboratory standard deviation S_r:

$$S_r = \sqrt{\frac{0.055}{2 \times 4}} = \sqrt{0.0006875} = 0.0262$$

Table 7–6 Results of interlaboratory trial of diacetyl measurement and calculation of d_i^2

	Diacetyl content mg/liter			
Laboratory	a	b	d_i	d_i^2
1	0.15	0.10	+0.05	0.0025
2	0.10	0.09	+0.01	0.0001
3	0.20	0.15	+0.05	0.0025
4	0.08	0.10	−0.02	0.0004

The between laboratory standard deviation S_L is calculated using the formula:

$$S_L = \sqrt{\frac{\Sigma\,(y_i - \bar{y})^2}{(p-1)} - \frac{S_r^2}{n}} \qquad \text{[Equation 7]}$$

Where y_i are the average values of each pair of measurements and $\bar{y}$ is the average of all the values (0.12125). Returning to our example (Table 7–7): Entering the sum of $(y_i - \bar{y})^2$ (0.005531), $n = 2$ and $p = 4$ into Equation 7:

$$S_L = \sqrt{\frac{0.005531}{3} - \frac{0.0006875}{2}} = \sqrt{0.00149999} = 0.0387$$

It is expected that more variation is seen between rather than within laboratories ($S_L > S_r$; different analysts, different equipment), but they should be of a similar order. If S_L is several times larger than S_r, then there is a problem in the process requiring further investigation.

The overall standard deviation of the reproducibility of the method S_R takes into account **both** within and between laboratory variation and is calculated from the S_r and S_L values:

$$S_R^2 = S_L^2 + S_r^2 \qquad \text{[Equation 8]}$$

Table 7–7 Results of interlaboratory trial of dicetyl measurement and calculation of $(y_i - y)^2$

	Diacetyl content mg/liter		y_i	$y_i - \bar{y}$	$(y_i - \bar{y})^2$
Laboratory	a	b		(a + b)/2	
1	0.15	0.10	0.125	+0.00375	0.000976
2	0.10	0.09	0.095	−0.02625	0.000689
3	0.20	0.15	0.175	+0.05375	0.002889
4	0.08	0.10	0.09	−0.03125	0.000977

For the example:

$$S_R^2 = 0.00149999 + 0.0006875 = 0.0021874$$

$$S_R = \sqrt{S_R^2} = \sqrt{0.0021874} = 0.047$$

For most practical purposes S_R is the value to use.
Following convention, CV_R and CV_r are calculated as:

$$CV_R = 100 \times \frac{S_R}{\overline{x}} \text{ and } CV_r = 100 \times \frac{S_r}{\overline{x}} \qquad \text{[Equations 9a and 9b]}$$

In Example 3, the CV_r and CV_R for the diacetyl determination are 21.6 and 38.8, rather too large for comfort!

An alternative approach to presenting variation in terms of repeatability and reproducibility is to give them at the 95% confidence level, r_{95}, and R_{95} respectively.

These are obtained by multiplying S_r and S_R by the factor 2.8.

Accordingly for the analysis of diacetyl, the repeatability value (r_{95}) is $2.8 \times 0.0262 = 0.07$ and the reproducibility value (R_{95}) is $2.8 \times 0.047 = 0.13$

The r_{95} is most easily understood as the maximum acceptable difference between repeat determinations at the 95% confidence level (on average not expected to occur more than once in 20 measurements). R_{95} is the maximum acceptable difference between single determinations in 2 different laboratories at the 95% confidence level.

CV_r and CV_R for the diacetyl determination are 21.6 and 38.8, much too large for comfort!

In measurements of chemical constituents of beer (e.g., alcohol, diacetyl), the statistical procedure also tests whether or not the values obtained are dependent or independent of the concentration of the constituent. If independent of concentration, the results are reported as a single r_{95} or R_{95} values as in the example above. Where there is a dependence on concentration, this is reported with the results and the nature of the dependence in the form of an appropriate equation is given.

For example, $r_{95} = 0.3m$ means that there is a linear dependence of r_{95} on the concentration (m) of the substance measured. This means that the graph of r_{95} versus concentration has a slope of 0.3 and passes through the origin (point 0,0). The appropriate r_{95} for the assay is therefore obtained by multiplying m by 0.3.

So, if the r_{95} referred to assay of ethanol in the range of concentrations 30 to 50 mg per liter, then the r_{95} for a concentration of 30 mg per liter would be 9 (0.3×30). For an alcohol concentration of 50 mg per liter, r_{95} would be 15. Other relationships may also be reported so that the appropriate r_{95} can be determined.

7.4.2 Control Charts

The standard deviation and mean of a normal distribution can be applied to the development of control charts for production processes. The most commonly used quality control chart is the mean and range chart. This type of chart was originally designed for process control in manufacturing engineering products. In such manufacturing processes, it is a simple matter to take a sample of product of five items (e.g., bolts) at regular intervals and measure some property such as length or diameter. In brewing, equivalent processes would be bottles of beer from a packaging line (e.g., for measuring fill height).

For many brewery operations only single samples are available. Under these circumstances an individuals chart is used.

Mean and range charts

The layout of a mean and range chart is shown in Figure 7–4. The relationship of the mean chart to the normal distribution is shown by includ-

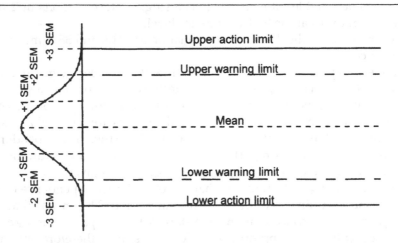

Figure 7–4 The parameters of a Process Control Mean Chart. See text for discussion.

ing the normal curve at the left-hand side (this would not be present on an actual chart). It is important to remember that this is the distribution

$$\left(\text{SEM} = \frac{S}{\sqrt{n}}\right)$$

of the means of samples, **not** that (S) of individual values. Action and Warning Limits are placed by drawing lines at ± 3 and ± 2 SEM about the mean value (Figure 7–4). Above the mean line are the upper warning limit and upper action limit, UWL and UAL (also known as upper control limit, UCL); these are set at +2 SEM and +3 SEM respectively. Below the mean line are two additional lines representing the lower warning limit (LWL) and lower action limit (LAL) at −2 SEM and −3 SEM respectively.

To aid calculation of the action and warning limits the following equations may be used

$$\text{UAL} = \bar{x} + \left(\frac{(3 \times S)}{d\sqrt{n}}\right) \text{ and LAL} = \bar{x} - \left(\frac{(3 \times S)}{d\sqrt{n}}\right) \qquad \text{[Equations 10a and 10b]}$$

To enable the range to be used to estimate S, these are commonly abbreviated to

$$\text{UAL/LAL} = \bar{x} \pm A_2\bar{R} \qquad \text{[Equation 11]}$$

Where for each of m samples of n measurements,

$$\text{The mean range} = \bar{R} = \sum_{i=1}^{i=m} \frac{r^i}{m} \qquad \text{[Equation 12]}$$

r_i is the range (difference between the highest and lowest value) within sample i and $i = 1$ to m; and A_2 is

$$\frac{3}{d\sqrt{n}}$$

d can be obtained from Table 7–5 but calculated values of A_2 are also given in the table.

Example 4

Consider 4 samples each of 4 measurements, i=1 to 4 and m=4 (Table 7–8):

Table 7–8 Calculation of grand mean and mean range for four samples each of four measurements. The grand mean is the sum of each of the sample means divided by the number of measurements (4) in each sample. The mean range is the sum of each of the sample ranges divided by the number (4) of measurements in each sample

Sample Number (i)	Measurements				Means	Ranges (r_i)
1	2.1	2.8	2.5	2.9	2.575	0.8
2	2.2	2.3	2.6	2.5	2.4	0.4
3	1.9	2.5	2.6	2.4	2.35	0.6
4	2.3	2.6	2.4	2.1	2.35	0.5
				Sum of means	9.675	Sum of ranges 2.3
				Grand Mean	2.41875	Mean Range 0.575

The Grand Mean is 2.42 (rounded up) and the Mean Range is 2.3/4 = 0.575

The UAL/LAL, for A_2 n = 4, 0.729 (Table 7–5) are:

0.729 × 0.575 = 0.42. On the mean chart, the Mean Line would be placed at 2.42, and the UAL at (2.42 + 0.42) 2.84, with the LAL at 2.00.

In some charts, the action limits are placed at 3.09 × SEM and the warning limits 1.96 × SEM; 99.8% and 95.0% of all measurements respectively are expected to lie within these lines. Accordingly, 0.2% (100–99.8) of all values are expected by chance to lie outside the UAL and LAL. On average, half of these (0.1%) will be above UAL and half below. Similarly for UWL and LWL, by chance we expect 5% (100–95) of all values will lie outside the limits with half of these (2.5%) being above the UWL and the other half below the LWL.

The action and warning limits for a range chart are calculated from the mean range ($\bar{R}$; Equation 12) by multiplying by a factor $D_{0.001}$ and $D_{0.025}$ for LAL and LWL respectively, and $D_{0.975}$ and $D_{0.999}$ for UWL and UAL respectively. The values of D are given in Table 7–5.

Example 5

Using the data in Example 4 and taking the values of $D_{0.999}$ and $D_{0.001}$ from Table 7–5, the UAL is 0.575 × 2.57 = 1.48. On a range chart then, the Mean Range line would be at 0.575 and the upper action limit line 1.48. The LAL is at 0.0575 (0.1 × 0.575).

Examples of a mean and range chart are presented in Figure 7–5 and Figure 7–6 respectively and are discussed below.

Why monitor both mean and range? Mean without range shows changes in the parameter measured. Range without mean shows change in distribution (increased variability around the same mean). Plotting both gives more information on the process under observation.

The data for calculating the mean, range, and warning and action limits are obtained by analyzing a process assumed (expected) to be operating normally showing only random variation. Typically, 25 samples each made up of 4 to 6 measurements will be taken. Plotting the data on the mean and range chart created will indicate whether or not all data points lie within the expected limits, that is, whether or not the process was under "statistical control." If the process is not in control, then further analysis must be undertaken. In many brewing operations, obtaining a process operating under control may not be as easy to achieve as one might think.

The process of selecting the data and choosing the most appropriate control chart is the province of a statistician. Badly designed control charts lead to confusion, dissatisfaction, and eventually the discrediting of Statistical Process Control techniques and this does nothing to improve process and product quality assurance.

Using the mean and range chart, a process is in statistical control when all of the following conditions apply on the mean and range charts:

1. No mean or range values lie outside the action limits
2. No more than about one value in 40 lies between either the upper warning and action limits or the lower ones
3. There is no instance of two consecutive mean or range values lying outside the same warning limit
4. There will be no run or trend of five or more that infringe a warning limit
5. No run of more than 6 sample means lies above (or below) the grand mean
6. There are no trends of more than six values of the sample means which are either rising or falling.

Inspection of Figure 7–5 (a mean chart) shows that the process monitored is under control for the first 9 samples (even though sample 7 falls below the Lower Warning Limit). The rising trend of values from samples 7 to 11 does not warrant any action because the UWL is not exceeded. Action must be taken at sample 10 because this is the seventh consecutive value below the mean and at sample 15 because a value exceeds the

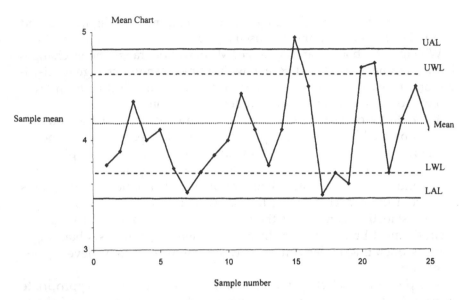

Figure 7–5 Examples of process control mean. See text for interpretation.

Upper Action Limit. Action is also needed at sample 22 where two consecutive values have exceeded the Upper Warning Limit. Also of concern are the observations that three values in 25 lie between LWL and LAL and two in 25 between UWL and UAL (exceeding the requirements of condition 2 above). Generally speaking this process is not under good statistical control.

The corresponding range chart (Figure 7–6) shows that action must be taken at sample 19 where the upper warning limit is exceeded. Note that the mean chart showed no problem with this sample. This type of situation indicates that the spread of samples (range) is unacceptable even though the mean is fine and illustrates the advantages of plotting both types of chart. A comparison of both charts clearly shows that, for the same sample data, the mean chart is much more sensitive to process variation than the range chart.

The mean and range charts enable the brewer to establish that a process is under statistical control That is to say, the only variation is that due to non-subscribable causes (random chance variations). This does not, however, show whether or not the process is running within specification unless the specification has the same mean and standard error as the process. In such a case, the process mean coincides with the midpoint of

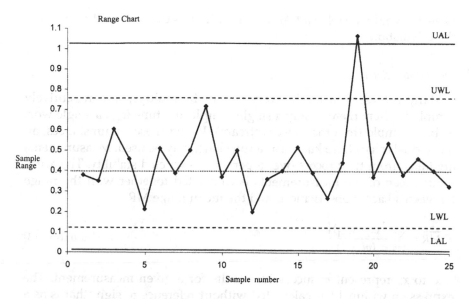

Figure 7–6 Examples of process control range. See text for interpretation.

a desired specification range, so setting the specification at mean $+/-$ UAL is feasible. However, it would be more efficacious to set them at some value slightly larger. How much larger should it be? The concept of the Capability Index (CpK), a measure of the degree to which the bandwidth of the specification covers that of the process, is used to resolve this issue.

CpK = Specification Range/UAL-LAL. This has a value of 1 when the specification range exactly matches the process range (UAL to LAL). For example, if the process had a mean of 100 and SEM of 2, then the UAL is 106 and the LAL 94. So, a CpK of 1 means that the Specification Range is 12 (mean $\pm$ 3 SEM). In some processes, brewers set the CpK at a value higher than 1, ensuring that the specification is protected from normal process variation.

Obviously, there is no point in setting specifications that cannot be achieved in practice (band width less than 1).

When the process mean is not in the midpoint of the desired Specification Range, then either UAL or LAL will be closest to the specification limits. In this situation, then the closest value is used to set the specification. If the UAL is closest, then Specification Limit = process mean − UAL / 3 × process SEM. In situations of one-sided specifications, e.g., maximum haze, then only the relevant Action Limit (in this case, the UAL) is used in the calculation.

Individuals chart

Many processes in the brewery cannot be multiply and repetitively sampled. Often, there is only a single sample at a time (eg., a single wort or beer sample from the process stream). Under these circumstances, an individuals chart (also known as a run chart) may be used. Measurements are taken from the process over a period of expected stability. The arithmetic mean of the measurements is calculated together with the range between adjacent observations and the mean range $\overline{MR}$.

$$\overline{MR} = \sum_{i=2}^{i=m} \frac{|(x_i - x_{i-1})|}{(m-1)} \qquad \text{[Equation 13]}$$

x_i to x_m represent m successive results for a given measurement. The expression within $||$ is calculated without reference to sign (that is as a difference, not an absolute amount).

Example 6

Consider 5 successive measurements of a beer parameter (Table 7–9).

So $m = 5$, $|x_i - x_{i-1}| = 2.7$ and therefore from equation 13, $\overline{MR} = \dfrac{2.7}{4} = 0.675$

Action lines are placed at $\pm 3 \times \overline{MR} \times \dfrac{1}{d}$ and warning lines at $\pm 2 \times \overline{MR} \times \dfrac{1}{d}$.

Table 7–9 Table of values of five successive measurements of a beer parameter and calculation of the differences ($|(x_i - x_{i-1})|$) in successive measurements

x_i	Measurement	$|x_i - x_{i-1}|$
x_1	5.2	
x_2	4.1	1.1
x_3	3.8	0.3
x_4	4.2	0.6
x_5	3.5	0.7
		Sum = 2.7

d is taken from Table 7–5 for a sample size of 2 (1.880). The mean is calculated in the usual way. It is very important to note that unlike the mean and range chart (discussed above), the action and warning limits are based on S not SEM. Accordingly, it is less sensitive and not as good at detecting small changes in process centering. On the other hand, it is simple and does indicate changes in the mean level (accuracy). The layout of the chart is the same as Figure 7–5.

The "rules" dictate that action is taken if any points lie outside the 3S limits, two out of three successive points lie outside the 2S limits or there are eight points in a run on one side of the mean. Because of the relative insensitivity of these charts, lines at ± 1S are also often included.

The preceding discussion has dealt with methods applicable to analyses giving results as continuous variables, eg, pH, color, haze, etc. Some analyses consider attributes, for example, flavor, properties that a product either has or doesn't have. Different statistical process control methods must be applied under these circumstances.

Statistical analysis has application in other areas of brewing to evaluate principal factors affecting beer quality and in the detailed analysis of flavor (see Chapter 6). Techniques such as ANOVA (analysis of variance) and analysis of principal coordinates used to assign the relative importance of different factors may be used. These are outside the scope of this introductory chapter but can be found in any advanced statistical text.

CHAPTER 8

Cleaning and sanitation

8.1 INTRODUCTION

In almost all circumstances, **cleaning** and sanitation must be maintained as separate and different (though complementary) processes and technologies. Cleaning precedes sanitation and prepares the way for sanitary treatment. This is the case whether cleaning is by simple manual methods using bucket and brush or by sophisticated CIP (cleaning-in-place) methods. In a few cases, a combined cleaner-sanitizer approach might be possible or even desirable, but it is essential to correctly analyze such circumstances to avoid costly mistakes.

Modern CIP systems are preferred to manual cleaning in all but the smallest and simplest breweries because aggressive cleaning regimes can be used in relative safety and CIP can be set up to conserve water, energy, and cleaning chemicals. Such a system is assumed in the following discussion. One notes in passing that where manual cleaning is done, ready access to the equipment, ease of dismantling and reassembly, and ruggedness are necessary features of plant design.

The word sanitation has a wider sense than that used above. It also implies a healthful (Latin: sanitas) or wholesome and holistic approach to processing. In this wider context, sanitation applies to the whole spectrum of food plant operations encompassing even the location, orientation, and landscape choices of the plant. It includes control of rodents and other mammals, insects and birds that might enter the building and compromise the raw materials, process, or product. It includes proper choices for layout and construction of the plant, ship-shape organization of the brewery and its cleanliness, training of operatives, and quality of raw material (especially as a vector for contamination). The term "good housekeeping" is often applied to these sorts of concerns. The need for this kind of control in a brewery is most clearly seen around grain and spent-grain handling facilities, and indeed sanitation inspectors might well spend most of their time at these locations. This broader context of sanitation is vital but outside the scope of this short chapter.

8.2 THE CLEANING SYSTEM

8.2.1 The CIP system

A CIP system comprises a set of tanks that holds cleaning and sanitizing chemicals, can maintain them at required strength and temperature, and delivers these solutions in necessary volume through pipes and pumps on an automatic cycle in a desired sequence when required. A system might also be designed to recover rinse waters for reuse, e.g., post-cleaning rinse water for pre-rinse use.

CIP systems work on one of two general principles: either high volume/low pressure or high pressure/low volume. The latter is popular because of the effectiveness of mechanical energy and the smaller volumes involved; however, the delivery system (a rotating gun, for example) is more complex and costly than a spray ball. Note that neither system works well without proper maintenance. In some cases, non-CIP parts must be removed and cleaned separately. Some equipment must be more completely disassembled for cleaning-out-of-place (COP) or manually.

Though the specifics of time, cleaner concentration and kind, and temperature (see below) vary from application to application, the **sequence** of an ordinary cleaning cycle is universal.

1. As soon as practicable after emptying, the tank is first opened and made safe for cleaning, e.g., by exhausting CO_2 in the case of alkaline cleaning and removing non-CIP parts.
2. Surfaces are then rinsed with warm water to remove the bulk of loose soil present and the rinse-water is flushed to drain.
3. This is followed by the cleaner, commonly an hot alkaline cleaner applied for a prescribed time, which is recovered for reconditioning and reuse.
4. The surface is then rinsed (at the same temperature as the cleaning cycle) to remove cleaner and suspended soil and can be recovered for re-use. The process can end here or:
5. This can be followed by an acid rinse to neutralize remaining alkali and to control stone; alternatively, acid might be used, e.g., every third clean or on a weekly basis depending on the number of uses of the vessel. Then final rinse.

Cleaning is followed by sanitizing. Generally, cleaning is done as soon as possible after emptying and sanitizing is done immediately before filling. There may be a gap of some time between the two events. If the gap is large (24 hours, for example), the tank should be recleaned and then sanitized before use.

Note that if a tank is not vented and alkaline cleaners are sprayed into a CO_2 atmosphere or (alternatively) cold solutions are sprayed into a hot tank, the vessel will implode.

8.2.2 Beer contact surfaces

Cleaning success depends in large part on a cleanable surface, i.e., one that is fully accessible to cleaning solutions, smooth, and resistant to aggressive cleaning agents. In most modern settings, this requirement is met by fabrication of pipe-systems and tanks and other beer-contact surfaces with stainless steel with an appropriate finish. Pipe systems must be welded and/or assembled with sanitary fittings, be free of sags and dead-ends (short sections of pipe with no outlet), and tanks must be free of cleaning shadows, e.g., below manways or probes. The CIP system must be able to deliver cleaning solutions with sufficient volume and velocity to clean the surfaces and suspend and carry away the soil.

Beer contact surfaces must be cleaned to eliminate soil as a potential beer contaminant and as a harborage for microbes and to prevent the transfer of flavors and colors, etc., among batches of product that flow through the same system. Cleaning also assures plant efficiency, e.g., the effectiveness of heat transfer surfaces by removing soil or fouling that impedes flow of heat. Cleaning also creates a surface that is capable of being sanitized because it is free of soil that would protect microbes and otherwise prevent proper action of the sanitizer. Clean food contact surfaces are required by law.

8.3 FACTORS THAT AFFECT CLEANING

8.3.1 Water and soil

Water is the solvent for the cleaner and the soil, and is the medium for pre-clean rinse and post-clean rinse. It must be potable. Salts dissolved in the water affect cleaning; hardness (Ca^{++} and Mg^{++}) causes precipitates with alkaline cleaners (reducing their effectiveness) and help to deposit "stones" (beerstone) on surfaces especially during alkaline detergency. Iron salts in water can cause equipment to stain. In hard water areas, cleaners need to be formulated to deal with such salts, e.g., with sequestering or chelating agents, or water needs to be softened for cleaning, or an acid cleaning cycle should follow alkaline cleaning at suitable intervals. In any case, "stones" should not be permitted to build up as they become more unsightly, a greater hazard and more difficult to remove with time.

In addition to mineral ("stone") deposits, soils in brewing are various mixtures of carbohydrates, including simple sugars, and proteins. Although sugars are water soluble, proteinaceous soils are most easily removed by alkaline cleaners, and that is the standard practice of the industry. Soils should be removed quickly when the tank becomes available for cleaning because cleaning becomes more difficult with time, especially if the surface is hot and the soil can dry out or polymerize or bake on. Because of the amount, kind, and quality of the soil, some cleaning tasks are more difficult than others and so require more aggressive cleaning.

8.3.2 Energy input and cleaners

Soil is not held to a surface by serendipity or magic but by energetic bonds. To remove soil, therefore, energy must be put into the system in three forms: as heat, chemical energy, or mechanical force and applied for sufficient time. These are the four interdependent variables that affect cleaning. Hot solutions are generally more effective than cooler ones and can therefore operate with a lower concentration of cleaners; similarly, for example, vigorous mechanical action can substantially reduce the amount of time required to clean. Some suitable blend of all four variables is used in virtually all cleaning applications, taking into account the cleaning task at hand, safety, and cost.

Cleaners dissolve in water (hardness affects this) and reduce its surface tension and that of the soil-surface-solution interface. Water-soluble soils dissolve (temperature affects this). Surface and soil are wetted, penetrated, and dispersed in suspension (cleaner concentration and wetting agents affect this). Some soils react chemically with a cleaner, e.g., hydrolysis or "peptizing" of proteins, saponification of fats (temperature, alkalinity, and concentration affect this). Finally, the soil in solution/suspension is fully rinsed away (surfactants and agitation affect rinsing properties). Not one chemical has all the cleaning qualities needed for effective and economic cleaning in all applications, and mixtures of chemicals for particular uses are common and are usually desirable. Only one or a few cleaning agents should be used in a plant. Although cleaner formulations vary considerably for different applications, they all contain the same basic components (see below).

Note that cleaning removes far more microbes than does sanitizing, but incompletely. The removal of soil and substantial reduction of the microbial population by cleaning markedly increases the effectiveness of sanitizers.

8.4 CHEMICAL COMPONENTS OF CLEANERS

8.4.1 Alkaline components

Alkaline sodium salts are the most common components of cleaners. **Caustic soda** (NaOH, sodium hydroxide or lye or KOH) is very strongly alkaline, dangerously so, and must be used with great caution. The extraordinary dissolving power of caustic soda makes it dangerous to human tissue and to soft metals. It dissolves oily and proteinaceous soils readily and is an indispensable part of cleaning agents for heavily soiled applications (e.g., brew kettles, heat exchangers, fermenters, returnable bottles). However, it forms a sludge with hard water and does not rinse well and so is best used as a component of a "built" or mixed cleaner containing water softeners and wetting agents, for example.

Soda ash (sodium carbonate) provides high alkalinity and softens water by precipitating calcium and magnesium as their carbonate salts (at high pH). **Sodium silicates** also soften water, suspend soil well, have wetting and emulsifying properties, and good buffering action (resist pH change). They tend to inhibit corrosion of soft metals. **Sodium phosphates** are either orthophosphates, such as trisodium phosphate or chlorinated TSP (chlorine is often added in alkaline cleaning where oxidizing power is useful, e.g., for cleaning protein soils) or complex phosphates such as tetrasodium pyrophosphate, sodium tripolyphosphate, sodium tetraphosphate, or sodium hexametaphosphate. Orthophosphates soften water by precipitation as Ca^{++} salts, and complex phosphates by sequestering metal ions, especially hexametaphosphate. These materials also aid in deflocculation and suspension of soils. The complex phosphates tend to revert to the simpler form in concentrated solutions of caustic soda or at high temperature. Phosphates are a troubling part of the composition of effluent streams from cleaning operations as they are difficult to remove.

8.4.2 Other components

Surfactants (synthetic detergents) have hydrophilic and hydrophobic parts on the same molecule so they can dissolve in water and soil (especially fat-based soils), but their effect on surface tension (e.g., penetration, dispersion, suspension, and rinsability) make them useful as wetting agents, generally in low concentration, in brewing applications too. The most used are anionic (high foaming) and non-ionic (low foaming) molecules and mixtures of them. Cationic molecules (e.g., "quats" or quaternary ammonium compounds) are more used as santizers than as cleaners.

Phosphoric acid is commonly used as an acid cleaner especially for controlling "stone" deposits. Other useful acid cleaners include mineral acids such as sulfuric and nitric acids and sulfamic acid, and organic acids such as gluconic acid, acetic, hydroxy acetic, and citric acids. Acids can be used in the presence of CO_2 and so often find application in finishing cellars where light soil is present in a CO_2 atmosphere. Added wetting agents promote effectiveness of acid cleaners.

Chelating agents EDTA (ethylene diamine tetra acetate), NTA (nitrillo triacetate), and sodium gluconate soften water by chelating (binding) metal ions particularly Ca^{++} and Mg^{++} and so find application in hard water areas. Gluconate works well in very alkaline cleaners and so can partially substitute for complex phosphates.

8.5 SANITATION OF CLEAN SURFACES

8.5.1 Sanitation and sanitizers

Sanitation is used to prepare the surface for beer contact by reducing the load of microorganisms present to acceptable levels. This prevents microbial contamination of the process and product and so the undesirable flavor changes (and possibly hazes) associated with growth of microbes. In brewing, wort contamination by microbes remaining in the equipment downstream from the kettle and in the fermenter are particularly critical because wort is an excellent nutrient for most microbes to grow in. Yeast is also a repository of contaminating microbes (see Chapter 16).

Sanitizing works best on thoroughly cleaned surfaces and all good sanitation begins with that premise. This is because (a) cleaning substanially reduces the microbial load to be killed, (b) removes harborages where microbes might survive sanitation procedures, and (c) because soil reacts to neutralize sanitizing chemicals.

There are two techniques of sanitation: by application of wet heat (as hot water or steam) and by chemicals. Sanitizers should be non-toxic, non-corrosive, and non-tainting, be widely and rapidly biocidal at low use-concentrations, and be easily measured and free rinsing. Hot water or steam easily meet all of these criteria, but is not appropriate in all applications, and can be dangerous and expensive. The range of chemical sanitizers that meet these criteria is limited in practice to **halogens** (chlorine-releasing compounds and the iodophors), the quaternary ammonium compounds, and anionic/acid detergent formulations. In the U.S., the Environmental Protection Agency (EPA) registers sanitizers and dictates how they should be used in food plants.

8.6 CHEMICAL SANITIZERS

8.6.1 Halogens

Sodium hypochlorite solution (e.g., dilutions of common bleach) is extremely useful, readily available, and cheap and such a solution is fast-acting and effective against a broad range of microbes. The active agent is hypochlorous acid (HOCl), which is a powerful oxidizing agent. It is somewhat unstable at high temperature and in the presence of organic soils (which is the reason for cleaning before sanitizing). Hypochlorite should be used at neutral or slightly alkaline pH because, in acid conditions, it can rapidly corrode even stainless steel and release noxious chlorine gas. It is commonly used at up to 200 ppm for 10 seconds exposure at room temperature. A much lower concentration is effective with longer exposure times.

Other chlorine yielding compounds probably also release hypochlorous acid as the active agent and so are similar, though not necessarily identical, to sodium hypochlorite in their effects and limitations. Chlorine dioxide is more stable than hypochlorite in the presence of organic soils (which incidentally should not be present when sanitizing), though similarly dangerous in acid conditions. Chlorine gas is commonly used for in-plant chlorination of all water in food plants at 5 to 25 ppm depending on application.

Iodophors are iodine formulated with surface active agents plus acids (e.g., phosphoric acid) as a carrier medium. They are sensitive to high temperature and tend to foam, but are fast-acting and effective biocides at 15 to 25 ppm. Their amber-yellow color is a useful indicator of concentration. Iodine is effective because it reacts with microbial proteins.

8.6.2 Other sanitizers

Anionic/acid surfactants (e.g., alkyl aryl suphonates) plus acids such as phosphoric, are effective biocides at pH less than 3.0. They can advantageously provide an acidifying and sanitizing rinse following alkaline cleaning or they might be useful in CO_2 atmospheres. They disrupt microbial surfaces. Peracetic acid, being a powerful oxidizing agent, is an effective biocide; it degrades to water and acetic acid (non-toxic and volatile) after use.

8.7 SANITATION QUALITY CONTROL

8.7.1 Sanitation quality control plan

A sanitation quality control plan is a **preventative** program (to guard against product failure) rather than a trouble shooting program (to identify sources of problems after they happen). It must be designed to measure and control those factors that affect product quality, and must have an operational objective and approach, criteria and methods of measurement, records and reports, and safety standards.

The plan must be reasonably detailed and specific as to the areas of application (i.e., pieces of equipment or vessels), the objective, the cleaning/sanitizing method, and the methods and frequency of monitoring. For each piece of equipment, the plan should define what is to be achieved and how it is to be achieved. For example, specify timing (when to clean/ sanitize); use concentration and temperature of chemicals to be used for cleaning and sanitation; removal or dismantling and handling of non-CIPed parts; and opening of cocks, manways, and valves, etc. The plan should take account of the soil type and amount to be removed and the location of the equipment, e.g., daily aggressive, hot, caustic cleaning of a beer storage tank in a cold cellar might not be a good choice!

Standards for the number of organisms remaining on beer contact surfaces should be set that can be achieved and measured using prescribed methods, such as swabbing a known area of the sanitized surface, or contact plates, or plating of rinse water, etc. Visual inspection for soil is useful if done on dry, well-lighted surfaces; a surface that looks, smells, and feels clean probably is clean, though residual microbes are invisible, of course. Standards of identity, concentration, and temperature must also be established for cleaner solutions, especially if they are reused, for example, in a CIP system. Such standards protect the product, process, equipment, and employees. Records must be kept and reports issued regularly, most conveniently in the form of charts, to assure management and work staff that sanitation is satisfactory.

8.8 SAFETY

Safety in cleaning and sanitation is a prime consideration because most cleaning and sanitizing chemicals are hazardous in some form, and spillage plus water can cause unsafe and slippery floors. No cleaning/sanitizing regime should be established without adequate training of employees in the proper use of chemicals and without providing adequate protective equipment and clothing. For example, caustic materials are dangerous if inhaled and heat up and can boil and splash when added to water; hypochlorites with acids release chlorine gas. Hot water and solutions, boiling wort, and steam are extreme hazards. In breweries, CO_2 is a deadly heavy gas that workers must learn to recognize and avoid. All workers deserve a safe place of employment and none should be allowed into a brewery without adequate safety training and equipment.

PART II

Processes

CHAPTER 9

Barley

9.1 INTRODUCTION

Barley is an unsuitable material for making beer. It lacks the necessary enzymes for brewing, it lacks friability for easy milling, it produces a highly viscous extract that is deficient in amino acids and lacks the color and flavor required for making beer. Malting changes all these properties in crucial ways. The collective word for these changes is **modification**, which encompasses the sum of changes wrought in the physical, chemical, and biological properties of barley by the process of controlled germination called malting. Malting comprises **steeping, germination**, and **kilning**.

9.2 OVERVIEW OF MALTING

To make malt, selected and prepared barley is steeped (soaked) in cold, aerated water for 40–50 hours; this is followed by 3–5 days of cool, aerated germination, when the shoot and rootlet grow and important enzymes form and act. The grain is then dried by kilning with warm air, which fixes the properties of the malt and imbues malt with its unique flavor. The process takes place in a **maltings** or malt house. The properties of the barley selected for malting and the processes used, determine the kind of malt produced and the economics of the enterprise.

Malting may be thought of as a contest between the maltster and the grain. Germination permits the living embryo of the barley grain to produce a new plant, utilizing the stored reserves (mostly starch) of the grain for that purpose. The maltster intends to sell those same stored reserves (potential extract) of the grain to the brewer, and so maltsters carefully control the action of the growing embryo to achieve a desired level of modification, with minimum loss of reserves. Some loss always occurs however; this is called **malting loss** and is measured as the loss of dry weight as barley is converted into malt. Malting loss is the price the maltster must pay the embryo for its cooperation. In a well-controlled opera-

149

tion, there is a good correlation between malting loss and degree of modification. Lack of control of processing parameters yields high malting loss, reduced malt quality, and low plant efficiency. The progress and control of germination concerns four central themes: the penetration of water into the grain during steeping, the stimulatory action of the embryo, the resulting onset of enzyme synthesis, and the action of the enzymes synthesized. These are fully dealt with in the next chapter.

Brewing industry professionals around the world have agreed to standard methods of analysis by which malt quality is reported. The measurements fall into three general categories: (a) those that imply the amount of beer that can be made from a malt (e.g., extract yield), (b) those that imply the quality of the extract yielded [e.g., free amino nitrogen (FAN)], and (c) those measures that imply the ease of obtaining the extract required (e.g., diastatic power). Brewers use these sorts of measures to specify their requirements. Maltsters can rarely meet a brewer's specification with a single batch of malt, but must blend batches of different qualities to meet the overall quality (specification) demanded. This is an important feature of practical malting.

9.3 BARLEY FOR MALTING

Barley, along with all other cereal grains, belongs to the huge botanical family of grasses or the *Graminae*. Malting barleys, as opposed to feed barleys, are specially bred or specially identified varieties of barley that, experience has shown, malt well. That is, they yield malts with satisfactory and, in some cases, superior brewing qualities. In the U.S., barleys specifically bred as malting varieties are tested and recommended by AMBA (American Malting Barley Association). Malting barleys do not yield as much grain per acre as do barleys for animal feed, and so American maltsters customarily pay farmers a premium to grow malting varieties and for meeting malting quality specifications. Barley breeders constantly attempt to improve the existing malting varieties to compete with the yield of feed varieties.

In Britain and Europe, similarly, maltsters strongly prefer certain varieties over others, which are approved malting varieties. These are assigned a malting quality number by NIAB (National Institute for Agricultural Botany) as they are developed, and maltsters pay a premium for malting quality barley.

Barley for malting has well-defined analytical, agronomic, and physiological properties. Analytically, the barley sample must be viable (at least

96% alive); dry (about 12 to 13% moisture); free of disease, infestation, and discoloration; and reasonably free of debris including dust, weeds, broken corns (collectively called dockage), and skinned corns. The grain should be low in nitrogen (protein) because this slows modification and lowers extract yield. Plump grain is preferred because it contains relatively less husk and more starch (extract) than thinner grain. Necessary barley physiological properties include vigorous germination, potential for rapid and complete modification, and the potential to yield high levels of brewing enzymes.

The farmer, of course, has a different list of desirable barley properties including the centrally important agronomic and economic ones of yield per acre, disease resistance, resistance to shattering of the ear in the field, and lodging (falling down) in winds. These properties determine whether or not farmers plant desirable varieties.

9.4 BARLEY EVALUATION

Maltsters inspect a sample of the barley they intend to buy. Samples are taken from the lot by carefully chosen methods, e.g., using a 6–12 ft. (2–4 m) long "trier" for box cars and hopper cars, and spout samplers or belt samplers for grain flowing in bulk. This large composite sample must then be split randomly to yield a representative sample of about 2 lb. (1 kg) for analysis. The value of barley is a function of its variety and its quality, and decisions to buy are based on these factors. Traditionally, barley growers, breeders, and maltsters identified the variety of a barley visually, by inspecting the morphology (shape) of the corns. Intensive crossing in breeding programs to the same gene pool have now rendered this technique of little practical value. Today, maltsters use sophisticated biochemical methods of identification (e.g., electrophoretic separation of proteins of the grain or the esterase enzymes of the germinating grain) to identify the variety being offered for purchase and to confirm that they receive the barley they buy.

By visual inspection and by laboratory count, they can estimate the amount of weed seeds, chaff and dust, and broken and discolored kernels present. This "clean out" material or dockage is bought at barley prices, but reduces the storage stability of the barley and must be removed before storage and malting. Skinned barley drowns in the steep, fails to malt, and reduces malt quality. A bright appearance of the grain suggests good conditions during harvest and storage. Maltsters also pay strict attention to the weather in the growing regions throughout the year, and are in touch

with farmers growing grain under contract, for example. They are, therefore, well informed about the likely qualities of the barley crop from
region to region long before it is harvested and offered for sale.

9.4.1 Kernel quality and analysis

When cut open lengthwise, pre-germinated kernels can be seen and the
appearance of the endosperm judged. A white, opaque or "mealy" endosperm, caused by a myriad of tiny air chambers throughout the endosperm, suggests a good quality barley that will malt well. On the other
hand, a grayish, slightly translucent "steely" or "glassy" endosperm suggests a poor quality barley, possibly high in nitrogen (protein). Such barley will modify less easily. The assortment of plump, medium, and thin
grains can also be estimated visually and measured on a sifting device.
Plump grains (over a 6/64-inch screen) yield more extract than thin ones
(through a 5/64-inch screen), and grains of different plumpness have to
be malted separately. Maltsters therefore demand a high proportion of
plump grains. The bushel weight (pounds of barley per bushel) provides
an additional measure of kernel size and density. Similarly, the 1,000-ker-
nel weight measures the average weight (size) of a kernel and can be used
to compare barley input to malt yield to estimate malting loss. By germinating a known number of grains spread on blotting paper on a covered wet plate, maltsters estimate the number of grains that germinate
promptly (count after 2 days), those that germinate more slowly (idlers,
count after 3 days), and those that fail to germinate (dormant, see below,
or dead kernels). Maltsters desire at least 96 to 98% viable kernels.
Because they have no means of separating dead kernels from live ones,
after malting, dead kernels (unmalted barley) will form part of the finished malt and compromise its quality. Barley moisture is measured simply by weighing a small sample of finely ground barley before and after
heating in an oven at 130°C. This value is typically 11 to 13% in American
barleys as harvested. Lower moisture values are quite possible but the
grain is then more prone to skinning and breakage during handling.
Maltsters must have less than 14% water in the grain because, above this
level, perceptible respiration (breathing) of the barley in storage releases
moisture and heat, which promotes more respiration. Over time, the accu-
mulating moisture causes germination in storage and the heat build-up
kills the embryo, which cannot then be malted. A desirable moisture content in barley may be the natural result of dry weather at harvest, or the
barley may be cut and wind-rowed, or the threshed barley may be air-
dried before storage. In Britain, barley is typically harvested at 20% mois-

ture or so and is dried for storage. The moisture content and nitrogen content of grains can be estimated by near infra-red reflectance.

The nitrogen content of barley, often reported as the "protein" content (nitrogen content×6.25), helps to predict the extract yield of the grain, the ease of modification, and the potential for enzyme synthesis. Generally, a high level of protein, i.e., much above 13%, compromises extract yield because the protein replaces starch in the kernel and may yield hazes in beer. Protein also inhibits the rate of modification because as the level of protein increases, it is deposited in the matrix surrounding the starch in the endosperm as less easily metabolized hordein. On the other hand, high-protein barleys tend to yield more enzymically-active malts than those low in nitrogen. Nitrogen content is measured by the Kjeldahl method in which a small sample of barley is digested with pure boiling sulfuric acid. The amount of ammonia formed is measured. The American Society of Brewing Chemists (ASBC) *Methods of Analysis* handbook also provides for the determination of potential extract and potential diastatic power of barley.

All else being equal, the quality of malt, as judged by its extract yield, is a function of barley variety, kernel size, and nitrogen content. Bishop expressed this idea in one of his "Equations of Regularity," which reads

$$E = A - 11.0N + 0.22W$$

where E is the extract yield, A a varietal constant, N the nitrogen content, and W the 1,000 corn weight. This equation does not literally work with today's methods of expressing extract yield, but Bishop's original equation usefully illustrates important relationships among the analytical values mentioned above. Predicting malt quality from barley analysis remains of great interest to barley breeders who need rapid methods for this.

9.4.2 Germinative properties

In Britain and Europe, barley at harvest may be water sensitive or dormant. This happens in the U.S. less commonly, primarily because different and generally more vigorous varieties are grown. Also, the weather at harvest is usually dry and hot and so the grain is harvested quite dry, which tends to relieve dormancy. Dormancy has advantages for stored grain especially if it is somewhat high in moisture, because dormancy prevents germination in storage.

A dormant kernel is one that is alive but will not germinate; thus it has the **capacity** to germinate (because it is viable), but does not do so. The **germinative capacity** is the percentage of grains capable of germinating when dormancy is relieved (total viability). The **germinative energy** of a barley sample is the percentage of grains that germinate at the time of the test (as described above). The difference between the two values is the percentage of dormant grains. A maltster about to purchase a lot of barley wishes to know whether those grains that fail to germinate are dead or dormant (i.e., expressing low germinative energy). There are several ways to do this. Kernels can be split in half and stained with sodium biselenite or 2,3,5-triphenyl tetrazolium chloride; the embryo and aleurone of live kernels stain red. This method depends on the action of dehydrogenase enzymes, which may remain active in dead corns (and even malt), and therefore the test can give false-positive results. Alternatively, kernels may be soaked in hydrogen peroxide (0.75% solution) and inspected for signs of germination after 2 or 4 days in this solution. Alternatively, kernels may be treated with sulfuric acid to remove the husk, washed, and put on a wet plate to germinate. These techniques are much slower, but more reliable than the staining methods.

In a well-stored dormant barley, the germinative energy progressively increases until it equals the germinative capacity (Figure 9–1). It is then suitable for malting. Some U.S. barleys may lose germinative energy during storage, sometimes as a result of pre-germination. If mild, such barley may be taken as soon as possible to the malt house, but if more severe, the material is sold off as feed barley.

Barley that is not dormant may still fail to germinate when it goes to the malt house because it is water sensitive. In a laboratory test for this property, water-sensitive barley germinates if the barley is partially covered with water, but not if immersed (Figure 9–2). This is referred to as the 4 ml and 8 ml test. This test is done routinely where water sensitivity is a common phenomenon, because it is a matter of importance in steeping.

Maltsters make laboratory analyses on a few hundred kernels or 1 to 2 ounces of barley, which must be a representative sample of the whole lot of many tons of grain. Analyses form the basis for purchasing decisions. Low viability would prevent a purchase, but other analyses may provide the basis for price negotiation. Excessive dockage, for example, may lower the value of an otherwise satisfactory barley lot.

9.5 THE BARLEY KERNEL

Barley grain is the fruit or seed of the barley plant in which only the embryo and the aleurone layer is alive. The vast bulk of the tissue is dead and functions either to protect or to feed the embryo.

Though naked or huskless barleys exist, in commercial barleys, the husk tightly adheres to the grain. In a barley kernel, the husk, pericarp,

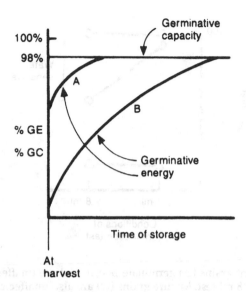

Figure 9–1 Effect of storage on the germinative capacity (GC = viability, assumed to be 98%) and germinative energy (GE = present ability to germinate) in two barleys A and B, which are partially dormant at harvest. (Dormancy = GC − GE.)

and testa or seedcoat, which were originally parts of the barley flower, fuse to each other and enclose the living embryo and aleurone tissues, which in turn surround the endosperm (Figure 9–3). The husk protects the endosperm during the mechanical rigors of malting and, being largely insoluble, forms the filter bed for lautering in the brewhouse. The pericarp/testa surrounds the entire kernel. Because the pericarp is waxy and rather waterproof, and the testa acts as a semi-permeable membrane; this layer effectively defines the exterior and the interior of the kernel. The testa is very thin or perhaps absent at the micropyle region close to the embryo, where water penetrates readily during steeping.

The endosperm is the part of the barley from which beer will be made. It contains primarily starch in the form of large (25 μm) or small (5 μm) crystalline granules, which are buried in a proteinaceous matrix. These two elements, in turn, are located inside cells with thin cell walls (Figure 9–4). These cell walls throughout the endosperm render barley difficult to crush and so barley is hard or not friable. The cell wall contains some protein and much β-glucan (about 4% of barley weight, though the value varies greatly), which has the potential to cause wort and beer viscosity, as well as to form hazes and gels. Germination of barley dissolves these cell

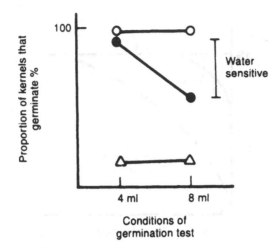

Figure 9–2 Dormant grains (△) germinate poorly and are unaffected by the conditions of the 4 ml/8 ml test. Mature grains (○) are also unaffected and their germinative energy equals their germinative capacity (i.e., virtually 100% germinate). Grains that are not profoundly dormant, but have residual dormancy in the form of water sensitivity (●), grow well in 4 ml of water, but poorly in 8 ml.

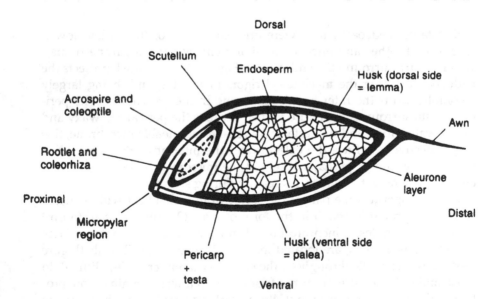

Figure 9–3 Schematic representation of a barley kernel in longitudinal section.

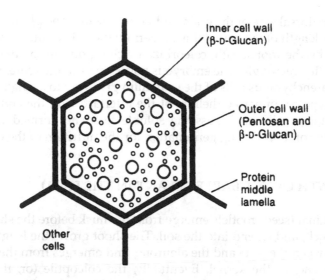

Inner cell wall
(β-D-Glucan)

Outer cell wall
(Pentosan and
β-D-Glucan)

Protein
middle
lamella

Other
cells

Figure 9–4 Schematic representation of a barley endosperm cell showing large and small starch granules embedded in a protein matrix surrounded by the cell wall and the middle lamella cementing cells together.

walls. This gives access to the starch and protein within the cells, renders barley friable, and substantially solves the problem of β-glucan viscosity.

Near the scutellum, the endosperm cells are empty and crushed. Just below the aleurone in a region called the sub-aleurone layer, the endosperm cells are smaller and tend to contain more protein than those toward the center of the endosperm. The aleurone layer is part of the endosperm. The aleurone lies just beneath the testa and its associated nucellar tissue. Like the testa, it envelops the entire kernel except at the ventral furrow and at the base of the kernel. The living aleurone layer is 2 to 4 cells thick and has thick cell walls. The cells are rich in reserve substances such as lipids and the aleurone grains, which contain a protein/polysaccharide/minerals/phytin complex, but starch is absent. These cells contain well-developed mitochondria, necessary for energy metabolism, and endoplasmic reticulum, necessary for protein synthesis and transport. During germination, the aleurone layer reacts to the plant hormone **gibberellic acid** to form and release into the endosperm enzymes that cause modification. Such enzymes are also released by the scutellum.

The embryo occupies the proximal base of the kernel. Within the embryo, the embryonic leaves protected by the leaf sheath or coleoptile (commonly called the shoot, acrospire or plumule in malting) are oriented toward the dorsal side of the grain and point distally. The plumule grows almost the length of the kernel during germination. The embryonic roots, protected by the root sheath, coleorhiza or chit, point to the base of the grain and the micropyle. The embryo is borne on or in the scutellum. The general parenchymous cells of the scutellum give way to a single layer of palisade-type columnar epithelium at the interface with the endosperm. Through these cells must pass the soluble nutrients formed from the endosperm contents during germination for the nutrition of the embryo.

9.6 GROWTH OF THE BARLEY PLANT IN THE FIELD

The seminal (seed) rootlets emerge from the husk before the shoot, and grow, branch, and extend into the soil. The shoot grows the length of the kernel between the testa and the aleurone, and emerges from the husk at the distal end of the kernel. Eventually, the coleoptile (or, if planted deeply enough, a stem extension of the coleoptile) reaches the surface and the first leaf emerges through an opening in the coleoptile tip. New leaves initiate and emerge from the tube formed by earlier leaves. They are numbered in order of appearance because farmers time various field treatments, such as adding fertilizer, by recognizing each successive stage of growth. Leaves form at nodes or joints. From the first node, the crown forms at or near the soil surface, and from this, develop the adventitious roots, which will sustain the growing plant to maturity. Here also, form the apical bud(s) from which new stems or tillers grow. This is called the tillering stage. The number of tillers is a varietal characteristic and, because each tiller carries no more than one ear of grain, an important characteristic relative to yield and kernel size. The stems elongate, a phase called jointing or shooting, to elevate the ear, which initiates at this time. The flag leaf (boot stage) and then the ear or spike emerges (heading).

The rachis or stem of the ear upon which the flowers (inflorescence), and later the seeds, are borne is a differentiated extension of the stem of each tiller. The morphology of the rachis determines to a large extent the kind of ear produced and a strong rachis is necessary to prevent the ear breaking or shattering during maturation and harvest. At each node or joint of the rachis emerges a triad of florets. Each floret is called a spikelet and each spikelet comprises (among many other parts) the lemma and the palea, which later enclose the flower and form the husk of the grain. The sexual parts of the spikelet develop: the stamens grow anthers, which will form and release pollen in anthesis. The ovary bears the style upon which the stigma rides to receive pollen. Barley is designed to be pollinated by wind, but the proximity of anthers and stigmas causes most spikelets to pollinate themselves.

Each triad comprises a central and two lateral spikelets. The central spikelet is fertile in all barleys; if only this flower develops into a seed, there is one seed per node of the rachis and the barley is called two-rowed (Figure 9–5) because the appearance is of two rows of kernels in the ear. However, if all three spikelets give rise to seed, there are three seeds per node and the barley is called six-rowed. The seeds of six-rowed barleys are generally somewhat smaller and, hence, tend to yield less extract than those of two-rowed types, and the lateral seeds are slightly twisted.

9.6.1 Growth rates

A barley plant requires about 100 days (North America) to about 140 days (Britain and Europe) to grow from sowing to maturity. The time depends on many factors including the variety, the vagaries of the weather, especially temperature and moisture, soil fertility, latitude, and the presence of weeds, pests, and disease. After ear emergence and anthesis, the kernel lengthens in about a week from about 2 mm to about

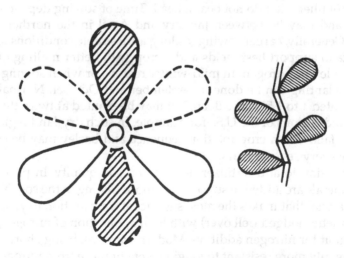

Figure 9–5 A triad of florets arise at each node of the rachis offset from each other above (solid outline) and below (dotted outline). In two-rowed barley, only the central floret of each triad is fertile and gives rise to one kernel at each node; this gives the impression of two rows in the ear (shown shaded). In six-rowed barley, all the florets are fertile, giving the impression of six rows in the ear.

10 mm. The grain then fills more slowly and increases in girth and dry weight over some 2 to 3 weeks. During this time, total simple sugars, especially reducing sugars such as glucose and fructose, decline as the total starch increases. In this period of some 20 to 30 days, the moisture content of the grain falls slowly at first and then rapidly during ripening. The entire plant dies shortly before harvest.

9.6.2 Yield

Yield has a number of meanings depending on local conditions, but most generally means the total weight of grain produced per unit area planted. Ultimately, of course, it is the economic value and profitability, which are functions of the marketplace and the efficiency of production, that govern whether a farmer is successful or not with the crop. Barley, of a suitable variety for the region that has been properly selected and treated as a seed barley, must be planted into well-prepared, weed-free soil containing adequate moisture. Adequate selection and preparation of seed assures a high survival of seedlings and, hence, a relatively low seeding rate can be used, say some 100 to 150 lbs/acre. The density of sowing, or more correctly the density of mature plants that develop, affects yield and quality. Crowding decreases survival, and increases the number of tillers that do not bear a head. Time of sowing depends on the locality and may be between January and April in the northern hemisphere. Generally, earlier sowing (as long as desirable conditions for sowing pertain) support best yields and a crop with better malting qualities including lower nitrogen. In mild-winter regions or when sowing winter barleys, planting can be done in September and October. Normally barley is planted 1 to 1.5 inches deep, but may be planted at twice this depth in dry-land regions of the U.S., for example. In such dry-land regions, the land lies fallow in a crop rotation sequence and barley may be cropped but once every several years.

The decision to add fertilizer depends on soil quality. In principle, at least, minerals are added to supplement those lacking in the soil. Nitrogen is the material that makes the most significant contribution to yield. Some older varieties lodged (fell over) with high application of nitrogen, which set the limit for nitrogen addition. Modern varieties, being short-strawed and generally more resistant to lodging, benefit more from nitrogen addition even at high levels. However, high nitrogen application causes high seed nitrogen, which increases feed value but decreases malting value. Early (even at sowing) addition of nitrogen, as part of a balanced fertilizer program, supports high grain yield with least increase of grain nitrogen and is the usual fertilizer practice in cultivation of barley for malting. Weeds must be controlled because they compete for soil nutrients, moisture and even light and effectively reduce grain yield.

Barley seedlings fail to thrive in water-logged soil that warms slowly, but adequate soil moisture as rainfall or irrigation is essential for good yields. Harvest conditions affect yield and crop quality. The weather at harvest time ideally should be dry and warm. Barley yields of 1 to 3 tons per acre (with double these amounts in very good years) depend on many factors including the weather, agronomic practices such as crop rotation and control of weeds and diseases, and harvesting conditions and practices, especially preventing grain loss by ear shattering.

9.6.3 Strain improvement

Plant breeders improve barley by one of several methods: the most traditional is the selection of the best plants from among a native population (land race) of plants. Repeated selection yields a more uniform and so less improvable population. The most powerful method of barley improvement is hybridization. Pollen from a selected "male" fertilizes a selected "female" (emasculated flower). The breeder then seeks the desired traits among the progeny of successive generations. Additional crosses and back-crosses may be made among selected progeny and parents to build the genetic base required, which, when expressed under local conditions, produces an agronomically superior barley. Most crosses to produce new malting varieties are made among barleys with impeccable malting pedigree, but because the malting performance is genetically unconnected to field performance, specific barleys with desired agronomic traits must be included in the breeding program.

CHAPTER 10

Malting technology: malt, specialized malts and non-malt adjuncts

10.1 OVERVIEW OF TECHNOLOGY

Some brewers make malt for their own use. This gives them a good sense of the proper price for malt and for the quality of malt that commercial maltsters should deliver from the current crop. But, by making less than all the malt they need, they assure that their own malting capacity always operates near full capacity. The vast bulk of the world's malt is made by independent malting companies.

Brewers write specifications for the malt they require; they minimally seek malt with a high extract yield, that is well modified and has a high complement of diastatic enzymes. But some specifications (e.g., high dextrinizing units (DU, a measure of α-amylase) **and** high extract) are internal contradictions, because good modification and a high enzyme content come at the expense of malting loss, which leads to a lower extract (all else being equal). To meet the posted specification for the lot of delivered malt, maltsters must blend malts from different batches and, in some cases, produce batches of malt specifically for blending. This means, of course, that kernels within the batch are different from each other and there may be *no* kernels in a blend that precisely meets the specification.

Brewers also expect malt to be constant in quality from batch to batch within each crop year and from year to year. This is rarely the case. Barley varieties differ from each other; the same variety grown in different regions or from a different crop year will likely have different properties; even within a variety and region and year the analysis for protein, for example, of different lots could differ sufficiently to affect malting performance. Barley properties change during storage, not always for the better. Thus, maltsters may satisfy brewers' specifications in different ways as time progresses, so that brewers occasionally observe changes in malt performance in the brewery, although the malt has ostensibly identical batch-to-batch composition and meets specification.

The maltster occupies a difficult place between the farmer, farm practice, and the vagaries of the weather, which control barley quality, and the brewer who demands a staring material of constant quality to feed an increasingly more efficient, but less flexible brewing process. Within limits, maltsters achieve what is expected of them by strictly selecting the barleys they use and segregating barley lots of different varieties, from different regions and of different nitrogen or moisture contents. Each of these materials is then malted separately by methods known (or at least intended) to get the best out of each barley.

The malting process comprises five stages: **preparation** of the barley and its **storage, steeping, germination, kilning**, and **blending** of the finished malt.

10.2 PREPARATION OF BARLEY FOR MALTING

Barley is stored in deep concrete silos of as much as 5,000 tonnes capacity for up to 15 months. The silos have conical bottoms designed to accommodate the angle of repose of the grain (26°+ for barley or 35°+ for malt). Barley is a living tissue and respires (breathes); hence, to prevent suffocation, it must be moved from time to time (e.g., every 3 to 4 months) to aerate it and to keep it cool. During such moves, maltsters take the opportunity to clean the barley and sometimes fumigate it to control insect pests. However, the first defense against pests, including insects, rodents, and birds, is scrupulous cleanliness in the malting apparatus and in the maltings and its environs.

Upon receipt at the maltings, barley is weighed and then cleaned and reweighed. Barley passes over vibrating or revolving screens or sieves of various sizes, and over air jets and magnets to remove non-barley material that is substantially larger or smaller, or lighter or heavier, than barley kernels. This is the dockage or clean-out. Such material includes stones, straw, leaves and twigs, pieces of metal, clods and dust, and indeed anything to which any raw agricultural product is prone. Stones and metal must be removed with great care because if they should strike a spark in a dusty atmosphere in the malt house or brewery, a violent explosion could follow. Coarse cleaning is relatively rapid and maltsters clean the grain as it passes from the fields into storage; additional cleaning is also performed each time barley is moved.

At the point of harvest, barley is stored in field elevators after coarse cleaning to remove the dockage, which can negatively influence storage stability of the grain. Barley often enters storage at field temperature, which could be as high as 30°C (86°F). It is necessary to cool it by blowing cool air through it. Generally, stored grain can be spoiled by insect infestation at high temperature or by microbial contamination at high moisture. Grain, therefore, should be stored cool and dry, e.g., 13°C (55°F) and 13% moisture, depending on the length of storage time anticipated (Figure 10–1).

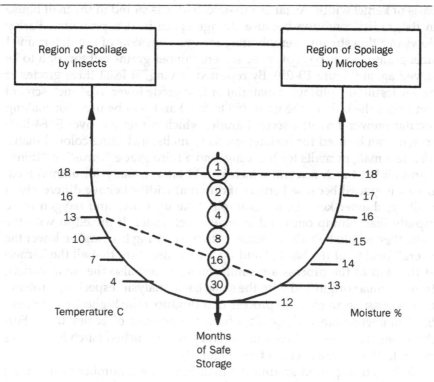

Figure 10–1 This schematic represents the idea that grain temperature and moisture both affect its storability. Even if the grain is very dry, it can be spoiled by insects if it is warm (top left side of the figure). If it is wet (top right side), even very cold storage will not protect the grain. Conditions of storage affect the length of time a grain can be satisfactorily stored: 18°C (61°F) and 18% moisture might suffice for a perhaps one month, but 13°C (55°F) and 13% moisture, for example, would be needed for the 15 months or so that malting barley is maximally normally stored.

After cleaning, the grain is first separated and then graded; these are relatively slow processes and are done at the malt house itself. Separation removes all kernels that are shorter than barley, such as most weed seeds and broken barley kernels or half-corns. For separation, barley passes through long rotating cylinders with pockets or indentations on interior surfaces into which half-corns or small spherical weed seeds fit. This lifts the short kernels out of the bulk of the grain and deposits them in a central trough (Figure 10–2a). Broken kernels are especially undesirable in stored grain because they foster insect infestation and moisture pick-up, and, of course, do not malt.

Grading of barley follows separation; the process grades barley on the basis of kernel width. As far as possible, barley is stored in the malt house in its graded condition because storage space is at a premium. Barley flows in a thin stream over vibrating screens with slots of pre-determined size; plump grains pass over the screen, thinner grains fall through to be sieved again (Figure 10–2b). By repeated sieving, at least three grades of barley result: the plumpest material or first grade (over 6/64-inch screen) comprises the bulk of the grain (80 to 85%) and will be used for making regular brewer's malt; a second grade, which is thinner (over 5/64-inch screen), can be used for making specialty malts, including colored malts, distillers malt, or malts for blending; and a third grade "thrus" or "thins" (through 5/64-inch screen), which is sold later, usually for animal feed. Barley is graded because kernels of different widths behave differently in malting; thinner kernels, for example, take up water and oxygen more rapidly than plump ones and so malt much faster. If mixed in with the bulk, they would bolt ahead, suffer extreme malting loss, and so lower the overall quality of the batch. Grading, by making sure that all the kernels at the start of the process are similar in size (and also the same variety, from a similar origin and with the same basic analysis, especially protein), makes possible an efficient process that produces the highest yield possible (commensurate with quality) from the amount of barley used. Furthermore, it assures that each malt kernel in the finished batch is, as far as possible, the same as every other (Figure 10–3).

Grain is transported around the malt house by a number of methods: bucket elevators can raise grain quickly and safely to great heights, and belt conveyors can transport grain in large volume over long distances more or less horizontally. Other methods of transport, including pneumatic systems, screw conveyors, or chain and flight conveyors, have advantages in particular applications, especially movements over short distances. Grain breakage (especially if the kernels are unusually dry), which can occur during transport, and the formation of dust, which is an

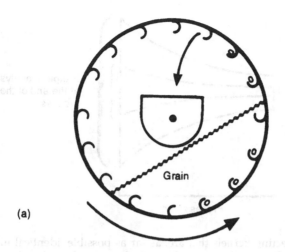

(a)

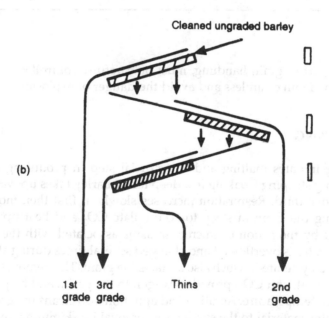

Cleaned ungraded barley

1st   3rd   Thins   2nd
grade grade   grade

Figure 10–2 Schematic representation of a grain separator (a) and a grader (b). Grains are separated by grain length (a relatively slow process) in which short grains and weed seeds fit into indentations on the inside of a revolving cylinder. Barley is graded by width when passing over vibrating screens of selected sizes.

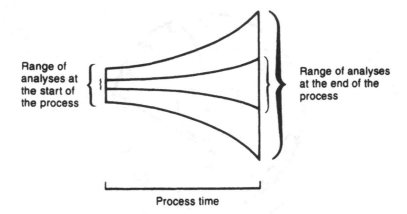

Range of
analyses at
the start of
the process

Range of analyses
at the end of the
process

Process time

Figure 10–3 By selecting kernels that are, as far as possible, identical in variety, origin, size, and analysis, maltsters can limit the kernel-to-kernel variation in the finished malt and control the process with some accuracy.

inevitable part of grain handling, must be carefully controlled in a malt house to maintain cleaniless and avoid the danger of explosion.

10.3 STEEPING

Steeping initiates malting and is a crucial step in producing quality malt. During steeping (soaking in a deep tank), barley takes up water and swells by one-third. Respiration increases slowly at first then more rapidly, causing the grain in steep to accumulate CO_2 and heat up; this is aggravated by the action of microorganisms associated with the barley. The steep water is overflowed and changed several times during steeping and vigorously aerated, which also agitates the grain. This removes microorganisms, heat, and CO_2, provides oxygen to the grain, and helps maintain an even temperature. Aerating and agitation also cleans the grain and brings lighter material to the surface for removal by skimming and overflowing the tank. Steep water leaches husk components that may impede germination and others that may affect beer flavor, and possibly some that prevent proper flocculation of yeast. The leached materials include sugars, amino acids, phenols and phenolic acids, tannins, and minerals including silicates and phosphates. Combined losses of skimmed and dissolved materials approaches 1.5% of the dry weight of grain and is part of malting loss.

Steep water is the main pollutant associated with malting, and much attention is now given to water conservation, especially spray steeping and water recirculation. In this technology, immersion of the grain is minimized. Nevertheless, sufficient steep water must be used to assure quality malt, and so most malt companies operate extensive water treatment facilities to improve water quality before discharge, typically to a municipal plant.

In practice, the deep conical-bottom steep tank (Figure 10-4) is partially filled with a cushion of water; barley (typically ≥40 000 lb.) is transported to the vessel by conveyor and falls into the vessel through a spray of water that wets all the grains and suppresses dust. The water should be potable, cool (10 to 15°C) and free of taint and excessive iron; otherwise, water quality for steeping is not crucial. However, lime or other alkalis may be added to the steep water to brighten weathered or stained barley.

The sequence of events that follows filling of the steep tank depends on individual malting practice and preference. After the initial flood stand, i.e., when the barley is buried in water, maltsters may simply change the steep water by overflowing, draining, and replacing it. This is done at least once but may be done three or four times. In addition, compressed air is forced through the steep water from nozzles located near the base of the vessel. Between water changes, the barley is allowed to "air rest" (also called "dry stand"), during which time air may be blown up (or drawn down) through the drained bed with or without a spray of water. More air is required in later air rests than in early ones. Such treatment is efficacious, especially with potentially water-sensitive barley and is necessary to prevent the grain overheating (i.e., exceeding 17 to 20°C). In a few cases, the barley-water-air mixture is pumped from the bottom to the top of the vessel for mixing and aeration. In modern practice, a good deal of air is used during steeping to meet the increasing demand for air of the respiring grain as steeping progresses and to avoid stifling it, to replace oxygen metabolized by microorganisms present, to combat water sensitivity, and to remove heat and CO_2. Generally, more aeration and less water is required as steeping progresses, and indeed the maltster need only maintain a film of water over each kernel to provide all its later needs for water. In practice, a rather regular routine of water and air application is used when the characteristics of the present crop and the varieties are known.

Moisture increases in steeping from about 10–12% to 42–46% and occasionally higher for special purposes. The exact final level of moisture

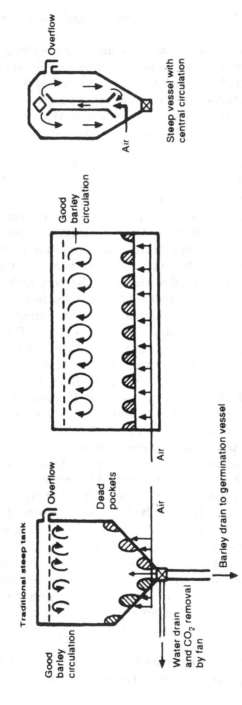

Figure 10–4 Schematic representation of designs for several types of steep tank. The traditional conical steep tank (left) has good barley circulation only in the topmost layer and significant dead pockets close to the air inlets. This problem also remains in the flat-bottomed steep (center), but barley circulation is better. In the steep vessel with central circulation (right), the barley circulates throughout the entire vessel.

depends on the variety being malted and its quality and the kind of malt being made; generally, barley varieties that are less easily modified or batches intended to yield more highly modified malt (including specialty or colored malts) are more fully steeped than a more vigorous barley or a batch intended for a regular pale malt. Steeping typically takes 40 to 50 hours, or longer in a few cases, at a water temperature of 10 to 15°C. The rate of water uptake is more rapid in warmer water and with damaged or thinner kernels, but slower with high protein (steely) barleys. Steep-out moisture is an important determinant of germination and needs to be within 1% of target.

The end of steeping is signaled when the coleorhiza ("chit" or root sheath) penetrates the husk and becomes visible as a white dot. Even chitting is essential. Chitting causes a leap in the requirement for oxygen for respiration that cannot be met by the relatively primitive aeration equipment of the steep tank. If the grain chits under water, it takes up too much water and drowns. The grain is therefore transferred to an environment where its changing demands can be more efficiently satisfied. This is called a **germination chamber**. Though there are many designs, all mechanical (or "pneumatic") germination chambers permit cool, humid air to be forced through a relatively shallow bed of grain that is turned regularly. Mechanical or pneumatic maltings were developed to replace floor maltings, in which steeped grain was simply spread out on a floor to grow; such practice was inefficient and labor intensive, and few floors remain today.

10.4 GERMINATION

Germination chambers are of two main types: **drums** (e.g., with a 30,000 lb. batch size) or, much more commonly, **boxes,** which might be three to ten times larger than drums). Drums revolve slowly and continuously to turn the malt and the air enters through longitudinal tubes that pass through the malt or from a perforated deck (Figure 10–5). In boxes, a bank of turners, supported on the side walls of the chamber, traverse the long bed to turn it, and air enters through a perforated floor (Figure 10–6). Drums provide more continuous turning and aeration than boxes, but are more damaging to the grain and more clumsy to fill and empty. Boxes are of two general types: rectangular ones such as the Saladin box, in which the turners travel up and down the bed; and circular ones (increasing in popularity), in which either the turner or the floor turns about its own axis. Circular vessels are particularly useful in cylindrical tower maltings, which are typically built on restricted sites.

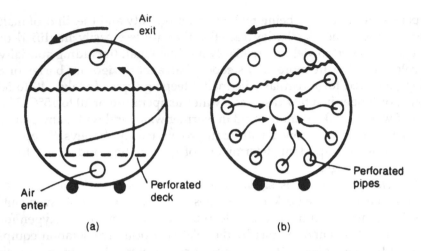

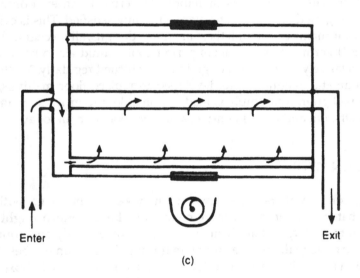

Figure 10–5 Transverse section of a decked drum (a) and a Galland drum (b) and longitudinal section of a Galland drum (c). The Galland drum turns slowly but continuously; the decked drum turns only as necessary. Arrows indicate the direction of air flow.

In some box maltings, the germination chamber may also serve as the kiln after germination is complete because the mechanical requirements for air flow and turning the grain are quite similar in both processes. These are called germination and kilning vessels (GKVs). There are also

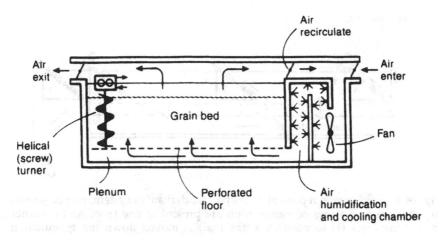

Figure 10–6 Schematic representation of a box or compartment germination chamber. Arrows show the typical direction of air flow and recirculation. In some boxes, however, with a different fan configuration, air is drawn down through the grain bed.

continuous (Domalt system) and semi-continuous (Wanderhaufen system; Figure 10–7) maltings in operation.

After steeping is complete, conveyors transport the wet, chitted grain to the germinating chamber, either mixed with water or, more commonly, after draining. Several steep tanks service one germination vessel. In many countries (though in the U.S. only for distillers malt) the grain may be sprayed during transport with a solution of gibberellic acid (0.1–0.5 µg/g based on dry barley weight), a plant growth hormone that promotes modification; potassium bromate may also be added to counter excessive protein breakdown and rootlet growth. The turners spread out the grain evenly in the germination chamber by passing up and down the bed. The turners thereafter run on a regular program (e.g., two or three times a day). The helical screw turners are designed to mix the malt, which achieves even germination, and to lighten the bed for more easy air flow and hence efficient cooling and CO_2 removal. Turning also prevents the rootlets, which appear during germination, from entangling and forming clumps of grain, which comprise a microclimate, which leads to uneven malt.

In all mechanical maltings, large volumes of air (e.g., 6 to 10 cubic feet per minute per bushel) are forced by fan through a water-spray chamber to clean, cool, and humidify the air to at least 98% relative humidity. The air then passes under the grain bed and up through it to exit at the top of

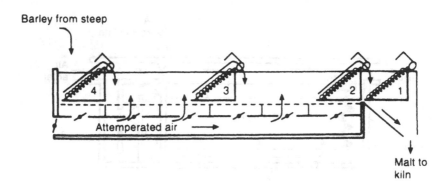

Figure 10–7 Schematic representation of a Wanderhaufen system, one of several in which various stages of germination are present at one time. As the turner moves from rest (1) to position 4, the malt is moved down the germination "street." At position 2, the malt is moved off the street to the kiln and at position 4 a space is cleared for grain from the steeps.

the chamber. Down-draft chambers exist, but are less common. The exhaust air can be partially recirculated. In dry climates, the air cools sufficiently in the spray chamber due to the latent heat of water evaporation, but in warm humid locales refrigeration may have to be applied. The air supply supports grain respiration by supplying O_2 and removing CO_2, helps maintain adequate moisture, and cools the bed by removing the heat formed by respiration.

The upward direction of air flow in the grain bed implies that gradients can arise through the 4–6 ft. depth of the bed. Thus cool, moist air, which is low in CO_2, enters at the bottom of the grain bed (the **"air-on"**) and exits at the top of the grain bed (the **"air-off"**) up to 3°C warmer (which causes it to pick up moisture and to dry the bed), and carrying more CO_2. This means that grain towards the top of the bed germinates under slightly warmer, drier conditions in an atmosphere containing less O_2 and more CO_2 than the barley at the bottom of the bed. Running the turners and sprinkling with water to maintain or raise the malt moisture content ameliorates these gradients, but does not eliminate the problem. There may be a temperature differential of 1°C at the start of germination rising to 3°C in later stages between the air-on and air-off in a grain bed. In drums, the malt mixes continuously and there is much less problem with gradients; moreover, the small batch size allows rather delicate control. Germination continues typically for 3 to 5 days, at about 15–20°C in most modern maltings, rising to 20–22°C towards the end of the process. Within these ranges, the times are somewhat shorter and the temperatures somewhat higher in the U.S. than in the U.K. and Europe. The processing parameters depend on the barley variety being malted, the malting objective and the peculiarities of the process and equipment.

Towards the end of germination, the barley respires vigorously. Temperature control becomes more difficult, and there is some danger that the maltster may lose control if depending on cooling by air flow alone. To combat this, the grain is allowed to dry out somewhat towards the end of germination ("withering"), and the exit air with its enriched CO_2 content is recirculated through the bed of grain to stifle it. These actions tend to slow germination and the drying prepares the malt for the kiln.

To limit barley growth, barley is malted for the shortest time and at the lowest temperature commensurate with quality and efficiency. Excessive barley growth unnecessarily consumes endosperm matter, which provides the substance of the rootlet (removed) and the shoot (insoluble) and fuels grain respiration (conversion of starch to CO_2 and water). Endosperm material, primarily starch, consumed in this way is not then available to yield extract in mashing in the brewery. Losses during germination are the most important and largest part of overall malting loss and may be 4–8% of barley dry weight depending on many factors.

As germination time advances, the barley modifies sufficiently under the influence of enzymes made in and released from the aleurone; while greater modification, achieved by longer time in germination, may improve the malt in some ways, it causes unacceptably high malting loss. Thus, eventually, germination must be stopped and the malt properties fixed. Kilning (drying) the green malt achieves this and yields a dry, storable, and friable product with sufficient enzyme content to meet specification and with desirable malt flavor. As mentioned above, some designs permit the kilning of green malt in the germination vessel itself (a GKV), but more usually grain is transferred to a separate kiln. Germination chambers are usually unloaded or "stripped" by dropping the end wall of the chamber and ploughing the malt into, e.g., a screw conveyor system so revealed or by pneumatic transfer.

10.5 KILNING

Kilning reduces the moisture content of malt from about 45–50% to about 3–5% with a current of heated air. This fixes malt properties. Kilning must achieve low moisture and desirable flavor in the malt. This requires intense heat, yet the malt enzymes, which heat tends to inactivate, must be conserved. These objectives seem antithetical. However, enzymes are much more heat stable in dry environments than in wet ones. The objectives of kilning can therefore be achieved by a drying program that removes most of the moisture at relatively low temperature, and then applies intense heat to achieve final moisture reduction and flavor change when the grain is rather dry; this conserves sufficient enzyme. Kilning is highly energy intensive; some 75 to 80% or more of the total energy used in malting is consumed at this stage.

If the air exits the kiln carrying its maximum load of moisture, the drying process is at its most efficient, and that is what kilns are designed to do. Kilns may have one, commonly two, or even three floors for drying malt efficiently. Each design achieves the same end, but the process is most easily understood in a single-floor kiln (Figure 10–8) and that process will be described first. A kiln comprises a heat source below the grain bed, devices in the grain bed for turning it, a fan that draws air through the bed (usually in an upward direction), and ducts for recirculating the air.

In the first stage of kilning, called **free drying**, the malt is moist and water can be easily removed from the surface and outer regions of the grain by a high volume of air flow, which is the main determinant of rate of water removal, at relatively low temperature (50–65°C). The volume and temperature of the air defines its water-carrying capacity, which must be sufficient to prevent "drip back" in the kiln or "stewing" the malt. These conditions arise when the air carries excess moisture as might happen, for example, when saturated air leaves the grain bed and hits the cold interior of the kiln above the bed (drip back) or when the air cools too much in the bed of grain itself (stewing). The condition is corrected by increasing the water-carrying capacity of the air by making it warmer and/or increasing its flow volume. During free drying, evaporation of water from the grain cools it due to the latent heat of evaporation; this cooling effect means that the malt enzymes are not as sensitive to initial kilning temperature as might be expected. In fact, in some kilning practices, quite high air-on temperatures can be applied without undue enzyme destruction. Typically, the air-on temperature is increased either stepwise (Figure 10–9) or a few degrees each hour for the first 12 hours or so. During free drying, the moisture content of the malt decreases to about 23–25%. Free drying is characterized by (a) a wide difference between air-on and air-off temperature (about 30°C, Figure 10–9), (b) the temperature of the malt itself is less than the air-off temperature, and (c) the exit air approaches 95+% humidity. Along with removal of water, undesirable flavor volatiles also evaporate, especially those responsible for grainy or grassy aromas and this continues throughout kilning.

Kilning

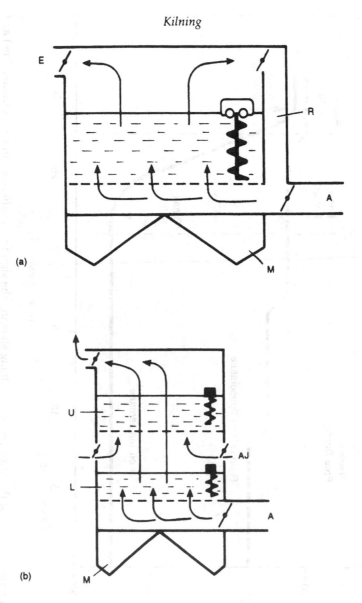

(a)

(b)

Figure 10–8 Schematic representation of the essential features of (a) single floor and (b) two-floor malt kilns. Key to (a): A, air from furnace; R, recirculated air; E, exhaust air or to a second kiln; M, malt out. Key to (b): A, air from furnace at volume and temperature suitable for lower floor; AJ, air to adjust volume and temperature of air suitable for upper kiln; U, upper grain bed in I and early II phase of drying; L, lower grain bed in late phase II and phase III of drying; M, malt out.

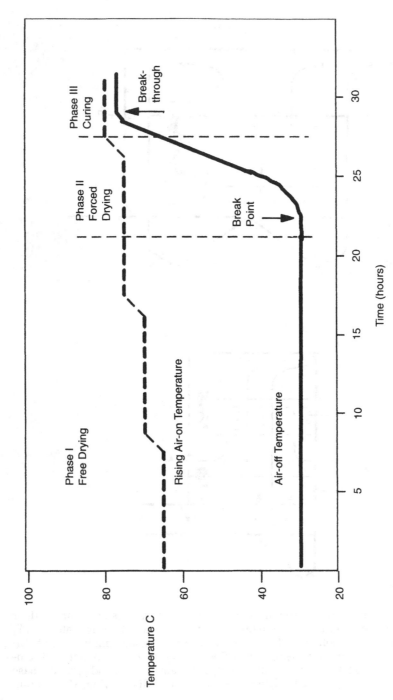

Figure 10–9 A representation of the three phases of kilning showing the significant difference in air-on (dotted line) and air-off temperatures associated with the three phases.

Free drying soon depletes surface moisture and that in the outermost layers of the barley kernels. Thereafter, moisture must diffuse to the surface from deeper layers of each kernel before removal; this is the second stage drying, called **forced drying**, during which time malt moisture decreases to about 10–12%. The grain shrinks as it dries. This shortens the distance that moisture must diffuse, but reduces the surface area from which it can be evaporated. If the temperature and volume of air remain the same as in stage one, the air exits the kiln carrying less than its maximum load of water. To partly correct this inefficiency, the air is warmed (to about 70–75°C) to speed moisture diffusion, air-flow volume is decreased, and, especially toward the end of this stage, some of the air is recirculated through the grain bed. Although there is some cooling effect as a result of water evaporation, especially at the start of this phase, it is much less than in stage one, and as drying progresses, the air-on/air-off/malt temperatures eventually become about equal (Figure 10–9). The "break point" is the point at which the temperature of the air-off begins to rise, which signals the end of free drying.

Below about 10–12% moisture, only so-called "bound" water, associated with the macromolecules of the kernel, remains in the malt. To remove this water, the air temperature must be further increased and the volume of air reduced to the lowest practical level that assures even drying. The air is substantially re-circulated through the kiln. Eventually, as the grain approaches 5% moisture, the malt is "cured." In this stage the air-on temperature reaches 80–85°C (e.g., American pale malt) or up to 110°C (e.g., British ale malts), and is maintained for the last 4–8 hours of kilning (curing stage). There is very little drying during curing and the air is almost entirely recirculated. The final moisture content reaches 3–5% or even lower in a few cases. Curing creates the desirable malty aromas and flavors by lightly toasting the grain (the Maillard reaction) and evaporates the last of the unwanted volatile aroma characters. Because of the disproportionate amount of energy required to remove the last few pounds of water from malt, there is a modern tendency to somewhat higher moisture values in some markets, e.g., in Europe, some malts may have up to perhaps 6% moisture.

The description above applies to single-floor kilns in which the malt is turned regularly during drying and, therefore, the malt in the bed is relatively constant in moisture and other properties from top to bottom. In such case, an analogy may be drawn with germination. Some modern kilns, however, apply heat to rather deep beds at a relatively high initial constant temperature, say 80°C or more, without turning the malt during drying. Such systems depend on evaporative cooling of the malt to conserve enzymes. Even so, malt at the bottom of the bed is drier and hotter than grain in the upper layers and such a differential can hardly contribute to evenness in the finished malt. In such a kilning program, a stage arises called "breakthrough" (Figure 10–9), when the air-off temperature closely approaches the air-on temperature. The malt is then turned and cured by raising the air-on temperature.

Many kilns have two floors (Figure 10–8b), which achieve efficient utilization of air flow without recirculation of air through the bed. Similarly, many single-floor kilns are run in tandem, operating effectively as a two-floor kiln. In a two-floor kiln, the malt on the upper kiln is one-half of the kilning cycle behind that on the lower kiln, and the air-off of the lower kiln is the air-on of the upper kiln. The air is used twice. The air-off the lower kiln is always too hot and too low in volume to enter the upper kiln, which is loaded with more moist malt. Therefore, between floors, the air is diluted with additional air to increase its volume and lower its temperature to that suitable for air-on in the upper kiln. This assures efficient use of air, which exits from the top of the kiln at 95 + % humidity.

In practice, in a two-floor kiln, after stripping the lower kiln, the contents of the upper kiln drop to the lower floor; the top kiln is then loaded with a fresh batch of green malt and the process continues. The floors usually tip in sections, rather like a large horizontal venetian blind. In some kilns, the entire floor tips in one direction to discharge the malt.

Thermal efficiency is important in kilning because this process consumes 75–80% of the energy used in malting. In some kilns, entering air is pre-heated by exchange with exit air, for useful savings. In a few places, cogeneration is possible; in this process, the maltster burns oil or gas to generate electricity, which is sold to the national grid and the "waste" heat from the generator is used for kilning.

Indirect heating of air for kilns is less efficient than direct firing in which the flue gases of the fuel pass through the malt. However, oxides of nitrogen (NO_x) are formed in flue gases, even with low-NO_x burners, and can react with organic amines of malt rootlets and shoot to form nitrosamines, which are potentially carcinogenic. Therefore, indirect heating of kilns, in which flue gases exchange heat with a supply of fresh air, is used in most places to control nitrosamines. Even so, it may be necessary to introduce some SO_2 into the air stream, especially in urban locations with ambient NO_x in the atmosphere. SO_2 also serves to preserve the pale color of malt, where that character is prized.

Kilned malt retains the rootlets, which must be removed immediately after kilning before they pick up moisture and while they are still brittle. The hot malt is dropped from the kiln, cooled rapidly with dry air, and beaten and screened to separate the rootlets. The cool malt is transported to storage silos where it is commonly retained for up to a month or more before use. This allows the moisture content to even out within each kernel and among kernels and the malt is said to mature. Mature malt brews better. The malt is analyzed and blended to meet brewers' specifications and, finally, loaded into rail cars or trucks for transportation to the breweries.

10.6 MALT QUALITY, ANALYSIS AND SPECIFICATION

Given the key role of malt as the primary source of enzymes, carbohydrates, and nitrogenous compounds, which ultimately form the brewers' wort, it is not surprising that brewers lay great stress on malt quality.

Much can be learned about a malt from its appearance, odor, and taste. Visual inspection reveals whether or not it is of a uniform blend (same-colored grains, same-sized grains), and that dust and debris such as stones or metal (hazardous in milling) and weed seeds (no brewing value) are absent. The aroma of malt should be "clean"; no mustiness (earthy, moldy aroma), indicative of mold contamination, should be present. The malt should be dry. Biting through a kernel of malt reveals the degree of modification. Well-modified malt is crunchy (friable) throughout and sweet to the taste, whereas less well-modified malt will contain hard ends, which can be easily discerned by their resistance to biting or chewing.

With the increase in understanding of the chemistry and biochemistry that underlies the role of malt in brewing, has come the definition of more precise specifications. This, in turn, has been made possible only by the development of reliable methods of analysis. New methods are the subject of continuing research. Quality assurance of malt is achieved by setting precise specifications for those parameters of importance to the brewer. A description of individual methods is beyond the scope of this text. However, compendia of methods are produced by the American Society of Brewing Chemists, the Institute of Brewing, and the European Brewing Convention.

One consequence of the existence of different sets of analytical methods is that, very often, it is impossible for students of brewing to make accurate comparisons of different malts from published data. This is because the analytical techniques described in the several compendia all differ somewhat in methodology and results are often reported in different units! There are some parameters, such as hot water extract (HWE), which, because of their economic implications in terms of yield of extract, are important to all. Other parameters, such as the amount of total nitrogen, level of precursors of dimethyl sulfide (DMS), amount of β-glucan, or level of saccharifying enzyme are more important to some brewers than others. This is because of the different beers produced or brewing practices adopted. Thus, high nitrogen (protein content) is considered undesirable by brewers of traditional ales. Their beers are not processed by long-term cold storage, which facilitates precipitation of protein and, as a result, protein can precipitate in finished beer, lowering its shelf-life. DMS precursors are necessary for brewers whose beers contain characteristically high levels of this flavor compound, but otherwise are undesirable. High levels of saccharifying enzymes are essential for higher temperature mashing processes and mashes containing adjuncts.

Some common values for malt analyses are shown in Table 10–1. Every attempt has been made to put them on the same basis to enable comparison. Hot water extract by a fine grinding process is a measure of the maximum potential extract available from the malt. This is then the target figure for the brewer. A 100% efficient wort production process would yield this amount. Of course, this is unlikely to be achieved in practice. The U.S. six-row malt (Table 10–1) shows lower extract potential than the other malts because the kernels tend to be small; this results in a high proportion of husk and, consequently, less of the total weight is starch (extract). This malt is prized for its extraordinary diastatic power, which permits mashing with non-enzymic adjuncts. Total protein and nitrogen are lowest in the U.K. malt. This is because, where possible, varieties of malting barley containing low nitrogen are selected. Various indices of modification are included in the table:

- **Fine/coarse extract difference** (f/c difference): well-modified malt shows a value of about 2% or less. The greater the degree of modification, the easier it is to extract all the material by coarse milling. The fine/coarse difference is therefore smaller in better modified malt.
- **Soluble nitrogen ratio** (SNR or S/T or Kolbach Index): well-modified malts of ordinary nitrogen content, have values of 39–41%. A high ratio of soluble nitrogen to total nitrogen implies greater breakdown of proteinaceous material during malting, and so better modification.
- **Cold water extract** (CWE): well-modified malt shows a value of 18–20%. As modification proceeds, relatively insoluble high molecular weight molecules tend to be broken down to more soluble smaller molecules. High values of CWE, therefore, imply good modification. This measure is less popular now than it once was.

Other indicators of modification include a measure of **friability** (about 85% is desirable) or **homogeneity** (96 to 100% preferred), estimated by

Table 10–1 Analyses of different malts

Measurement	U.K. pale ale	German lager	U.S. 1	U.S. 2
Hot water extract fine (%)[a]	81	81	80	77
Fine/coarse extract				
difference ND	2.5	2.5	2.0	1.7
Total protein (%)	9	11	12	13
Total nitrogen (%)	1.5	1.8	1.9	2.1
Soluble nitrogen (%)[b]	0.58	0.76	0.78	0.82
Soluble nitrogen ratio[c]	39	42	41	39
Cold water extract (%)	19	ND	ND	ND
Color (°Lovibond)[d]	3	1.5	1.6	1.7
Dextrinizing units	ND	46	37	40
Diastatic power[e]	45	85	115	156
Moisture (%)	2.5	4.5	3.9	4.1

U.S. 1 and U.S. 2, North American malts from two- and six-row barleys, respectively.
[a]U.K. pale ale value calculated from 301 1°/kg determined by Institute of Brewing method; values for German and U.S. malts produced by EBC and ASBC methods respectively.
[b]Values for German and U.S. malts calculated from total nitrogen and soluble nitrogen ratio.
[c]This value is often referred to as the Kolbach index of modification.
[d]U.K. and German malt data converted from EBC measurement using the relationship 1° Lovibond = 0.5 EBC units.
[e]Value for German malt converted from EBC Windisch–Kolbach units.
ND, not determined.

---

staining corns cut in half longitudinally with e.g., calcofluor or methylene blue.

On the basis of the analyses in Table 10–1, it can be concluded that the German lager malt is better modified than the U.K. ale malt. This, of course, goes against tradition where the opposite is said to be the case! However, it would be a brave, if not to say foolhardy, brewer who used the German malt in a British infusion mash system or the U.K. malt in a German decoction system, or either malt in an American double mash. Vital though malt analyses and specifications are, they do not give a complete picture. Malts of the same specification produced in different maltings often give different brewing performance. The first time a new season's malt is used is always a stressful time in breweries, even though the analytical specifications are the same as those of the malt of the season before.

Two measures of enzyme activity are shown in Table 10–1: namely dextrinizing units (DU), which is largely a measure of α-amylase activity; and diastatic power (DP), a measure of the concerted activity of both α- and β-amylases. α-Amylase activity is at similar levels in German and U.S.

malts. Diastatic power, however, is much greater in U.S. malts. This indicates a higher content of β-amylase so necessary in high temperature adjunct mashes if satisfactory saccharification is to be achieved (Chapter 13). The low temperature thick mash used by ale brewers can readily tolerate the lower amounts of enzyme present in ale malt. The lower enzyme content of the U.K. malt is largely the result of using different barley varieties and high kilning temperature, which is also reflected in the lower moisture content and high color of this malt.

10.7 SPECIAL MALTS

Special malts are very important modifiers of the color and flavor of beers. Most contain no active enzymes. They do, however, contain extractable material. In some products, this is in the form of simple sugars, but starchy material is also present. Special malts are therefore readily extracted during mashing and contribute to total and fermentable extract. However, special malts always contain less extractable material and usually more color than regular pale malt (Table 10–2). In general terms it is said that increasing the color of a malt by a factor of two increases the flavor intensity by at least a factor of four.

Unfortunately, there is no consistent nomenclature for these special materials and brewers and suppliers should always be sure they are using a common language when discussing special malts. Furthermore, it is virtually impossible reliably to compare one product with another without a detailed knowledge of the malt used to make them, of the exact processing conditions used, and the analytical parameters of the products.

Table 10–2 Typical extract values (l°/kg dry basis) for special malts and roasted barley

Product	Extract	Color
Crystal	268	200
Carapils	260	30
Amber	275	50
Chocolate	270	1000
Roasted	265	1200
Roasted barley	270	1350

On the same basis a typical pale ale or lager malt would have an extract value of 302 and color of 3.

Special malts are used in relatively small quantities and often mixed and milled together with pale malt to form the grist. Where very large or very small amounts are used, the brewer may choose to mill them separately. Special malts and roasted materials can be extracted with hot water away from the brewhouse and that extract added at some convenient stage of processing, even to finished beers. The formulation of a grist in terms of its content of special malts is often a guarded secret.

A modern plant for making colored malts (Figure 10–10) comprises a heated horizontal revolving drum akin to a coffee roaster. The malt used is either (a) normal dry malt, (b) green (unkilned) malt or (c) kilned malt that has been re-wetted. The choice depends on the product being made. This malt can be "stewed" or dried/roasted. Stewing is heating without drying, and is achieved by heating the drum externally with no air passing through it. Drying and roasting occur when hot air is passed through the drum. The final product is discharged onto a cooling floor where it is turned and cooled with a flow of ambient air.

Thus, there are two product lines: (a) those represented by malts with unconverted endosperm (i.e., pale malt through the various high kiln, brown, and roasted malts), and (b) those with partially converted endosperm (achieved during stewing), such as carastan or cara-pils malt or

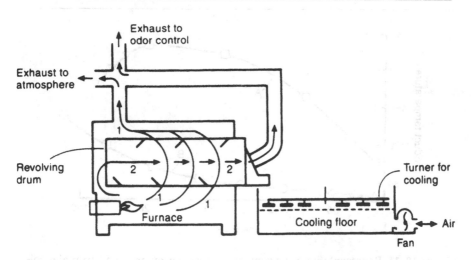

Figure 10–10 Schematic representation of a roasting drum for manufacture of some specialty malts and roasted barley. 1. Air passes around the drum; malt heats without drying ("stews"). 2. Air passes through the drum; malt is dried and roasted.

dextrine malt and crystal malts. In both product lines, higher color and more intense flavor go together and they are likely to contain little or no enzyme activity. We suggest that the term **caramel malts** be used to describe the first group of malts and the term **crystal malts** be reserved for the second type, which have a glassy or crystalline interior. Roasted barley fits with neither of these categories, being unmalted. Examples of the types of temperature profiles used are shown in Figure 10–11. Table 10–2 provides some typical analyses of some special malts.

To produce roasted products (caramel malts in our nomenclature), such as chocolate malt, pale dry malt (though usually not of premium quality) is dry-roasted in closed rotating drums at temperatures up to 225°C for 2 hours (profile A, Figure 10–11). This malt is used in stouts and porters at 10–20% or more of the total grist to give intense black color and roasted and bitter flavors in the beers.

Crystal malts (profile B, Figure 10–11) are made by wetting pale malt to 50% moisture, heating to 65°C without drying (stewing), and then drying and roasting the material as shown. In effect, the malt endosperm of each individual grain undergoes a mashing process followed by intense heating to encourage color formation (Maillard reaction). Such malts do contain unconverted starch. When dried at high temperature, the endosperm

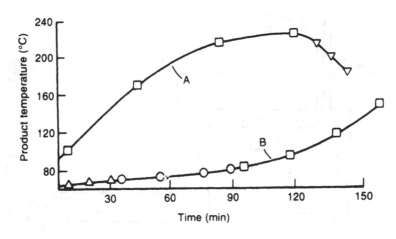

Figure 10–11 Representative programs of heating barley and malt to make colored products. Profile A: roasted barley and colored and roasted malts. Profile B: carastan and crystal malts etc. Actual colors and flavors are determined by modifying the temperature and time of treatment. The colors and flavors of any particular product require specific profiles. △, stewing; ○, drying; □, roasting; ▽, quenching.

hardens to a glassy mass within each kernel. The color and flavor of the product reflects the time and intensity of heating used during stewing, drying, and roasting, up to 140°C. A wide range of products is possible. Crystal malts are usually very dry, about 2% moisture, and quite free of enzymes. This type of malt is typically used at levels of 2–3% of the grist in English ales, for example, and higher levels in darker beers.

Amber malts are produced from dry malt of 4% moisture by heating to 93°C for 20 minutes, then very slowly heating up to 140°C and holding until the desired color develops. This malt gives some color and a biscuit or nutty flavor to beers when used at levels of 2–5% of the grist.

10.8 NON-MALT SOURCES OF EXTRACTABLE CARBOHYDRATE (ADJUNCTS)

Adjunct materials are used by brewers to replace malt for economic reasons and/or to modify the flavor of a beer. If malt is replaced with non-malt starch, e.g., from rice or corn grits or sugar syrups, a lighter-flavored, lighter colored and less satiating product results. Alternatively, the objective may be to increase the flavor of a beer and in these cases, non-malt-based specialized adjuncts (e.g., roasted products) could be used in place of, or in addition to, special malts.

Adjuncts are added to the process at the mashing stage (i.e., in the grist), at the kettle boil or occasionally in the fermentation or the beer before maturation.

10.8.1 Mash adjuncts

Many mash adjuncts, such as rice or corn grits, contain ungelatinized starch of high gelatinization temperature and must be boiled for 20 minutes or more in a cereal cooker, before entering the mash. On the other hand, if an adjunct contains starch with a sufficiently low temperature of gelatinization, it can be added directly to the grist (e.g., some wheat and barley products). Alternatively, the adjunct might contain gelatinized starch as a result of processing (e.g., flaking), and so can also be added directly to the mash. Adjuncts that have converted (hydrolyzed) starch such as corn syrups can be added directly to the kettle.

Mash adjuncts are generally processed cereal products and supply extract to the wort, and usually contain (pound for pound dry weight) more extract than malt (Table 10–3). They are primarily sources of starch derived from corn (maize), rice, barley, or wheat. The starch, when gelatinized, is readily hydrolyzed by malt amylases in the same manner as

Table 10–3 Typical extract values (l°/kg dry basis) for various cereal adjuncts

Cereal	Product type	Extract
Maize (corn)[a]	Grits	340
	Flakes	340
Rice	Flakes	360
Wheat	Flakes	300
	Flour	340
	Torrefied	300
Barley[b]	Flakes	280
	Torrefied	270

[a]May contain oils that influence beer quality.
[b]When used at high levels, increases wort viscosity by 50% (torrefied material), or two- to fourfold (flakes).

malt starch during mashing to provide a functionally identical spectrum of sugars. The cereals may be used as grits, which require cooking, or as flakes or torrefied products (sometimes referred to as micronized).

Grits are derived from grain endosperms by coarse milling. Flakes are grits that have been heated and rolled to compress the material. Torrefied products are treated with dry heat (often direct infra-red), which causes the grains to "pop" or micronize. (More recently extruded cereal products have become available but have not yet been adopted by brewers because their relatively low bulk density makes them expensive to transport and store and less convenient to use.) When products are heated during preparation, gelatinization and other changes to the endosperm structure occur, which increase the amount of material extracted in the mashing process. This may be of sufficient economic advantage to offset the cost of using the processed product.

Problems may be experienced with some adjuncts. For example, maize (corn) or rice grits must be low in fat and not rancid to prevent their undesirable flavor from entering the process; materials prepared from raw barley contain solubilized β-glucans, which can cause run-off and filtration difficulties. Grits of rice and maize must be cooked to gelatinize the starch (Chapters 12 and 13). In each case, these products are quite bulky and thus assist in keeping an open mash. Some adjuncts, such as grits and micronized products (but not flakes), are usually milled before use.

10.8.2 Kettle adjuncts

Adjuncts added to the kettle boil include invert sugar (hydrolyzed cane sugar), glucose syrups or solids (derived from maize starch), other cereal

Table 10–4 Typical extract values (l°/kg) for sugars and syrups

Product	Extract
Invert sugar (from cane)	320
Solid sugar (from maize)	310
Maize (corn) syrup	300
Malt extract	300

syrups, and malt extracts. These are sometimes called "wort extenders." The extract they contain is, of course, not subject to the action of malt enzymes and thus must be ready to enter the wort. Typical values for extract are shown in Table 10–4.

Brewing syrups produced by the acid hydrolysis of starch low in nitrogen (protein) contain D-glucose as the main fermentable component and are about 50% fermentable. Additional processing with enzymes increases the fermentability to 80%. Syrups produced by enzyme activity alone can be made with a sugar spectrum similar to that of malt wort (Table 10–5). Usually microbial enzymes are used: one with α-amylase activity at high temperature and then, at lower temperature, an enzyme with activity similar to β-amylase.

10.8.3 Other specialized adjuncts

Roasted raw barley is produced by a similar process to chocolate malt (see above) and is a cheaper alternative. This product may cause run-off problems in mashing or difficulties with beer filtration if β-glucans are extracted in soluble form. This problem may be circumvented by extracting the coloring and flavoring constituents from the roasted barley by using water at slightly alkaline pH. A concentrated extract can then be added to the mash or kettle boil or even beer in process.

Table 10–5 A comparison of the starch-derived carbohydrate (% total) of syrups and malt wort

Carbohydrate	Acid enzyme syrup	All enzyme syrup	Malt wort
Glucose	42	4	9[a]
Maltose	30	60	46[b]
Matotriose	8	16	20[c]
Oligosaccharides and dextrins	20	20	25

[a]Includes fructose.
[b]Includes 5% sucrose.
[c]Includes 6% maltotetraose.

Caramel is a substance produced by heating invert sugar or acid-hydrolyzed starches with ammonium ions. This produces colored compounds formed by the Maillard reaction (Chapter 15). The darker the color, the less sweet the product. Concerns have been expressed about the safety of caramel in foods. However, in defense of the product, it can be said that the substrates and conditions needed for the Maillard reaction are normally present in a kettle boil and in the roasting drums used to produce special malts (Figure 10–10). The International Technical Caramel Association has drafted detailed specifications for a range of products. These have been approved by the World Health Organization and the European Community. A 100% liquid caramel would give a color of 60,000° Lovibond (30,000 EBC units) and an extract of 250 l°/kg. A 0.1% solution would give a color of about 60 Lovibond. Caramels have found use in many dark beers and have characteristic and highly valued flavors.

Maltodextrin produced by partial acid hydrolysis of purified starch (negligible nitrogen content) contains little fermentable material. However, the dextrins in it have been used to give "body" to beers. Recent research suggests that dextrins do not affect beer viscosity, but may impart sweetness following their hydrolysis by amylase enzymes in saliva.

Lactose (milk sugar) is not fermented by brewer's yeast but has a characteristic sweetness and very smooth mouth-feel. Milk stouts traditionally contained 7–12% lactose. In a former age, this product was highly recommended for mothers who breast-fed their babies!

Finally, priming sugar is used by brewers to provide fermentable sugars in cask beers (traditional draught or live beers) and to adjust the residual sweetness of other beers. This material, which is invert sugar often with added caramel, may be used in solid or liquid form. Addition of primings is made to support "conditioning" whereby fermentation in cask (or bottle) carbonates the beer. Residual unfermented primings also give some sweetness to the product. Typical additions increase the specific gravity of the beer by one or two degrees that, when fermented, gives the requisite amount of carbon dioxide in the product.

CHAPTER 11

Malting biochemistry

11.1 INTRODUCTION

The previous chapter on malting technology viewed barley and malt as a population of kernels and used collective terms such as grain bed to describe this population. The measurements used and the concerns addressed dealt with larger issues, both in principle and in dimension, than malting biochemistry deals with. Malting can be effectively and usefully thought of on the larger scale, for example, for engineering new plant, economics of production, blending for specification and so on. However, germination, and the hydrating and dehydrating activities that precede and follow it, are not collective events but occur in each individual kernel, and it is only at that level that the biology of malting can be addressed and understood, because **modification** (the term for all those changes that accrue in malt as it develops from barley) occurs in each separate kernel. Moreover, the extent to which each kernel behaves in the same way as all other kernels determines the evenness of the malt. With the possible exception of acrospire growth and some staining techniques, which are measured on relatively few kernels, there is no practical measure for kernel-to-kernel variation in a batch of malt. All measures used in practice enumerate the quality of a population of kernels.

11.2 THE SINGLE KERNEL

In a single kernel, modification takes place as a series of well-orchestrated, overlapping and complementary events, which have a single purpose in nature: to assure that the viable embryo is adequately protected and nurtured until it is a plant sufficiently established to derive all its needs from its surroundings. The reserves of starch and protein laid down in the barley storage tissue, the endosperm, exists for the nutrition of the embryo, not for making beer, and those reserves only become a convenient source of brewer's extract if the embryo's drive to obtain short-term nutrition is intelligently harnessed and controlled. This is the maltster's task and dilemma; in practice, maltsters must achieve full modification with minimum malting loss, i.e., excellent brewing qualities and high extract yield.

Malting loss is the loss of dry weight as barley is converted into malt. It mainly results from the mobilization of starch of the endosperm to provide the mass of the growing embryo and the biochemical energy (through respiration) needed for that growth. Modification also originates in embryo growth. It is no surprise, therefore, that high malting loss and good modification tend to go hand in hand. The objective of control in malting is to restrict embryo growth (the seat of malting loss) while endosperm breakdown (modification) progresses over an economically short time.

11.3 BARLEY MODIFICATION

Barley modification follows an easily discerned pattern: it progresses from the embryo or proximal end of the kernel towards the distal end, and from the periphery, especially the dorsal periphery, towards the center of the endosperm. The least modified area therefore is always close to the distal tip and perhaps skewed to the ventral side. If the malt is significantly undermodified, this region is called a "steely tip" or "hard end." This overall pattern of modification results because the four events that comprise germination also have a strong proximal-to-distal pattern, which derives from the morphology and physiology of the kernel. These events are: (a) the entrance of water and the hydration of the kernel, (b) the formation and release of gibberellic acid by the embryo, (c) the stimulation of the aleurone layer and the formation of enzymes of modification, and (d) the release of these enzymes into the endosperm and their action.

11.3.1 Steeping

The state of suspended animation of the embryo and aleurone, brought on by dehydration of the grain as it ripened in the field, is broken by the advent of moisture. Steeping irrevocably sets malting in motion and is a crucial first step towards quality malt. Rapid and even water uptake is required with abundant aeration. Water provides a medium in which biochemical reactions can take place. Therefore the steep-out moisture content of the grain strongly influences the rate and extent of modification (Figure 11–1) and is hence a matter of close control in processing. The husk wets rapidly in steeping and within a few hours the moisture content approaches the high level of moisture (about 50%) that these

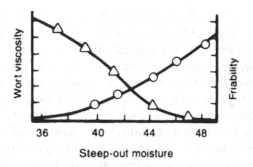

Figure 11-1 Relation of wort viscosity (△) and malt friability (○) to steep-out moisture (all else being equal in the process).

outer layers will retain throughout the process. The barley, thus, retains a surface film of water, made possible by the wettable husk, which is the first source of water uptake. This film should be maintained at all times during steeping and in the early stages (at least) of germination, because without this supply, the embryo will withdraw water from the endosperm, which impedes modification. Excessive use of spray steeping for water conservation tends to promote uneven modification perhaps for this reason.

The intact pericarp-testa in contrast, is relatively impermeable to water, having a cuticular or waxy composition, and water enters the kernel primarily through the micropyle region at the base of the embryo where the testa is thin or absent. This means that the embryonic axis hydrates and swells before the endosperm is wetted; it also dries first in dehumidifying conditions in germination or kilning. The embryo always has a higher moisture content (55–60%) than the endosperm (30–40%), which takes up water much more slowly. Water penetrates the starchy endosperm from the scutellum relatively slowly and in a proximal to distal direction; mealy kernels hydrate more evenly and faster than steely ones. In both cases, however, water penetrates along the dorsal plane ahead of the ventral, with the space between hydrating last. Water uptake in steeping is faster with warmer water and thinner barley kernels.

11.3.2 Germination

The germinating embryo produces gibberellins and abscisic acid. For malting this is an event of central importance, because these plant hormones stimulate and modulate the production of enzymes of modification (including α-amylase) by the aleurone layer and their release into the endosperm. From the maltster's point of view, release of gibberellins (Figure 11–2) is the only desirable function of the embryo (although residual levels of S-methyl methionine may be important in some lager malts). Maltsters in many parts of the world, but not in the U.S., apply one gibberellin, gibberellic acid (GA_3), to barley usually after steeping to bolster the natural hormone and to accelerate modification. Gibberellins are synthesized in the embryonic axis possibly from pre-formed precursors including *ent*-kaurene in the scutellum, and are released to stimulate the aleurone layer to synthesize enzymes. This formation and release of gibberellins takes place in the first two days of embryo growth, i.e. during steeping, and the significant proximal-to-distal flow of water can carry the hormones with it through the endosperm to the aleurone.

An alternative and more specific view reports that exogenous gibberellic acid, or gibberellins formed in the embryonic axis, are transported through the vascular tissue of the scutellum to the junction of the scutellum and aleurone layer. This connection is best developed on the dorsal side of the grain and serves to explain the somewhat asymmetric progress of modification (Figure 11–3). The scutellum is mounted askew to the endosperm and the presence of the ventral furrow may also impede ventral advance of modification (Figure 11–4). The gibberellins then travel through the aleurone from the proximal to the distal end.

The aleurone releases the bulk of the enzymes of modification, with the scutellum making an early and useful contribution; this happens strictly in response to gibberellins. Modification of the barley endosperm must await the formation and release of these enzymes. However, because the endosperm hydrates more slowly than the embryo, there is a delay before the products of enzyme action on the reserves of the endosperm become available to the embryo. The embryo is well supplied with sugars of low molecular weight to sustain it in this period. As sugars from the endosperm become available the supply of new gibberellins declines, because they are no longer needed. However, the pattern of gibberellin formation and release reinforces the water uptake pattern discussed above, providing for a strong proximal-to-distal influence on modification. This is reinforced again by enzyme synthesis, release, and action.

When the aleurone layer receives the chemical message delivered from the embryo in the form of the gibberellin hormones, it responds with a massive increase of enzyme (protein) synthesis at the expense of its reserve substances. The enzymes formed, including α-amylase, endo-β-glucanases and proteases, all have significant roles in endosperm degradation. Once

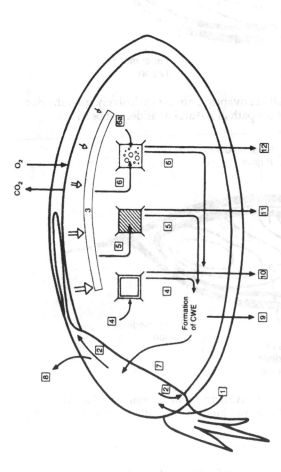

Figure 11–2 Summary of events occurring during grain germination that affect malt properties and are under the control of the maltster. Modification proceeds generally from proximal to distal end (from left to right) and from periphery to center of the endosperm. 1, Entrance of water through the micropyle (in general, flow is proximal to distal); 2, Formation and release of gibberellins, stimulating the aleurone; 3, Progressive stimulation of the aleurone to form and release enzymes of modification; 4, β-glucan solubilase and β-glucanases dissolve cell-wall β-glucans; 5, Proteases and peptidases partly break down proteins and release FAN; 6, α-Amylases attack small starch granules; 6a, Latent β-amylase becomes free β-amylase; 7, Embryo nutrition; 8, Malting loss from metabolism of CWE to CO_2 (respiration) and formation of acrospire (insoluble) and rootlets (removed); 9, CWE available in malts as an index of modification contributes to malt color; 10, Physical modification: (a) increase in friability; (b) lowering of wort and beer viscosity (factors correlated as indices of modification); 11, Index of modification as Kolbach Index (soluble nitrogen ratio, SNR or total soluble nitrogen/total nitrogen). This is important in beer foam/haze, beer flavor, and amino acids for yeast nutrition; 12, Formation of α-amylase (dextrinizing units, DU) effective in mashing; release of β-amylase (diastatic power, DP) for wort fermentability. These contribute to CWE in the major part. CWE = cold water extract.

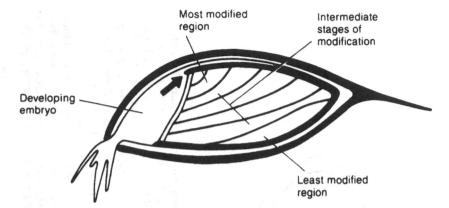

Figure 11–3 Progress of modification advancing almost exclusively from the dorsal aleurone. The arrow indicates the path of natural or added gibberellins.

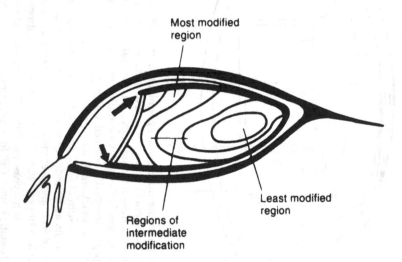

Figure 11–4 Progress of modification advancing mainly from dorsal aleurone but also from the ventral aleurone. The arrows indicate the path of natural and added gibberellins.

begun, enzyme manufacture continues unabated through the normal period of germination in the malting box or drum, and the level of these enzymes in the endosperm steadily increases. After the enzymes are released into the endosperm from the aleurone, they become effective in catalyzing modification; the release mechanism appears to require energy.

Enzymes move through the endosperm tissue by diffusion in the water present. The temperature of germination, the mealiness of the endosperm (i.e., the presence of a myriad of microscopic air passages that fill with water during steeping), the moisture content, the cell walls of the endosperm, and, of course, the amount of enzymes released all affect the rate and extent of enzyme diffusion and so influence the evenness and speed of modification. The temperature also affects the rate of enzyme action. The enzymes digest the endosperm cell walls, and the proteins and starch granules contained therein, by partially hydrolyzing them to release low-molecular weight products; these can diffuse and/or be transported back to the scutellum, the cells of which elongate to aid absorption, and thence to the developing embryo. Since the starch and protein is contained within cells, the walls of these cells must be hydrolyzed first. In well modified regions of the endosperm, cell walls are completely dissolved.

11.4 ENZYMES AND SUBSTRATES

11.4.1 β-Glucans

Endosperm cells contain the essential supplies required by the embryo for its nutrition. The cell walls of the endosperm into which the starch and protein is packed, render these materials unavailable to the embryo and to the brewer alike, and so the breakdown of these cell walls is crucial. Adequate degradation of cell walls causes good malt friability, and low values for fine/coarse extract difference, wort viscosity, and β-glucan.

The endosperm cell walls are complex structures comprising hemicelluloses and gums. These are chemically similar compounds that differ primarily in molecular weight and solubility: hemicelluloses dissolve in dilute alkali and gums in warm water. Depending on barley variety, the walls comprise about 70–75% β-D-glucan (glucan), 20–23% arabinoxylan (pentosan) about 5% protein, possibly present as a middle lamella between cell walls, and small amounts of glucomannan and other carbohydrates. The β-glucans, with a molecular weight of between 200,000 and 4×10^6 Daltons form a viscous solution in water; hot water extracts larger more viscous molecules than cold water. The water-soluble molecules are unbranched polymers of β-1-3- and β-1-4-linked glucose; the β-1-4 link dominates (70%). In larger molecules, there are extended sections of β-1-3-linked glucose units that do not exist in lower molecular weight compounds. This affects the susceptibility of the molecules to enzyme action, most crucially of course the β-glucanases.

The β-glucans are possibly linked by ester bonds, possibly involving ferulic acid, to peptides or proteins of the cell wall. Enzymatic degradation of β-glucans follows rupture of these ester bonds by enzymes that have been given the trivial name β-glucan solubilases which include a carboxypeptidase acting as an esterase. Ample amounts of this enzyme survive kilning and appear in malt where it is stable under mash conditions. In contrast, the β-glucanases are substantially degraded on the kiln and those that survive are quickly heat-inactivated in mashing. Therefore, sufficient release and breakdown of β-glucans must occur in malting; if not, β-glucans might be released during mashing in the absence of enzymes that can degrade them. This causes viscous wort that lauters slowly and beer that is difficult to filter.

Several enzymes called β-glucanases degrade these substrates; the enzymes include endo-β-1,3:1,4 glucanases, which hydrolyze β-1–4 links adjacent to β-1–3 links, endo-β-1–3 glucanase and exo-β-1–4-glucanase. These enzymes acting together are capable ultimately of hydrolyzing β-glucans of any molecular weight to cellobiose (glucose-β-1–4-glucose) or laminarobiose (glucose-β-1–3-glucose); these disaccharides yield glucose by the action of cellobiase and laminarobiase respectively.

The action of the glucanase group of enzymes is incomplete in ordinary malting and some β-glucan survives into malt to an extent that depends on the variety of barley and the conditions of malting (especially short duration and low moisture). The molecular weight and solubility of these remaining glucans, under the specific conditions of mashing, determines whether or not sufficient modification of the cell wall β-glucans has been achieved. Generally, β-glucans decrease to 10 to 20% or less of that present in barley. Brewers may add the commercially available (microbial) enzyme, endo-β-1–4-glucanase, which is fairly heat stable, to wort or beer to adjust viscosity if necessary. The treatment improves lautering and beer filtration under appropriate conditions of use.

The other major component of endosperm cell walls, the pentosans, comprises a backbone chain of β-1–4 linked xylose units to which short arabinose chains are linked at the 2- or 3- position of the xylose. The pentosans are generally not very soluble, and so remain in malt, but those that do dissolve are readily broken down in malting by the pentosanases present. Pentosans apparently play no significant role in beer properties although some hazes may arise from this source when brewing with wheat; some workers have suggested they could have effects similar to β-glucans.

11.4.2 Proteins

Proteins make up the matrix within each endosperm cell in which the starch granules are embedded. During malting, some modification of this proteinaceous ground substance is necessary to make starch more easily degraded in mashing, and also to form low-molecular weight nitrogenous compounds, especially amino acids. These will support the growth of the embryo and also the yeast during fermentation. Protein degradation in malting (but probably not mashing) may also promote colloidal stability in beer. In a well-modified malt, more than 40% of the protein is broken down to soluble components. The most auspicious time to degrade protein is during malting, when a full complement of proteases and peptidases is present. In contrast, extensive proteolysis is unlikely in mashing because many of the protease (endo-peptidase) enzymes are inactivated in kilning and the low temperature hold (protein rest) of the mash is short. In contrast, the exo-peptidases tolerate better the heat of kilning and in mashing could serve to form amino acids from peptides. Nevertheless the vast bulk of amino acids are formed in malting and the "protein" rest is likely a misnomer.

Solubility characteristics define barley proteins: albumins and globulins dissolve in water or dilute salt solutions, while hordein and glutelin dissolve, for example, in alkaline or hot alcoholic solutions containing reducing agents or detergents. The less easily dissolved proteins dominate in high-protein barley (Figure 11–5) and are the storage proteins. They are likely the origin of materials in beer foam, breaks, and chill haze, for example. Protein fractions isolated by solubility are by no means pure, and contain many components separable by modern methods such as electrophoresis. The protein matrix of the endosperm cells are packed together with the large and small starch granules. Under the electron microscope, the packing appears loose in a mealy endosperm and tight in a steely one. It may be the quality of this packing that influences proteolysis as much as the enzymes present: mealy endosperms modify more easily than steely ones.

The aleurone of germinating barley releases at least five endo-peptidases (proteinases, i.e., enzymes capable of producing peptides or polypeptides from proteins); two of them are sulfhydryl enzymes that form 90% of the proteolytic power. Endopeptidases are relatively heat labile and easily inactivated on the kiln. In addition a spectrum of peptidases exist; these exo-enzymes produce small peptides and amino acids from larger peptides. Exo-peptidases, including the carboxypeptidases, of which β-glucan solubilase is one, are present in barley endosperm and increase during steeping and germination. They are rather heat stable and persist in the endosperm after kilning. These enzymes produce the amino acids of malt and wort by cleaving single amino acids from the carboxyl end of peptides primarily during malting.

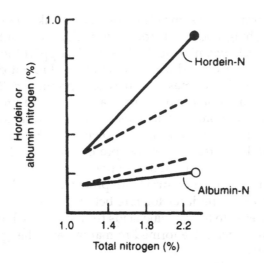

Figure 11–5 Schematic representation of the fate of nitrogen in barley (g/100g = %w/w). As the total nitrogen of barley rises, the proportion of nitrogen deposited as albumin (○) and hordein (●) does not remain constant (if it did it would follow the course indicated by the dotted lines). The extra nitrogen is disproportionately laid down as the storage protein hordein at the expense of albumin (solid lines).

11.4.3 Starch

Malting barley contains about 65% starch by weight, though poorer barleys with, for example, a thick husk, high protein or a thin corn, contain less. Starch occurs in barley in small (1–5 μm) and large (10–25 μm) spherical or egg-shaped granules. The large granules contain the vast bulk of barley starch, though they are much less numerous than the small granules. Granules have partly crystalline (ordered) and partly amorphous sections, giving them a lamellar or layered structure. This structure accounts for the characteristic black cross, which results from birefringence (interference) when granules are viewed under magnification in polarized light. The two main components of starch are amylose (about 25%), a straight-chain polymer of glucose linked α-1–4; and amylopectin, a branched glucose polymer with α-1–4 links in the chain and α-1–6 links at the branch points.

The granule surface associates with lipid material such as phospholipid and with matrix protein, which slows starch digestion. During malting, only about 10% of the barley starch degrades to feed the embryo. This mostly comes from the small starch granules, which degrade much faster than large ones and substantially disappear in malting. Although this represents malting loss, it is fortuitous that the small granules are preferentially used because they have a higher gelatinization temperature than large granules and so would dissolve more slowly in mashing. They may, therefore, be the cause of starchy worts and lautering problems when associated with other particulate material such as husk, cell walls, and precipitated protein, perhaps as teig material or top-dough. In contrast, degradation of large granules, evidenced by erosion ("pitting") of the granule surface, is undesirable; digestion of the lipid and proteinaceous material associated with and surrounding the large granules frees them for rapid degradation in mashing.

The enzymes of starch digestion include α-amylase(s), β-amylases, and limit dextrinase(s). Barley contains no α-amylase, and synthesis of this enzyme in the aleurone layer is among the central events of germination. α-Amylase can slowly attack ungelatinized granules, though β-amylase cannot. Generally, malt is well supplied with α-amylase. It is heat stable, especially in the presence of starch and so survives kilning quite well.

β-Amylase completes the breakdown of starch to produce maltose. In malt, α-glucosidase degrades maltose to glucose, which the embryo can use for energy metabolism. Barley contains β-amylase bound to another protein that renders it inactive or "latent." Action of endo-protease enzymes, or reducing substances, frees the β-amylase. Heat rapidly inactivates β-amylase; it is significantly inactivated in kilning, and speedily denatured at the conversion temperature of mashing. This influences the ratio of fermentable to non-fermentable extract in wort.

Limit dextrinase cleaves α-1–6 links of branched dextrins to produce glucose. The enzyme arises during germination, especially prolonged germination, but heat readily denatures it; it has some limited functions in practical malting, but none in brewing.

Starch represents the primary asset of malt, because it is the main substance from which brewers derive wort components. That the large starch granules are attacked only slowly, therefore, is a matter of great practical consequence; the reason for slow attack may be their structure, especially large-sized and non-gelatinized (crystalline) forms and the surrounding of protein and lipid. Rapid disappearance of small granules relative to large ones may be related to these factors or simply be a function of granule surface area/volume: a granule 2.5 μm in radius has 25-fold less surface area and 125-fold less volume than a granule of 12.5 μm radius.

11.4.4 Cold water extract

The major enzyme systems that develop in barley during germination are the β-glucanases, the proteases and peptidases, and the amylases. The aleurone layer and the embryo are alive, and enzyme levels and actions are controlled and organized within those tissues. This is not the case in the dead endosperm, where the extent of enzyme action is simply a function of enzyme concentration, its access to the substrate (including the availability of moisture), and the temperature. The only thing that is important in nature is that the supply of nutrients from the endosperm exceeds the embryo's need for them. Thus, there is a tendency for large molecules of the endosperm to be broken down somewhat faster than the embryo can use them. Maltsters call this accumulation of low-molecular weight materials soluble in cold water (or dilute ammonia) the Cold Water Extract (CWE); it is a useful indicator of the progress of modification, and essential in the manufacture of satisfactory malt. The CWE includes amino acids and reducing sugars that, when heated together on the kiln, form compounds that are colored and have desirable flavor and aroma typical of malt. This is called the Maillard reaction (refer to Figure 15–6; section 15.5.1). Manufacture of more highly colored malts may incorporate a longer germination and/or a long hold at relatively low temperature on the kiln to encourage formation of CWE before curing the malt at high temperature. Opposite practice preserves pale color where pale malts are prized. The amino acids of the CWE, i.e., those broken down in the endosperm but not utilized by the embryo, are the amino acids available from malt that will be extracted in mashing. In wort, these are essential for yeast nutrition and relate to the formation of beer flavor compounds (see Chapter 18).

The accumulation of soluble amino nitrogen during malting is also an index of modification expressed as the Kolbach Index or the soluble nitrogen ratio (soluble/total nitrogen or "S over T"). High CWE and S/T are linked to high malting loss and, of course, good modification.

11.5 CONTROL OF MALTING LOSS

The embryo is strictly necessary only for release of gibberellic acid to stimulate the aleurone to produce enzymes of modification. Further embryo action (growth and respiration) is undesirable and must be controlled by low germination temperature. Low temperature also tends to slow synthesis of enzymes, their diffusion through the endosperm and their action. Towards the end of germination, the malt may be allowed to warm up to foster the modifying action of released enzymes while stifling the embryo with accumulation of CO_2 (withering). Thus, long and cool processes produce thoroughly well-modified malts. Modern practices aspire to this ideal, but efficient production demands relatively short germination times. Modern malts tend, therefore, to be somewhat under-modified, primarily because they are malted a little too warm and consequently germinated not quite long enough for superior quality. The brewhouse process in which they are used are adapted to deal with this however.

Several modern technologies have emerged with the objective of producing well-modified malts in short processes. Commercially available gibberellic acid (GA_3) added to malt after steeping at about 0.1–0.5 mg/kg of original barley, rapidly enters at the micropyle and boosts the naturally produced gibberellins. It ensures quicker and more even germination and lowers malting loss by effectively advancing development of the aleurone ahead of the embryo. This favors enzyme production over embryo growth. GA_3 treatment tends to produce malt with high color, partly because the CWE is relatively high in nitrogenous compounds as a result of the selective stimulation of proteolytic activity, and the under-utilization of CWE by the embryo. Maltsters control this tendency with potassium bromate (100–200 µg/kg dry barley), which slows protein breakdown and embryo respiration.

A process called **abrasion** also accelerates modification. This involves deliberate damage to the barley kernel, probably at the distal end, such that added GA_3 is more effectual. Other techniques of grain damage have been tried, both to promote entrance of water and GA_3 and to inactivate the embryo. For example, embryo respiration is commonly slowed toward the end of germination by recirculating exhaust air from the grain bed to accumulate CO_2 or the germinating grain is under-aerated. Experiments found that grain can be stifled by re-steeping (flooding) the germinating grain with warm water or inhibited by gamma radiation. Inhibitory substances can be used: formaldehyde, sulfur dioxide, and acid are effective and have other advantages. Formaldehyde, for example, produces haze-stable beer. Acid steeping, followed by gibberellic acid, produces high extract (low malting loss) malt. All of these techniques have as their rationale inhibition of embryo growth and respiration (reduction of malting loss) while stimulating the aleurone layer (acceleration of modification through more rapid enzyme production). Note well that these experimental treatments do not find common practical application.

11.6 KILNING

The intensity of kilning can substantially alter what has been achieved during germination by its effect on malt flavor and color and on survival of enzymes. Pale malts are dried rather rapidly and finished at low curing temperatures. They are quite low in flavor and high in diastatic enzymes. More colored and more flavorful "high-kiln" malts are made by two strategies: first, prolonged low-temperature kilning to increase the CWE and particularly the amount of amino acids and low-molecular weight sugars in the malt; and second, high curing temperature. These two factors drive the Maillard reaction in which reducing sugars react with amino acids to form various high- and low-molecular weight compounds with intense flavors. These include condensation products of the reactants to form brown and flavorful pigments called melanoidins, low-molecular weight breakdown products including volatile aldehydes such as furfural, and a commonly isolated product of sugars with amines called maltol and iso-maltol (refer to Figure 15–6). These reactions are carried to extremes in the case of roasted malt, such as chocolate malt and black patent malt or roasted barley (Chapter 10). These products are heated to some 200–250°C to generate intensely colored and flavored products.

Various forms of crystal malt (Chapter 10) take the formation of CWE to extremes. In making these malts, green malt or re-wetted pale malt is "stewed" (heated without air-flow) on the kiln or more usually in a drum at about 70°C (Chapter 10) and then dried at high temperature.

CHAPTER 12

Mashing technology

12.1 INTRODUCTION

Brewers have three concerns when they extract wort from malt in the brewhouse: to produce the extract, to recover it, and to stabilize it.

- They achieve extract **production** during the mashing process, by dissolving the starch and other substrates and converting them, e.g., by action of the enzymes present, to a desirable spectrum of wort components.
- Extract is **recovered** in lautering or mash filtration, which is strictly a physical process of separation of solids from liquid.
- Boiling the wort achieves chemical, physical, and microbiological **stability** of the extract.

Around the world, brewhouse vessels vary considerably in size, number, and kind because of different brewing practices and traditions and different raw materials and products, but all brewhouse designs strive to attain the three objectives listed above economically and efficiently. Producing wort of the required characteristics concerns three related factors: (a) the raw materials used (including adjuncts), especially their extract and enzyme content and degree of modification, (b) the particle size of the materials, and (c) the times and temperatures of the processes used, especially mashing.

The brewhouse processes are **milling, mashing, wort separation, boiling, hop and trub separation,** and **wort cooling.**

12.2 MALT STORAGE AND MILLING

Malt contains everything for the successful manufacture of beer, except water and energy, which brewers must supply. Brewers merely have to open the package of malt, add water, and stir. Opening the package is called **milling**.

12.2.1 Storage

Breweries rarely store more than a few days' (or in some locations a few weeks') supply of malt on premise, because storage is expensive. However, the malt has been previously stored at the malthouse for a month or more to mature it. This assists proper brewhouse performance, especially wort separation. Malt is transported to the brewery by road or rail, usually in hopper-bottom transporters. These are unloaded at bays that, ideally, are protected from birds and other pests, and the malt is moved promptly through cleaners to storage. Like barley, malt requires cool, dry storage with control of pests, sparks, and dust at all times. Malts of different kinds and origins are stored separately until required when they are blended to the malt hopper above the mill commanding the brewhouse. In modern breweries, mill settings cannot be adjusted for exceptional malt lots and only those lots meeting stringent size specification can be processed. If large amounts of adjunct raw materials are used (e.g., rice or corn grits), special mills are installed for their reduction.

At the brewery, malt to be milled is batched to a special hopper commanding the mill and usually mounted on load cells or provided with some other device for measuring the weight of materials in process. The grain is dropped from the hopper to the mill and hence to a second temporary storage hopper, commonly called the grist case, which commands the mash vessel. In some small breweries, the malt falls directly from the mill into the **mash tun**; in most large breweries mill capacity is too small to permit such an arrangement.

12.2.2 Milling

Milling of any substrate must achieve two objectives: particle size **reduction** and particle size **control**. Roll mills are used for milling of malt for brewing (with one exception, see below), because this best suits one primary objective of milling: to leave the malt husk as intact as possible. An intact husk, including absence of a shredded husk, helps wort separation in lautering and may reduce extraction of tannins and other undesirable components. Roll mills crush the grain as it is drawn through the space between the rolls. Mill efficiency and capacity depend on the length and diameter of the rolls and their speed, the gap between them, and the friction between the malt and the roll surface. Rolls are commonly "fluted" to increase friction. Malt milling depends on two factors: compression and shear. Compression depends on the width of the gap and shear depends on the difference in peripheral speed between the two rolls of a pair. High shear results in mills in which one roll of a pair is driven and the other, undriven, follows it. In effect, the power of the first roll is transmitted to its partner through the agency of the malt.

The objective of milling is to produce crushed malt with the ideal spectrum of particles for extract production and recovery. Small particles more easily yield extract, but large ones allow faster wort separation. The way malt is milled, therefore, depends on the expectations of the brewer, the quality of the malt (especially its modification), and the equipment in which it will be processed. Wort separation in most breweries is done in either mash filters or lauter vessels or, in traditional ale breweries and many micro-breweries, in mash tuns. Mash filters permit a finer grist than lauters, primarily because of the shallow filter bed and the relatively tightly woven cloths. Indeed, the latest mash filter technology operates with hammer-milled (virtually powdered) malt. Mash tuns, with their deep grain bed, require coarse milling of malt.

Particle size distribution in milled malt can be analyzed by shaking a sample on a stack of sieves. This divides particles into husks, grits, coarse and fine grits, and flour. These particles tend to arise from different parts of the malt kernel and have a different composition. Husks are the largest particles and contain the outer layers and parts of the embryo, and yield the least extract. Fine grits and flour originate mostly in the well-modified portions of the endosperm because those parts break up most easily under the impact of the mill; they yield some 50–60% of the weight of milled malt but 80–90% of the extract. Coarse grits contain some under-modified endosperm and fractions of husk; at 20–25% of the milled malt they yield some 10–12% of the extract. Malt milled for mash filters may contain all of the endosperm material as fines and flour.

When malt passes through a single pair of rolls in a simple two-roll mill, for example, the malt kernel breaks up according to the friability of the individual parts of the kernel and the impact they receive. Although there is much transverse breakage, the husk tends to split longitudinally, releasing the endosperm contents and retaining the embryo. The endosperm itself breaks up according to the extent of modification. The least well-modified portions contain the original barley cell walls, which act as an adhesive to yield coarse grits. The extent to which this happens depends on the mill gap, but always the least modified regions will most resist milling. Brewers cannot simply reduce the mill gap of a two-roll mill to process poorly modified malt, because this tends to shred the husk. Two-roll mills are, therefore, best suited to well-modified malts, and/or to small breweries where their relatively low cost is advantageous.

However, undermodified malts benefit from a more selective milling, which is possible in multi-roll mills. Mills with four, five, or six rolls with screens are common. In a six-roll mill (i.e., with three pairs of rolls) portions of the grain may pass through only one, or two, or all three pairs of rolls. Particles are separated by vibrating screens between the pairs of rolls. In such a way, brewers produce grist with a relatively intact husk, yet with the endosperm of even under-modified malt reduced to fine grits and flour. In this sense, mill energy substitutes for modification. One method for measuring the degree of modification, called the fine/coarse (extract) difference, depends on the difference in extract yield of a coarsely or finely milled sample of a malt.

There is a tendency these days to mill all malts to the same particle size specification dominated by fine grits and flour to achieve maximum extract. Indeed that is the purpose of a multi-roll mill (Figure 12–1). However, excessive reduction of the under-modified portions of the grain also promotes the release of β-glucans and hence more viscous worts. Even though the capacity may exist to reduce malt to husk and flour, it should be resisted for this reason, except in those malts that are sufficiently low in β-glucans.

Milling sets the scene for what happens in the processes that follow and correct milling is important. Therefore, mill rolls should be routinely inspected and regularly adjusted to gap specification along their whole length, because they wear and vibration causes them to lose proper adjustment. Mill adjustment may also be necessary with change in malt supply, for example, the new season's malt. Mill screens must be intact and clean to maintain full efficiency.

An important form of milling, that is common, except in the U.S., is wet milling (Figure 12–2). This technique seeks to minimize damage to the husk and maximize reduction of the endosperm by wetting the grain before it enters the mill. This "toughens" the husk or, more precisely, makes it pliable. The husk can then deform as it passes through the rolls, which have a small gap to reduce the endosperm. In practice, the malt is steeped in water (the temperature of which may vary from cold to mash temperature) or steamed to raise the moisture to about 30%. The steep water is drained and discarded or drained to the mash mixer or it may simply pass through the mill with the malt into the mash. After passing through the rolls, the milled malt is mixed with water at mash temperature and pumped to the mash mixer. The benefits of wet milling are that the larger husk particles permit rapid run-off and the smaller endosperm particles allow higher extract recovery. These dual benefits seem to accrue in inefficient brewhouses that replace a dry mill with a wet mill. If the brewhouse is already efficient however, faster run-off results in lower yield, as might be expected. Impeccable sanitation is required.

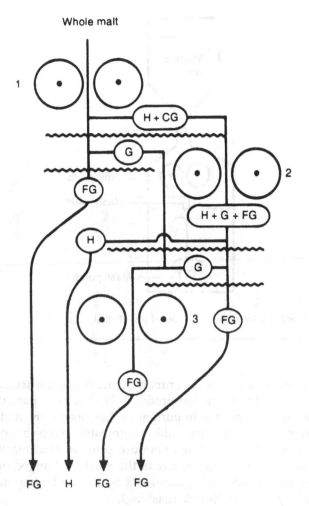

Figure 12–1 Schematic representation of the operation of a six-roll (i.e., three pairs of rolls: 1, 2, and 3) mill. The product from each pair of rolls is sieved through vibrating screens and only those particles requiring further reduction are passed on to the next appropriate set of rolls. Excellent milling of even under-modified malt can be achieved. H, husk; CG, coarse grits; G, grits; FG, fine grits and flour.

12.3 MASHING TECHNOLOGY

Brewers mash by one of three classical methods, which, with minor regional variations, describe most breweries: **infusion** mashing, **decoction** mashing, and **double** mashing. For good technical reasons, these

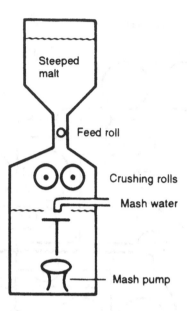

Figure 12–2 Schematic representation of a wet mill.

---

mashing regimes developed over many centuries in particular regions of the world for making particular products. For the most part, these historical reasons are no longer as important as they once were, and traditional methods of mashing, although still recognizable, have changed in modern times. A fourth method, **temperature programmed** mashing is now common, in which the temperature of the mash is changed by heat from steam coils (or even steam injection), not by adding boiling mash (decoction) or boiling adjunct (double mashing).

12.3.1 Infusion mashing

Infusion mashing is classically a British method of brewing. It requires a single, unstirred vessel (**mash tun**; Figure 12–3), in which brewers produce and recover the extract at a single mash temperature called the **conversion temperature**. The method requires very well-modified malt, which (in the British setting at least) is often quite low in enzymes. The method is unsuited for conversion of adjuncts that require gelatinization. Infusion mashing is particularly attractive in small breweries because the vessels are simple and cheap and easily scaled to their low output. However, demand for high throughput makes the technology unsuitable

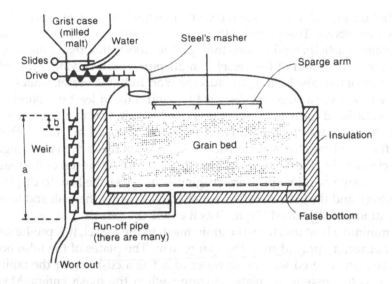

Figure 12–3 Schematic representation of an infusion mash tun. In such a device, extract is produced and recovered. The milled grain and mash water are intimately but gently mixed in the Steel's masher and delivered to the tun itself, which is pre-heated and in which the false bottom plates are covered with foundation water. After mashing, the wort is run off through a device designed to control the hydrostatic head applied to the bed; this is usually small at first (b) and can be much greater later (a). After the initial run-off, the bed is sparged to recover the maximum practical yield of extract.

---

in many large breweries. Thus, even in ale breweries, mash mixers and separate lauters are now common to help achieve high production capacity. Nevertheless, the infusion mash tun technology remains extant. A study of infusion mashing is instructive, because the factors in mashing that influence wort properties are most easily researched, explained, and understood in infusion mashing models. Indeed, much of the information about mashing was derived from such systems.

Mash tun operation

The infusion mash tun comprises a heavily insulated, covered vessel (Figure 12–3) about 6–8 feet (1.8 to 2.4 meters) deep, but usually 4 feet (1.2 meters) deep in small breweries, and up to about 30 feet (9.0 meters) in diameter, though most are smaller. It is equipped with a false bottom (which supports the grain bed) made of gun-metal or stainless steel and supported a few inches above the true bottom. The slots in the plates of the false bottom are 0.7–1 mm wide and several inches long and typically wider below than above. This provides the plates with some self-cleaning capacity because particles tend to pass through the slots rather than plug them. From the true bottom of the vessel, run-off pipes, each serving a roughly equal area of the false bottom, conduct the wort from the vessel, usually by gravity flow, over a structure designed to control (or at least measure) the hydrostatic head (or negative pressure) on the grain bed to a wort safe or grant. At the grant, the run-off (wort flow) is controlled by "taps." From here, the wort flows to the underback or kettle or, in the early stages, might be re-circulated to the mash vessel until it runs clear. At the top of the mash tun is a sparge arm for spraying sparging water over the bed, to displace the extract, and the Steel's masher, which folds together the malt and mash water (traditionally called "liquor") as it enters the vessel.

To minimize heat loss from the grain mash to the vessel, it is pre-heated with hot water sprayed from the sparge arm. The plates of the false bottom are then covered with mash water to act as a cushion for the falling mash and to prevent the plates clogging when the mash enters. Mash enters the vessel from the Steel's masher which intimately mixes brewing water and malt ("doughing in") to produce a thick mash at the mash conversion temperature of about 62–65°C (144 to 149°F). The mash temperature is fixed by the temperature of the malt and water, the amount of each being mixed, and by loss of heat to the vessel itself. Only by bringing the malt together evenly with water of the right temperature ("striking heat"), can a brewer lay an infusion mash with even composition and required temperature ("initial heat") throughout. This requires considerable skill. In a traditional mash, the grain is some 4–6 feet deep (1.2 to 1.8 meters) and the bed floats, although it is perhaps better described as buoyant. The air entrained by the Steel's masher and in the mash water and endosperm particles and the high density of the first worts contribute to this. After a stand of 20–60 minutes, wort run-off begins by setting (opening) the taps or valves or U-tubes controlling the run-off pipes.

Wort is drawn slowly at first with minimum negative pressure, to avoid pulling down or compacting the bed. Wort is drawn more rapidly when most of the dense wort has been displaced by sparge water sprayed on the top of the grain. Initially, if the wort is cloudy, it may be returned to the tun until it runs clear, but this is less common in modern brewing than it once was. Run-off continues until the wort has a low gravity and/or the required volume has been collected. The bed is then drained and the spent grains removed either manually, as is common in small breweries, or mechanically. After cleaning, the vessel can be reused; the cycle time is about 4–6 hours.

Infusion mashes are thick. They contain about 2.5 hl of water per 100 kg of malt, or a mash-in ratio of 2.5 to 1 by weight. This means that if the malt is cold, there must be a large difference between the temperature of the mash water and the temperature of the malt. Very hot mashing water striking cold malt may inactivate some malt enzymes. Some brewers warm the malt with warm air, hot water pipes, or steam to reduce this temperature difference. Other advantages accrue in this treatment: the malt mill yields a more intact husk if wetted with steam, for example, and the heat may "aromatize" the malt. Pre-heating and pre-wetting the malt is, therefore, not restricted to infusion mashing.

Correct mash temperature, vital yet quite difficult to achieve in infusion mashing, determines the ratio of fermentable to unfermentable sugars and other wort properties. Infusion mash temperature cannot be changed much once established because the mash is thick and because the vessel is unstirred. Pumping water in below the plates of the mash tun (underletting) raises the mash off the plates, if it has compacted there, and can be used to increase the temperature if very hot water is used. Underletting is also used in lautering. Sparge water is always hot (up to 80°C, 176°F) and can be used to raise bed temperature during run-off. This is desirable for recovery of extract and easier wort flow through reduced viscosity of the warmer wort. However, excessive increase in bed temperature may be undesirable because it may extract unwanted materials such as tannins, β-glucans, proteins, starch, and silica and other minerals from spent grain.

Run-off

The run-off of a mash tun requires skill and patience, because the bed is deep and, in a traditional tun, there are no mechanical aids provided such as rakes or knives. When the infusion mash is first formed, the insoluble particles are buoyed in a dense sugar solution (wort) that may approach 25°P. This extract is among and within the mash particles. Run-off is designed to recover the maximum amount of this extract in the required wort volume, which is fixed by the original gravity of the beer being made and the capacity of the kettle. It is quite usual to recover somewhat more wort than strictly required because up to 10% of volume is lost in normal wort boiling.

Wort separation does not take place on the slotted false bottom of the tun. The false bottom is not a filter plate, it merely acts as a support for the grain bed. Wort filtration takes places in the grain bed itself. The slots in the false bottom are therefore of no consideration in the problem of extract recovery as long as they are large enough and numerous enough not to impede wort flow and do not clog. However, since the entire flow of wort from the vessel passes through these slots and the entire negative pressure generated by that flow is exerted there, they do tend to clog and could restrict flow. Over time, wort can form a silica-containing deposit on the slots, which reduces their performance. Poorly designed or poorly maintained slots can therefore affect run-off. For the main part however, extract recovery concerns the grain bed itself (Figure 12–4a).

 When the wort flow begins, the first wort is dense and, even though hot, flows through the tiny channels of the grain bed with difficulty. It may be forced through by layering water deeply on the bed surface, by opening the taps more, lowering the run-off weir, or by accelerating a pump, all of which effectively increase the pressure on the grain bed or

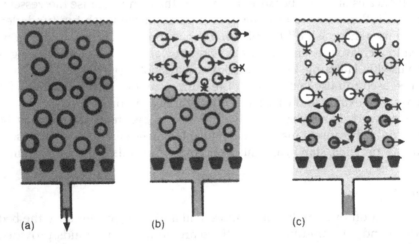

(a)   (b)   (c)

Figure 12–4 (a) At the start of mashing or lautering, there is strong wort throughout the bed and the bed is somewhat buoyant. In an infusion mash, the bed can float because of dense wort and entrained air. The particles are all saturated with strong wort, which exits the bed at high gravity. (b) At a mid-point in sparging, sparge water (lighter shading) follows the strong wort (darker shading) down through the bed, which becomes less buoyant; this and downward flow tend to compact the bed. Particles in the sparge water give up entrained wort (→) rather slowly from large particles (O) and more quickly from small ones (o). (c) Sparge water chases out the last of the strong wort and progressively weaker worts follow. Particles in the upper reaches of the bed have given up their content of strong wort (in b) and may now yield less desirable material (—x). The progress of these events explains why wort should be drawn slowly at first; why rapid run-off causes loss of extract; why later runnings can contain considerable extract, why last runnings contain undesirable materials, and why particle size is a critical factor in lautering.

the negative pressure below the bed. The flow will temporarily increase and then slow as the channels in the bed collapse (as when a child sucks a thick milk shake through a weak straw). The bed pulls down or compacts and decreases in depth. Flow through such a bed is restricted, leading to long run-off times and may even stop. A mash tun has no rakes or knives to cut and lift the bed to counter compaction, though these are provided in modern mash vessels and, of course, in lauters. While the dense wort is in the grain bed, it must therefore be drawn slowly.

As the dense wort is removed, the volume is made good with hot sparge water (Figure 12–4b). The sparge temperature is higher than the mash to help maintain mash temperature and reduce resistance to liquid flow. Although there is some mixing, the sparge water pushes the dense wort ahead of it, and so the gravity of recovered wort does not decline slowly and progressively over the entire run-off period but declines quite rapidly once the dense wort is pushed out (Figure 12–5a). As this point approaches, the speed of run-off can be increased because less-dense material is now flowing through the bed. However, the less-dense sparge makes the bed less buoyant and so the mash sinks and the bed compacts under any run-off regime. The key is to assure that the dense worts are substantially out of the bed before this collapse advances.

If there were no mixing between the dense wort and the sparge water, and if the bed were perfectly level and even in composition so that liquid flowed at the same speed through all parts of it (neither of which actually applies), the cut off between the flow of dense wort and sparge water would still not be sharp. The reason is that significant amounts of extract are trapped within the spent grain particles (Figure 12–4 b,c). This extract can be recovered only after the dense wort among the particles has been removed, because the extract must diffuse down a concentration gradient from the particle to the liquid among the particles. This diffusion takes time and therefore sparging cannot be too rapid, otherwise extract is lost. Diffusion is more rapid in a steep concentration gradient and when hot (part of the reason why sparge water generally raises the temperature of the bed), and if the mash particles are small. Each of these parameters has an optimum: if sparging/run-off is too fast and with large particles, extract recovery suffers; if too slow and with small particles, brewhouse throughput decreases. If sparging is too hot, undesirables may be extracted from the spent grain (Figure 12–4 b,c), especially unconverted starch at a time when amylases are exhausted, or tannic materials (these increase relative to the amount of sugar recovered in last worts, Figure 12–6 a,b). Brewers should note that extract recovered at the end of sparging is not simply diluted first wort (which would be truly worth recover-

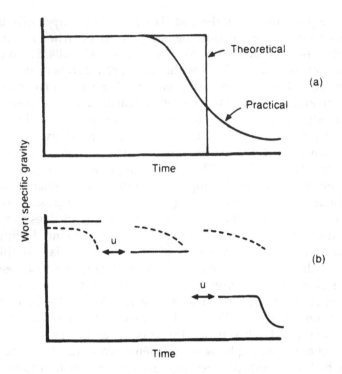

Figure 12–5 (a) Mash tuns are always run-off continuously to recover extract and lauters may be operated in this way too. (b) Alternatively, in lauters the grain bed is re-suspended and re-settled several times during run-off. The solid line represents extract recovered in each case and the dotted line in the lower drawing represents rate of wort flow. U, underletting.

ing); last runnings contain little of interest to brewers making quality beer.

If the mash particles are too small, run-off is inherently slow because the channels among them are narrow and may more easily clog. If particles are too large, recovery of extract trapped within them is poor.

Finally, the top-dough, or teig material, affects mash run-off. This is a deposit of protein precipitate plus fine particles from the grain, which settles on the surface of the mash. Water penetrates with difficulty through the top-dough, which acts as a plunger on top of the bed causing it to compact during wort run-off. It needs to be cut if present.

Excessive sparging, i.e., collecting a volume beyond the normal evaporation volume of the kettle, may be detrimental to wort quality and inefficient because the kettle boil must be extended to reach required gravity;

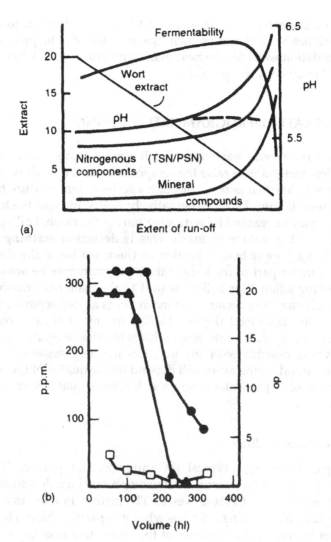

Figure 12–6 (a) Wort gravity falls rather sharply towards the end of run-off (*cf* Figure 12.5a, b). When gravity reaches about 5°P, wort pH rises rather significantly (except with calcareous water, ––––). Wort fermentability goes down and nitrogenous, fatty acid, mineral and tannic material increase dramatically *per unit of recovered extract*. TSN/PSN, total/permanently soluble nitrogen. (b) Run-off data (from a mash filter) in which the wort analyses are reported directly as concentration (ppm). ▲, Wort gravity; □, fatty acids; ●, polyphenols.

this uses excess energy and brewhouse time. It rarely pays to recover worts much below 1.5°P unless such "sweetwater" can be processed for use as foundation water for the next mash (sometimes called "weak-wort recycling") to avoid sewer penalties.

12.4 TEMPERATURE-PROGRAMMED MASHING

Brewers have several ways to establish a mash temperature program. The simplest method is to raise the temperature of the mash with steam jackets (or even steam injection), through a series of **temperature holds** or **stands** or **rests** (Figure 12–7a). Alternatively, one or more of the temperature stands may be reached by removing part of the mash, boiling it, and returning it to the main malt mash. This is **decoction mashing** (Figure 12–7c); when a brewer boils a portion of the mash he or she decocts it. Similarly, a major part of the temperature program may be achieved by adding boiling adjunct (or boiling water) to the main malt mash; this is called **double mashing** because the mash starts as two separate mashes, which are later combined (Figure 12–7b). In any of these three ways, brewers can manipulate mash temperatures to achieve results that would not always be possible with the infusion mashing system. The exact choice of time and temperature will depend on the quality of the malt and the objective of a particular operation for wort qualities such as fermentability.

12.4.1 Decoction mashing

This style of mashing is typical of German brewing practice. It is characterized by complex brewhouse arrangements that include a stirred mash mixer, a mash kettle and a lauter vessel. The method introduces a temperature program into mashing, which is wholly or partly achieved by boiling portions of the malt mash. Because of this, decoction mashing is particularly suited to dealing with (traditionally) under-modified malts. Modern German malts, however, are well-modified and the complex but traditional triple decoction system has been gradually simplified to double decoction (Figure 12–7c) or (more commonly) single decoction systems. Decoctions these days are, therefore, unnecessary for dealing with under-modification and likely have no real role to play in beer character or quality. Decoctions may now remain for reasons of tradition, though some claim possible impact on flavor and/or beer stability. The temperature program of the main mash, however, has some special functions.

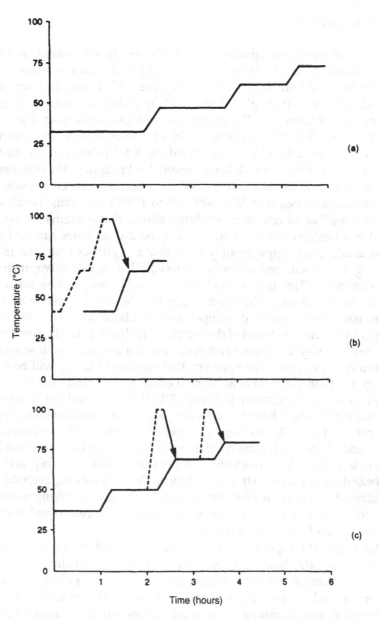

Figure 12–7 Example temperature-time programs for (a) temperature programmed, (b) double mashing, and (c) decoction mashing regimes. Solid lines refer to the main malt mash in all cases. In (b), the dashed line represents the adjunct (cereal mash) and in (c), represents the fractions of the mash taken for decoction in a two decoction process.

Equipment and technique

The mash mixer and smaller mash kettle are stirred, heated, and insulated vessels of simple design. Warm foundation water is placed in the mash mixer and with the stirrer running, the milled malt, usually from a six-roll mill, enters through a simple malt hydrator, which wets the grain. The mash is thinner than the equivalent infusion mash, with 3 to 5 hl of water/100 kg (3 to 5:1 weight ratio). In a traditional triple decoction, the first charge of mash, which is one-third the total volume of the mash, is soon transferred to the mash kettle where it is brought to the boil, usually with a short hold at about 70°C (158°F) for some starch conversion. The main mash however, is at 35 to 40°C (95 to 104°F); this stage is called the "protein rest" although recent evidence shows that proteins are not broken down here. At this low temperature, the β-glucanases can act (so the stand might more appropriately be called a β-glucan rest), the malt is thoroughly wetted, and all those materials that can dissolve, including proteins, do so. This initial stand is very much a preparation for further action. After boiling in the mash kettle, the decoction is returned to the main malt mash, raising its temperature to about 50°C (122°F). At this temperature, the substrates of the enzymes, including starch, dissolve further and significant enzyme action occurs. Certainly β-amylase and any peptidolytic or proteolytic enzymes that survived kilning will be active, though the α-amylase is not at its most efficient. The next decoction raises the main mash temperature to about 65°C (149°F), called the "conversion temperature," which favors rapid α-amylase action, but is too hot for other enzymes and even β-amylase action rapidly declines. The last decoction, when added back to the main mash, raises its temperature to about 75°C, which is called the "mash-off" temperature. This probably serves to squeeze the last part of extract out of the malt by dissolving even the most reluctant starch granules for degradation by the last surviving α-amylase molecules. Shortly, however, further enzyme action ceases and the mash properties can be said to be **fixed**.

The mash-off temperature also prepares the mash for lautering: it provides a little extra heat to counter that lost during transfer to the lauter, and, by making the mash warmer, renders the wort rather less viscous and more ready to flow through the grain bed during run-off. The extensive stirring and pumping of the mash drives out any entrained air and tends to break mash particles, forming fines, which reduce grain bed permeability. This is aggravated in low-calcium mashes typical of lager brewing. When the mash is finally transferred to the lauter vessel, the mash sinks to form a dense grain bed. Lauters therefore, unlike infusion mash tuns, spread the mash rather thinly over a wide area, so that the grain bed is relatively shallow, and the rakes run to maintain bed permeability.

In more modern decoction mashing regimes, the temperature program might follow a similar profile to that described, but incorporates only one or perhaps two decoctions. For example, the mash may be started at 50°C (122°F) or so instead of the first decoction; alternatively, the initial temperature rise could be done with steam jackets. In either case, the cycle could be completed with two decoctions (Figure 12–7C). If a single decoction were used, it could raise the mash to the conversion temperature *or* the mash-off temperature (as used for example in Japanese brewing).

12.4.2 Double mashing

Double mashing (Figure 12–7b), originally called the American double mashing system, may be viewed as a special case of the traditional decoction mashing method in which the under-modified material is treated separately from the malt by a single decoction. In double mashing, the under-modified material is the cereal adjunct (mainly corn (maize) or rice grits) and the decoction is the adjunct mash carried out in the cereal cooker (instead of the mash kettle). The main malt mash follows a similar temperature profile to that of the decoction method (Figure 12–7), although only one rise in temperature, to the conversion temperature, is driven by the addition of the boiling cereal.

In practice, the cereal (adjunct) mash is started first. It contains approximately 30 to 50% of the total extract required and 10% of the malt bill (or exogenous α-amylase). The adjunct mash is stirred and heated in the cereal cooker to about 70°C (158°F) with a short stand for amylase action. This thins the mash. Without this stand, at boiling temperatures the cereal mash would be a semi-solid paste and unpumpable. After this stand, the adjunct mash is raised to the boil and held for at least 20 minutes. Meanwhile the malt mash starts at about 35°C (95°F, the "protein/glucan rest"). Shortly thereafter it receives the cereal mash "pumped over" into it, which raises the temperature to the required conversion temperature.

The double mashing technique is quite flexible: the temperature ramp at pump-over may be steep or shallow depending on the speed of pumping, and the volume of the cereal mash determines the temperature reached at the end of pump-over. The ramp can easily contain one temperature stand, for example, at 55°C (131°F). This is sometimes call a β-amylase stand and is used to promote wort fermentability for example in making "dry" beers. The mash can then be taken to the conversion temperature, and later to mash-off temperature, by steam heat. Alternatively, the ramp can be steep and rise directly to the conversion temperature at 68 to 72°C (154 to 162°F) to restrict action of β-amylase and control fermentability. The reasons for each hold have been briefly reviewed above and are dealt with in some detail in Chapter 13.

12.5 WORT SEPARATION

There are two modern devices for wort separation: lauters and mash filters. Though there may be some local preference for one technology or the other, each has advantages, and neither has clearly displaced the other from the world-wide brewing scene.

12.5.1 Lauters

Lauters in general design are much like an infusion mash tun (section 12.3.1), but are wider and shallower and invariably equipped with rakes or knives for cutting the grain bed. The false bottom has generally rather narrower slots and fewer of them per square meter than a mash tun because the mash particles are smaller and the false bottom is considerably wider. Lauters receive the stirred mash from the mash mixer and spread it over the vessel bottom so that the grain bed is 25 to 30 centimeters (9–12 inches) deep, although up to 50 centimeters (20 inches) is possible. The particles have been extensively stirred and pumped during mashing and any entrained air by which the mash particles might have floated has long since been driven out. The mash therefore sinks onto the plates of the false bottom and forms the grain bed. The most dense particles sink first and are on the bottom and the lightest ones on top. The upper reaches of the bed are therefore quite prone to collapse and clog if wort is drawn too rapidly and the first cuts with the knives are typically in this top layer.

There are two main strategies for running-off a lauter (Figure 12–5 a,b). In the first, the bed initially formed is maintained and the rakes run constantly to cut and lift it and to keep the bed open for percolation of wort and sparge water. Rakes may be lowered more deeply into the bed as run-off proceeds and compaction (measured by an increase in pressure across the bed) rises. In the second technique, the same general approach is followed, but the wort is drawn more rapidly and the rate of run-off slows dramatically as run-off proceeds. When this happens, run-off is stopped, sparge water is pumped in under the plates (underletting), and the rakes run faster to resuspend the bed. After time to resettle the bed, the wort is again drawn until the need for rousing the bed again arises. Many modern lauters are operated by computer programs that decide how best to handle the lauter depending on such factors as wort flow rate, density, clarity, and the pressure differential across the grain bed.

In the U.S., lauter times are typically 90 minutes or so. Elsewhere, somewhat longer turn-around times seem usual, for reasons that are not clear. In modern breweries, the separation of wort from spent grain is the slowest step and limits the number of brews per day that a brewhouse can achieve. With a 90 minutes turn-around time, a brewhouse can make, at least theoretically, 16 brews in a 24-hour day per lauter vessel. It may be worth sacrificing some minor amounts of extract in last runnings to achieve this capacity in some situations.

12.5.2 Mash filters

The alternative common method of wort separation is the **mash filter**. The device comprises large rectangular **plates** with a metal grid that support the filter cloths (made of plastic material for easy mash discharge and cleaning). These plates are separated by much deeper **frames**, which form a cavity about 15 cm (6 inches) deep, and are bounded by the filter cloths. This is the chamber into which the mash is pumped. On one side of the frame, the plate forms a shallow chamber through which sparge water enters the mash and, on the other side, a shallow chamber through which the wort/spargings leave the mash.

A mash filter occupies a much smaller space than a lauter vessel and has fewer revolving and mechanical parts. However, mash filters are more labor intensive, require a rather predictable mash volume, and discharge wetter grains. The two devices produce wort of essentially equal quality and enjoy roughly the same turn-around time. Some modern mash filter designs (Figure 12–8) achieve high extract yield (in fact, virtually theoretical yield) rapidly and with high wort density, and discharge relatively dry spent grains without any sacrifice in wort quality. These devices depend on two strategies for their effect: hammer milling the grain to a fine powder and squeezing the grain bed in the mash filter with an inflatable bladder. This may have clear advantages over lauters in some settings.

12.5.3 Other wort separation techniques

Other technologies for wort separation have been developed from time to time, but have not survived. A technique of wort separation that is

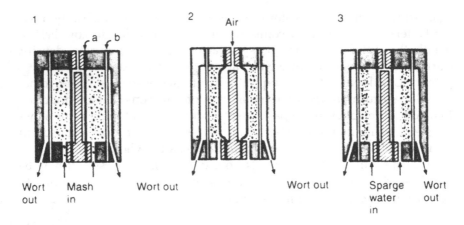

Figure 12–8 Contemporary mash filter designs may upset the traditional stand-off between mash filters and lauters, and may have particular advantages where lauters are rather small and run off quite slowly. This automated mash filter handles hammer-milled malt. The products are brilliant wort containing the theoretical (laboratory) yield of extract and rather dry spent grain at discharge. The filter comprises polypropylene plates (cross-hatched) with an inflatable rubber membrane (a) and frames (shaded) with polypropylene filter cloths (b). After filling the chamber (1), the first wort runs, the membranes then inflate to gently compress the bed (2), leaving a chamber for plug-flow sparging of the bed (3). Final wort is recovered by compressing the bed again (2) to recover end runnings with low volume and to yield a rather dry spent grain. The cycle time is 2 hours (12 brews per day).

extant but in decline is the Strainmaster. In this deep rectangular vessel, run-off tubes traverse the deep mash at many levels. Filtration takes place around each tube, which carries many filtration slots. This device is a modified lauter in which the slots of the lauter's false bottom are located on each of the many run-off tubes. The device is compact and effective but more difficult to clean than traditional vessels. Several have been replaced by lauter vessels in contemporary breweries.

12.6 PRACTICAL CALCULATIONS

Brewing calculations can be quite complicated or relatively simple. Simple approaches work best because the results of all brewers' calculations are usually modified based on experience. For instance, in many microbreweries the amount of malt used per brew will often be in fifty-pound increments, which is the weight of one bag of malt. In large breweries, the amount of malt used will almost always be in one hundred pound increments, because the accuracy of the malt scale is often plus or minus 100 pounds. All of the practical calculations described in this section can easily be loaded into a programmable calculator or a computer spreadsheet program.

12.6.1 Raw material requirements

The amount of malt and other raw materials required for a particular volume of wort at a desired specific gravity (°Plato) must be calculated for each type of wort produced. The calculations may be done in many ways, but all are based on the yield of extract available from the raw material(s), the efficiency of extraction, and the wort specific gravity desired.

Extract Calculations

The desired, or target, wort specific gravity can be converted into pounds of extract for ease of calculation. The American Society of Brewing Chemists provides a calculation that allows one to perform this conversion. This calculation uses the figure 259 pounds as the weight of one barrel of water at approximately 39°F. This approximation is close enough for practical use. (Actual figures are 258.7 lb. of water at 39.2°F, and 247.93 pounds of water at 212°F and the value used depends on whether the wort is in the kettle or at cellar temperature.) The extract figure may be calculated per barrel of wort, or per brew.

Example

ASBC Formula:

$$\frac{(259 \text{ lb./bbl} + \text{Plato}) \, (\text{Plato})}{100} = \text{pounds of extract/bbl}$$

For wort at 15.00 °Plato:

$$\frac{(259 \text{ lb./bbl} + 15.00°\text{Plato}) \, (15.00°\text{Plato})}{100} = 41.1 \text{ pounds of extract/bbl}$$

For 1000 bbls of wort at 15.00° Plato: = 41,100 pounds of extract/brew

Now that the brewer knows the total amount of extract required for the brew, the amount of extract required from each ingredient can be calculated. This is done by multiplying the total extract required by the percent of extract desired from each raw material.

Example

Calculate the amount of extract required from each raw material to produce 1000 barrels of 15.00°Plato wort. The material bill consists of 60% malt and 40% rice. From the preceding example, the brewer knows that this will require 41,100 pounds of extract/brew.

60% (60/100 = 0.6) malt:
(41,100 pounds of extract/brew)(0.6) = 24,660 pounds of extract/brew from malt.

40% rice (40/100 = 0.4):
(41,100 pounds of extract/brew)(0.4) = 16,440 pounds of extract/brew from rice.

Raw Materials Calculations

Once the amount of extract needed from each raw material is known, the brewer is able to calculate the theoretical weight of each raw materials required for the brew. These calculations assume that the brewer recovers every bit of extract that is available in the mash. This is an unrealistic assumption and almost certainly will never occur. The section on brewhouse efficiency below will put some practicality into this theoretical calculation.

The raw materials used in brewing vary in the content of extract available. The material specification sheet often states the amount of extract available to the brewer under specified conditions. If this information is not listed on the specification sheet, it can be easily obtained by contacting the vendor. In the following example, the malt supplied has 79% extract (coarse grind, as is), and the rice has 93% extract.

Example

Continuing from the previous example, the brewer knows the amount of extract per brew expected from each of the raw materials. Theoretically, how many pounds of each raw material will be required for the brew?

Malt extract = 79%, coarse grind, as is; Rice extract = 93%, as is

Theoretical Pounds of malt required:

$$\frac{(24{,}660 \text{ pounds of extract/brew})}{(0.79 \text{ pounds of extract/pound of malt})} = 31{,}215 \text{ pounds of malt/brew}$$

Theoretical Pounds of rice required:

$$\frac{(16{,}440 \text{ pounds of extract/brew})}{(0.93 \text{ pounds of extract/pound of rice})} = 17{,}677 \text{ pounds of rice/brew}$$

Brewhouse Efficiency Calculations

The theoretical weight of each raw material needs one final adjustment to yield a practical, working figure for each material. In large breweries, the brewer will recover, in the final wort, between 90 and 99 percent of the extract available in each raw material. That is, for each pound of extract per barrel available in the raw materials, the brewer expects to recover only 0.90 to 0.99 pounds per barrel of extract in the wort. The volume of wort collected and its gravity determines the practical value achieved.

These figures are called brewhouse efficiencies and will vary for each type of wort and for each individual brewhouse. Most large breweries calculate a brewhouse efficiency for each brew. This calculation provides an easy spot check of the brewhouse equipment. Metering problems, for example, may not be noticed until the brewhouse efficiency calculation yields unusual numbers or even unrealistically high figures such as 105% or 110%. Low brewhouse efficiencies may indicate problems such as inaccurate scales or mill gap settings that are out of adjustment.

Example

The brewer knows that the typical brewhouse efficiency for the brew in the preceding example is 95%. How much malt and how much rice needs to be weighed into the scale hopper in order to achieve the goal of 1000 barrels of 15.00 Plato wort?

Pounds of malt required:
$$\frac{31{,}215 \text{ pounds of malt/brew}}{0.95} = 32{,}858 \text{ pounds of malt}$$

Pounds of rice required:
$$\frac{17{,}677 \text{ pounds of rice/brew}}{0.95} = 18{,}607 \text{ pounds of rice}$$

The theoretical raw material and the brewhouse efficiency calculations are often combined into one calculation such as:

Pounds of material required per brew =

$$\frac{\text{(Theoretical Pounds of material required/brew)}}{\text{(Pounds of extract/Pound of material)(Brewhouse efficiency)}}$$

Calculations using liter degrees per kilogram (L°/kg)

Liter degrees per kilogram (L°/kg) is a useful and convenient way to express the extract yield potential of raw materials. The degrees used are specific gravity not° Plato. One liter degree per kilogram means that one kilogram of the material will yield one liter of wort with a specific gravity of 1.001. Specific gravity is the ratio of the weight of wort to the weight of the same volume of water under identical conditions and might have a value of say 1.048 (quoted as "ten-forty eight" in traditional parlance). Water has a value of 1.000, and so the degrees of gravity in one liter of such a wort are 1.048 − 1.000 = 0.048 × 1000 ml/L = 48 liter degrees. If we know the total volume of such wort required we can calculate the total number of liter degrees, and then knowing the number of liter degrees in each kilogram of material (L°/kg) easily derive the weight of material needed.

All of the other calculations illustrated above can also be done with this system using, of course, liters (hl×100), kilograms and specific gravity. It is, perhaps, high time this was the industry standard.

Note that we use the symbol L to represent liters in this section to avoid confusion with the number one.

Example

A brewer wants to produce 1500 hl of wort at specific gravity 1.048 from a material bill consisting of 75% malt and 25% adjunct. The malt has a material extract factor of 296 (L°/kg) and the adjunct has a material extract factor of 355 (L°/kg). Brewhouse efficiency is assumed to be 95%. What weight of malt and adjunct does the brewer need?

1. Calculate the total extract required for the brew:

   (1500 hl/brew)(48L°)(100L/hl) = 7,200,000 L°/brew.

2. Calculate the amount of extract from each raw material:

Extract from malt: (0.75)(7,200,000 L°/brew) = 5,400,000 L°/brew.

Extract from adjunct: (0.25)(7,200,000 L°/brew) = 1,800,000 L°/brew.

3. Calculate the weight of each material needed per brew:

Kilograms of malt required per brew considering brewhouse efficiency of 95%:

$$\frac{(5,400,000 \text{ L°/brew})}{(296 \text{ L°/kg})(0.95)} = 19,203 \text{ kg/brew}$$

Kilograms of adjunct required per brew:

$$\frac{(1,800,000 \text{ L°/brew})}{(355 \text{ L°/kg})(0.95)} = 5,337 \text{ kg/brew}$$

12.6.2 Mash Volume

The brewer will occasionally have a need to determine the actual volume of the mash. This figure may be used, for example, to determine if the mash for a new brand will fit into the mash vessel. Convenience, not accuracy, allows the brewer to convert the grain weight to an equivalent volume using the following conversions:

1 kilogram of grain in the mash occupies 0.7 liters.

1 pound of grain in the mash occupies 0.0027 barrels.

Example

A mash is composed of 170 bbls of water and 17,000 lb. of pale malt. What is the total volume of the mash?

(17,000 pounds of malt)(0.0027 bbls/pound grain) + 170 bbls of water
= 216 bbls

12.6.3 Mash Temperature

Whether the brewer is making an infusion mash, or a temperature programmed mash of some sort, at some point warm water and grains will be combined. When they are combined, the temperature of the resulting mash is important, and brewers take care to calculate the required temperature of the water that will be combined with the grain so that the desired target mash temperature can be achieved. The heat content of any material is its weight multiplied by its temperature and its specific heat. If the specific heat of water is taken as 1.0 kJ/kg°C, the specific heat of grain is assumed to be 0.4 kJ/kg°C.

Example

A brewer in a cool climate is targeting a mash-in temperature of 44°C after all the grain and water have been added to the mash vessel. The mash is composed of 173 hl of water and 4,680 kg of grain. The grain is stored outside and the outdoor temperature is 10°C. Assume the temperature of the grain is also 10°C. How warm should the 173 hl of water be in order to compensate for the cold grain?

Specific heats: Water = 1.0 kJ/kg°C; Grain = 0.4 kJ/kg°C.

Weight of water = 100 kg/hl.

(Assume that the mash vessel has been preheated and will not contribute to heat loss.)

$$\text{Mash temperature} = \frac{\begin{array}{c}(\text{Specific heat water})(\text{Water weight})\\(\text{Water temp.}) + (\text{Specific heat grain})\\(\text{Grain weight})(\text{Grain temp.})\end{array}}{\begin{array}{c}(\text{Specific heat water})(\text{Water weight})\\+ (\text{Specific heat grain})(\text{Grain weight})\end{array}}$$

$$44°C = \frac{\begin{array}{c}(1.0\ \text{kJ/kg°C})(173\ \text{hl} \times 100\text{kg/hl})(\text{Water temp.}) +\\(0.4\ \text{kJ/kg°C})(4680\ \text{kg})(10°C)\end{array}}{(1.0\ \text{kJ/kg°C})(173\ \text{hl } 3\ 100\ \text{kg/hl}) + (0.4\ \text{kJ/kg°C})(4{,}680\ \text{kg})}$$

$$44°C = \frac{(17{,}300\ \text{kJ/°C})(\text{Water temp.}) + 18{,}720\ \text{kJ}}{17{,}300\ \text{kJ/°C} + 1{,}872\ \text{kJ/°C}}$$

by rearranging:

$$\text{Water temperature required} = \frac{(44°C)(19{,}172\ \text{kJ/°C}) - 18{,}720\ \text{kJ}}{17{,}300\ \text{kJ/°C}} = 47.68°C$$

12.6.4 Hop utilization

The brewer often needs to calculate the amount of hops, or hop extracts, needed to supply a fixed number of international bittering units (IBUs) to the wort in the kettle, or to the beer. In either case, the calculation is the same. The difficult part is to decide a reasonable factor for hop utilization. This is based on experience.

$$\text{Weight of hops to be added} = \frac{\text{(Wort or beer volume)(Desired IBU)}}{\text{(\% alpha acids in the hops)(\% utilization)}}$$

Example

A brewer wants to formulate a beer with 25 IBUs using only Cascade hops (10.3% alpha acids). The expected hop utilization is 30%. The brew size is 500 bbls. How many pounds of hops will be added to the brew? This is now easily done by converting to liters.

Assume that 1 IBU = 1 mg/L (ppm) iso-alpha acid. One barrel = 117 liters.

$$\text{Weight of Cascade hops} = \frac{\text{(500 bbl)(117 L/bbl)(25 mg/L)}}{\text{(0.103 alpha acids)(0.30)}} = \begin{array}{c} 47{,}330{,}097 \text{ mg} \\ (47.3 \text{ kg}) \end{array}$$

To convert milligrams into pounds:

(47,330,097mg)(1 g/1000 mg)(1 pound/454 grams) = 104.25 pounds.

CHAPTER 13

Mashing biochemistry

13.1 INTRODUCTION

By the end of the nineteenth century brewers had a vast sum of empirical knowledge for which they lacked satisfactory explanations. The opening of the fields of microbiology and biochemistry at that time excited intense interest among brewers because these sciences were so clearly appropriate to their daily experiences and observations. These new sciences, though late-comers in brewing history, provided explanations for common observations and, perhaps more importantly, supplied ideas and information needed for intelligent development of brewing processes and for solving practical problems. Biochemistry and microbiology became the scientific partners of the brewers' art. That partnership flourished throughout the 20th century.

The biochemistry of **mashing** seeks to explain at the molecular level what brewers have observed in practice over many centuries and have come to regard as true. The amount and quality of extract in wort is about 90% the result of heat plus malt enzymes acting on their substrates in malt and adjuncts during mashing. The conditions of the mash, especially its temperature, therefore intimately affect wort properties.

Many of the arguments made here derive from what is known about the behavior of enzymes in model systems, i.e. idealized environments. These ideal environments include constant temperature, constant pH and ionic strength, excess substrate, a pure solution of a single enzyme at constant concentration, a defined and pure substrate, and so on. None of these conditions pertain in the heterogeneous mixture of a practical brewers' mash, which therefore does not always behave in expected ways. Assumptions and conclusions that biochemists have drawn about mashing have not always proved true. There is also a modern tendency to assume that traditional brewers developed certain arts and practices for real (i.e. biochemical!) reasons although they did not necessarily understand those arts in scientific terms. For example, in modern times the low temperature hold of a double mash has been ascribed the biochemical function of protein denaturation and digestion (the "protein rest"), when in fact it may have served the brewers of 150 years ago (who invented the procedure) the more prosaic but vitally important function of cooling the boiling cereal adjunct.

This chapter seeks to explain wort properties and composition as the end results of particular enzymes acting on particular substrates under particular sets of conditions. Each of these variables (**enzyme, substrate and conditions**) has a role to play in the outcome of the enterprise.

13.2 STARCH

The vast bulk of extract in wort arises from starch dissolution and digestion. Starch (Figure 13–1) is a glucose polymer comprising two kinds of molecules: (a) a straight chain of glucose residues linked α-1–4 called **amylose** which makes up 20–25% of the total native starch and (b) **amylopectin,** which is a similar molecule made of α-1–4 linked glucose residues but with branches linked at the α-1–6 position. On average branches occur every 27 glucose residues. Amylopectin has a so-called ramified tree structure and has a very large molecular weight. The nature of the glucose molecule (Chapter 2) explains why amylose has one non-reducing end and one reducing end, and why amylopectin has many non-reducing ends and just one reducing end.

Every time an α-1–4 glucosidic link is hydrolyzed one new reducing end (at the 1-carbon), and one new non-reducing end (at the 4-carbon) is formed. Therefore the extent of starch breakdown can be measured by the number of new reducing groups formed during digestion, for example, by the ability of the starch solution to reduce a ferricyanide salt solution to ferrocyanide. The change can be detected by titration with a standard solution of iodine. Alternatively, rate of formation of reducing groups can be used to estimate the efficacy of amylolytic enzymes in malt (e.g. for diastatic power determination).

Starch has a powerful color reaction with iodine solution: amylose yields an intense deep blue–black color because the iodine incorporates itself easily into the helical structure of amylose to form a chemical entity called a clathrate. Amylopectin gives a considerably less intense blue–violet color because it lacks extensive helical regions. Exhaustive action of β-amylase on amylopectin leaves β-limit dextrins which give a brick-red or brown color with iodine. The iodine color is commonly used to follow the course of starch breakdown in mashing, and lack of an iodine color is a presumptive test for starch-free wort. The iodine test must be done on cool samples to avoid a false result.

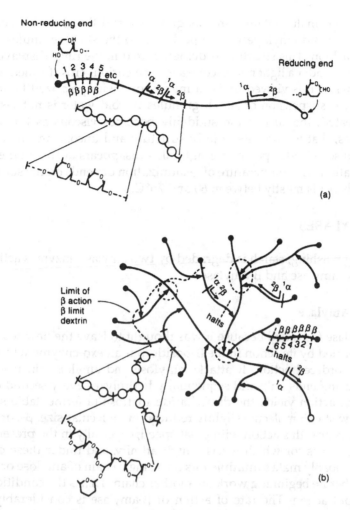

Figure 13–1 (a) Amylose: an unbranched chain of glucose molecules linked only by α-1–4 links. The structures at the reducing end (○) and non-reducing end (●) are shown. β-Amylase attacks this molecule sequentially at point 1 then point 2 and so on, releasing a disaccharide (maltose). β-Amylase activity can also initiate at points where α-amylase (1α) cleaves the amylose chain. (b) Amylopectin: a branched structure of glucose molecules joined by α-1–4 links in the main chain and α-1–6 links at the branch points, as shown by the arrow heads. β-Amylase acting alone cleaves maltose molecules starting at the non-reducing ends and halts when it approaches a branch-point. The remaining molecule is aptly called the β-limit dextrin. If α-amylase acts at 1α it gives access to the interior of the molecule and new β-amylase action initiates at the new non-reducing end revealed. Neither enzyme can cleave or approach the branch points, and these structures survive into beer as unfermentable dextrins, some with quite significant molecular size (dotted circles).

Barley and malt contain starch as granules, and the glucose chains of the amylose and amylopectin are packed into the starch granules in an orderly fashion. This structure is demonstrated in the rings displayed by the granule under a light microscope and by a curious cross formed on the granule when it is viewed under a microscope with polarized light illumination. A suspension of starch granules in cold water is not viscous. When heated, the suspension suddenly becomes viscous as the starch gelatinizes, that is loses its orderly structure and uncoils and hydrates. The cross seen under polarizing light also disappears at the moment of gelatinization. The temperature of gelatinization depends on the source of the starch but is mostly between 60 and 75°C.

13.3 AMYLASES

During mashing, starch is degraded by two amylase enzymes acting in concert: α-amylase and β-amylase.

13.3.1 β-Amylase

β-amylase (so called because it was thought to leave the new reducing group formed by its action in the β-position) is an exo-enzyme which has a rigidly ordered action. It attacks amylose and amylopectin from the non-reducing ends only and sequentially hydrolyzes every second α-1–4 link. Each action yields the disaccharide **maltose** (a fermentable sugar) and a new starch molecule slightly reduced in molecular size. β-Amylase is able to repeat this action with great speed, especially in the presence of large molecules for which it has a high affinity, and under these conditions it probably makes multiple hits on a single chain of amylose or amylopectin before beginning work on another chain. This is the condition for most rapid action. The rate of action of β-amylase is considerably less with small molecules, for which it has less affinity. β-Amylase, acting alone (and given sufficient time), is capable of converting amylose almost entirely to maltose, with complete loss of the iodine color.

β-Amylase cannot hydrolyze α-1–6 links or reach over them and action slows and then stops as it approaches the α-1–6 branch point of amylopectin. β-amylase therefore attacks only the outer regions of amylopectin (being blocked by the α-1–6 bond). It releases no more than 10–15% of the glucose residues as maltose and leaves β-limit dextrins. Thus, β-Amylase lacks access to the interior of the amylopectin structure, which, since this comprises 75–80% of the native starch, represents the bulk of the potential extract of starch in malt and adjuncts.

Because β-amylase releases the sweet-tasting sugar maltose from starch it has been traditionally called the **saccharifying enzyme**. More usefully it could be called the **fermentability enzyme** because its dominant practical role is to determine wort fermentability.

13.3.2 α-Amylase

α-Amylase (so called because it leaves the new reducing group formed as a result of its action in the α-position) is an endo-enzyme. Like β-amylase, it acts only on α-1-4 links, but it acts randomly. This means that any α-1-4 link in the starch molecule (except those in the immediate vicinity of an α-1-6 link branch point) is as likely to be hydrolyzed as any other. Thus, α-amylase does not produce significant amounts of fermentable sugars (glucose, maltose or maltotriose) except when it acts on relatively small molecules, i.e. towards the end of mashing. Because conditions are generally unfavorable for enzyme action at this stage the direct practical effect of α-amylase on wort fermentability is small. However, when α-amylase acts on starch molecules a few scissions by this enzyme causes rapid decrease in the size of starch molecules, and there is an accompanying dramatic decrease in the viscosity of a starch paste. For this reason, the enzyme is commonly called the **liquifying enzyme**. Although it is heat that renders starch soluble in mashing to form extract, it is often assumed that α-amylase action helps starch solubilization. This implies in practical terms that α-amylase is the extract-producing enzyme. For this reason, and perhaps because it is formed *de novo* during germination of barley, it has been given perhaps excessive attention by brewers and maltsters.

Most importantly, α-amylase opens up the large starch molecules (especially amylopectin) to β-amylase action. Every α-1-4 link hydrolyzed by α-amylase creates a new non-reducing end where multiple scissions by β-amylase action can ensue (Figure 13-1). However, since β-amylase works best on large molecules, there is an argument to be made that an excessive amount of α-amylase (which quickly breaks starch down to smaller dextrins) could be detrimental to high fermentability. For this and other reasons wort properties are strongly affected by the relative amounts of α- and β-amylase in malt and by conditions of mashing that might promote the action of one enzyme over the other.

13.3.3 Temperature effects

The two enzymes have one crucial difference that is central to control of wort properties: they react differently to temperature. α-Amylase tolerates higher temperatures (70°C, 158°F or even higher) than β-amylase, and works well at temperatures which rapidly inactivate β-amylase. The mash is a very supportive milieu for enzymes and they tolerate heat better in a mash than in water or a buffer solution, for example, but nevertheless β-amylase works best at 55 to 60° (130 to 140°F) and α-amylase about 10 to 15°C higher.

Enzymes survive better in thick mashes than in thin ones and in cool mashes better than in hot ones, but in all mashes β-amylase action declines faster than α-amylase action (Figure 13–2). This implies intuitively what will be argued in the following paragraphs, namely that mashing is a β-amylase-sensitive and therefore fermentability-sensitive event. This implies that any change in mashing method, particularly mash temperature, mash thickness and mash dilution (as malt–adjunct ratio) will show up in wort fermentability before any other measure, including extract yield. Wort fermentability is therefore a most useful means of monitoring satisfactory mashing.

13.4 MASH TEMPERATURE

The temperature of an infusion mash affects wort fermentability before extract yield because mash temperature can inactivate β-amylase without significant effect on α-amylase. The temperature effect is therefore more obvious if the total amount of enzyme is low and/or if the β-amylase : α-amylase ratio is low (Figure 13–3).

When experimental infusion mashes are made over a very wide range of temperatures (Figure 13–4), mash temperature dramatically affects the extract yield as well as the fermentability of wort. At low mash temperature the extract is low because the starch substrate dissolves poorly. The fermentability (%) of this relatively small amount of extract is high, however, because β-amylase is active even at low temperature and produces a good deal of maltose. At higher mash temperatures extract yield increases to a maximum and does not decrease much even at a very high mash temperature partly because α-amylase retains some action but also because heat alone dissolves starch. However, at such elevated mash temperatures the fermentability of the wort is low because the β-amylase is rapidly denatured. Somewhere between these two extremes there is a narrow range of temperature where extract yield and fermentability are both high. This can be called (for our purposes) the 'brewer's window' of opportunity. Above this quite narrow range of mash temperature wort fermentability decreases substantially (loss of β-amylase) and below this range extract yield decreases (incomplete starch solution).

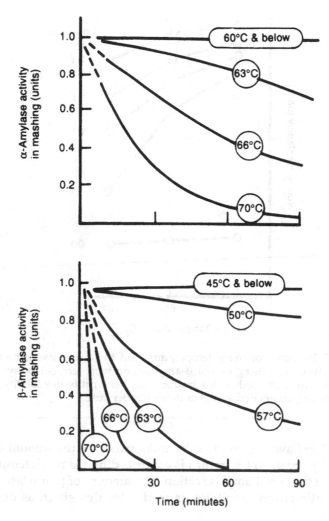

Figure 13–2 Survival of α-amylase and β-amylase in thin mashes held at the constant temperatures shown. β-amylase is clearly much more quickly inactivated than α-amylase at any given temperature.

This phenomenon provides a clear mechanism for the control of wort fermentability, and justifies, indeed demands, close control of mash temperature. In an infusion mash, highest wort fermentability accrues when mashing at the lowest temperature at which maximum extract can be achieved, and wort of progressively lower fermentability can be made by using higher mash temperatures up to the point where extract begins to

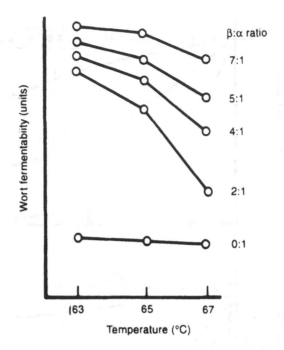

Figure 13–3 Influence of mash temperature and the β-amylase content of an enzyme mixture comprising α- and β-amylase, on wort fermentability. More β-amylase in the mixture (higher DP) produced higher wort fermentability and this effect was more exaggerated at higher mash temperature.

decrease. Beer flavor derives in substantial part from the amount of sugar processed by yeast, which in turn is controlled by the mash temperature through its effect on β-amylase action. The amount of β-amylase present (malt diastatic power) amplifies or moderates this effect, as does malt modification.

Malt modification primarily defines the temperature range of the 'brewer's window'. The lowest temperature at which maximum extract can be achieved is a function of malt modification because the extent of modification influences the rate of starch solution. The 'brewer's window' is lower on the temperature scale of Figure 13–4 when the malt is thoroughly well-modified and higher when the malt is poorly modified. Thus poorly modified malt must be mashed hotter than well-modified malt to achieve adequate extract yield. To produce the required level of fermentable sugar there must then be sufficient enzyme (especially β-amylase) to survive the higher mash temperature used.

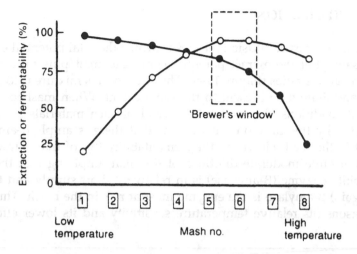

Figure 13–4 Extract yield (○) and fermentability (●) were measured. In a set of eight individual mashes (1–8), each held at a different but constant temperature. Only over a relatively narrow range was there a high fermentability **and** a high extract yield. This temperature range is called the 'brewer's window' of opportunity in the text, because, (a) if a mash falls much *below* this temperature, extract falls off and (b) *above* this temperature, fermentability falls off precipitously. The temperature range of the 'brewer's window' depends on malt quality and the mashing technique used.

Well-modified malts can be mashed successfully at low temperature because the starch easily dissolves and, if these malts are low in enzymes, the enzymes are conserved at low temperature and adequate fermentability results. Traditional British malts, for which the infusion mash was developed, are of this kind. Less well-modified malt must be mashed hotter to achieve adequate extract and must contain sufficient enzyme to tolerate this higher mash temperature. Thus, American malt mashed in a traditional infusion mash at relatively low temperature (say 62°C, 144°F) could give a rather poor extract yield; at 70°C (158°F) it yields well and the wort is sufficiently fermentable because of its very high enzyme content. Poorly modified malt, low in enzyme, is difficult to mash in any system, and impossible to mash successfully in infusion mashing systems for the reasons explained here.

The arguments made here demonstrates that the mash conditions must be chosen to accommodate the enzyme content and degree of modification of the malt. This is the basis of traditional styles of mashing.

13.5 MASH DILUTION

Imagine a series of infusion mashes in which the total potential extract is the same, but the extract potential comprises malt plus pure starch paste in various ratios (Figure 13–5). The effect of this mixture is to dilute the α- and β-amylase present to the same extent. When mashed under identical conditions, the total extract yield of such materials is almost unaffected by this dilution. This means that there is ample α-amylase present in the malt. However, the fermentability (%) of that extract decreases at quite moderate dilutions of the malt, implying that the fermentability enzyme (β-amylase) is in relatively short supply and therefore (again) β-amylase is the enzyme most at risk in the mash. Thus for two reasons (its relative temperature sensitivity and its lower effective

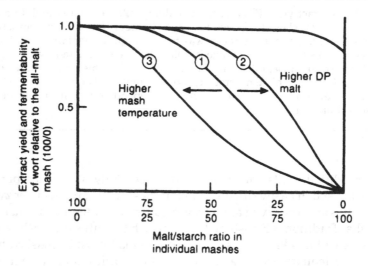

Figure 13–5 If pure starch and malt are mixed in the ratios shown so that the total potential extract remains the same and the mixture is then mashed, there is virtually no effect on extract (top line) presuming there is enough heat to dissolve the starch to create the extract. Fermentability is quite another matter however. Dilution of the malt with starch also dilutes the β-amylase so that fermentability falls off with dilution (line 1). Under exactly the same experimental conditions, malt with a higher DP (diastatic power) yields line 2 because it better resists dilution. Conducting the experiment at a higher mash temperature than 1, produces line 3 because the effect of dilution is magnified by heat inactivation of the enzyme. In such an experiment there is sufficient α-amylase present to discharge the starch–iodine blue color, even at quite extreme dilution. Thus mashes are always more at risk for fermentability than for extract.

concentration in malt) the β-amylase is the more difficult enzyme in malt to manage during mashing. Brewing errors that affect β-amylase activity will show up in lower fermentability before total extract (α-amylase) is affected by the same error.

13.5.1 Fermentable sugars

Fermentable sugars dominate the composition of malt wort. The maximum fermentability of wort that can be produced by malt enzymes in conventional mashing is 75–78%. This is mostly made up of maltose and maltotriose which are the result of α- and β-amylase acting together. Interestingly, if α-amylase action is favored during mashing (e.g. at relatively high mash temperature) maltotriose forms a higher proportion of the fermentable sugar fraction than if β-amylase action is favored. The reason is that maltotriose arises from β-amylase action on odd-numbered linear dextrins. Favoring α-amylase action produces more such dextrins for β-amylase action. This is important if a brewer's yeast has some difficulty handling maltotriose.

Less than 15% of the fermentable sugar is glucose, fructose and sucrose, which for the most part pre-exist in the malt. However, small amounts of glucose can be formed by α-amylase action on smaller dextrins towards the end of mashing; this also reduces the molecular weight of the residual dextrins in beer (Figure 13–6).

13.5.2 Non-fermentable sugars

The non-fermentable fraction of wort is at least 20% of the total extract in those cases where exhaustive amylase action occurs, and may be as much as 35 to 50%. The combined action of the amylases reduces the amylose portion of starch substantially to maltose (and maltotriose). There are some linear dextrins in wort but they are of relatively low molecular weight only, because larger ones would be degraded by β-amylase.

Most wort dextrins are branched molecules and contain the α-1–6 links of the starch which malt α- and β-amylases cannot attack. The degree of degradation by α-amylase, which cannot approach within several glucose units of a branch point, and the closeness of the branch points to each other in the amylopectin molecule, determine whether the dextrins contain one, two or several branch points. These factors determine the size of the wort dextrins (Figure 13–1). The enzyme limit dextrinase, as the name implies, is capable of hydrolyzing α-1–6 links but not much of this enzyme is produced in ordinary malting and the enzyme is sensitive to

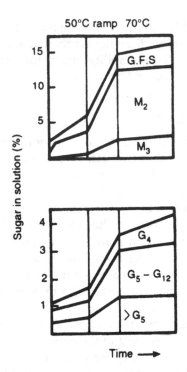

Figure 13–6 During mashing, maltose is mostly formed during the temperature ramp (upper figure). At the conversion temperature total extract continues to accumulate and lower-molecular weight dextrins form an increasing proportion of the total unfermentable extract (lower figure). At conversion temperature very little further increase in fermentable extract accrues. G, glucose; F, fructose; M_2, maltose; M_3, maltotriose; G_4–G_{12}, dextrins comprising 4–12 glucose units.

heating on the kiln and in mashing, and so its role, if any, is likely to be small.

Wort dextrins have no flavor of their own and are not viscous enough in solution to account for the perceived (sensory) viscosity or 'body' of beer. However, they probably contribute some texture to the beer and the hydrolysis of these dextrins by salivary amylases in the oral cavity does yield glucose. This could easily contribute to the aftertaste of beer and perhaps a sense of richness. It is always wise to remember that no-one ever tasted beer as it exists in the bottle, but only as modified in the oral cavity—that is warmed up, degassed, changed in pH and diluted and reacted with saliva (!).

13.6 TEMPERATURE-PROGRAMMED MASHES

The need to manage malt in more creative ways, e.g. to make good short-comings in the malt, accounts for temperature-programmed mashing. The biochemistry of such mashes is more complex than infusion mashing, because they provide the opportunity to adjust malt modification, to bring into play a wider spectrum of enzymes and to sequence enzyme action.

Temperature-programmed mashes start at low temperature and are heated up over a period of time either by adding a boiling mass of material (in double mashing or decoction mashing) and/or heating with steam jackets. At low temperature the malt is wetted, soluble substances dissolve and enzyme action starts. If an all-malt mash is steadily heated from e.g. 40 to 70°C (104 to 158°F), the aqueous portion of the mash is always starch-free, as judged by the iodine reaction. Thus the rate of starch dissolution, not the rate of enzyme action, is the rate-limiting step in such a circumstance. During heating, a significant amount of extract accumulates in solution, even at quite low temperatures although the mash particles continue to react with iodine until the conversion temperature is approached. Close to 70°C (158°F) or so the iodine color is finally discharged and the theoretical yield of extract accumulates. Thus, temperatures well below the conversion stand are immensely influential in determining wort properties.

The action of β-amylase is particularly crucial. At the conversion stand (about 70°C, 158°F) β-amylase survives for a short time only (Figure 13–2). A prolonged conversion stand therefore serves mainly to adjust the molecular size distribution of the dextrins by the action of α-amylase, rather than to affect fermentability. Thus, if good fermentability has not been achieved at lower temperature and by the *beginning* of the conversion stand, wort fermentability cannot be increased by prolonging the conversion stand. Some mashing procedures incorporate a temperature stand at about 55°C (130°F) specifically to promote β-amylase action and hence wort fermentability, e.g. in production of 'dry' beers where residual dextrins need to be minimized.

The situation is similar in double mashes or decoction mashes when a large volume of boiling adjunct or mash (containing solubilized and gelatinized starch) is progressively mixed into the main malt mash. This is because the starch is hydrolyzed rapidly and the soluble portion of the wort remains mainly starch free throughout normal pump-over. In such mashes it is relatively easy to introduce a 55°C stand for β-amylase action into the profile. The rate of pump-over determines the steepness of the temperature 'ramp' and whether or not there are any stands in this ramp.

This decision directly influences β-amylase action and the fermentability of the wort. Consistency therefore demands that the temperature profile of a mash be rigorously duplicated at each mash.

In summary, the low temperature stages of the temperature profile of a programmed mash promote β-amylase action with little, but probably sufficient, α-amylase action to open-up the amylopectin molecules that dissolve. As the temperature rises to about 55°C (130°F), β-amylase action and α-amylase action both accelerate substantially as does the dissolution of starch and the bulk of the extract yield and most of the fermentable sugar accumulates in solution. Above this temperature β-amylase action progressively declines but α-amylase action increases. At the onset of the conversion stand the action of β-amylase is in steep decline. At this point the last of the starch dissolves to yield the theoretical extract and this starch and the dextrins present are broken down further by α-amylase during the conversion stand.

The decoction method of mashing, in which part of the malt mash is boiled, may be thought of in the same way. The boiling portions of the malt mash added to the main mash raise its temperature through a series of stands and promotes the action of amylases, as discussed above. Also important in this mashing method is that the decoctions serve to solubilize malt starch, proteins (which are also precipitated) and β-glucans and so can function as an extension of modification. Decoction mashing, especially the traditional triple decoction mash, therefore has been interpreted as a means of handling under-modified malt.

13.7 OTHER ENZYMES

Germinating malt contains the full and balanced spectrum of enzymes necessary for the total reduction of endosperm polymers to low-molecular weight compounds suitable for feeding the barley embryo. Kilning radically alters this balance by extensive destruction of many enzymes so that the amount and kinds of enzymes in finished malt fall well short of those in green malt. This means that the opportunities for adjustment of modification in mashing are at best restricted. However, low-temperature stands, and (in the case of decoction mashing) boiling part of the malt do provide some opportunity for enzymes and substrates other than starch and amylases to react. It must be stressed that these actions are severely muted by enzyme destruction in kilning.

If the digestion of proteins during the mash were crucially important to brewers doubtless a method for endo-peptidase measurement in malt would have been developed. There is no such method. Most, if not all, proteolysis is a function of malting. This is important because proteins cause hazes in beers, contribute to foam stability and are ultimately the source of wort amino acids for yeast nutrition. Kilning substantially inactivates proteolytic enzymes (endo-peptidases, mostly sulfhydryl-dependent enzymes) and therefore extensive modification of proteins in solution during the short low-temperature stand at about 45°C (113°F) commonly called the 'protein rest') is unlikely. However, proteins do dissolve during this phase and are then precipitated in the mash with some tannins as the temperature approaches the conversion stand. Thus, mashing affects wort proteins but not significantly by enzyme action, and precipitation (not proteolysis) accounts for the disappearance of proteins during mashing. This proteinaceous precipitate may contribute to the teig material or top-dough which inhibits lautering. Incidentally, considerably more malt protein is precipitated in mashing than in kettle-boiling.

In contrast, exo-peptidases (especially the carboxypeptidases which cleave amino acids from the carboxyl end of peptides and polypeptides) are more heat stable. In those cases where formation of amino acids in mashing can be demonstrated, these enzymes are undoubtedly responsible. In American mashing systems which tend to be dilute, short and hot, less than 5 to 10% of the amino acids of wort arise during mashing. That is, at least 90% of wort amino acids are pre-formed in the malt. It is claimed that considerably greater levels of amino acids (30% to even 50% of the total) arise in other mashing systems. In particular, thick all-malt mashes with extended low-temperature stands would favor amino acid formation. The correct amount, and possibly kinds, of amino acids in wort contribute to satisfactory yeast performance (rate of growth and fermentation) and could influence the spectrum of flavor compounds produced. Furthermore, if excessive amounts of amino acids survive into beer they could foster the growth of spoilage organisms in unpasteurized products.

Some carboxypeptidases, probably acting as esterases, can free β-glucans from the endosperm cell walls during malting which makes the glucans available for the action of endo-β-glucanases (Figure 13–7) during malting. This substantially reduces their viscosity in malt. Carboxypeptidases survive kilning but endo-β-glucanases are much more heat labile and mostly do not. There is a danger therefore that if β-glucanolysis is incomplete in malting, β-glucans could dissolve in mashing under the influence of a carboxypeptidase (called solubilase) with insufficient endo-β-glucanase to reduce its viscosity. This would be especially true in high-temperature mashes, which would favor β-glucan solution but not the action of β-glucanases.

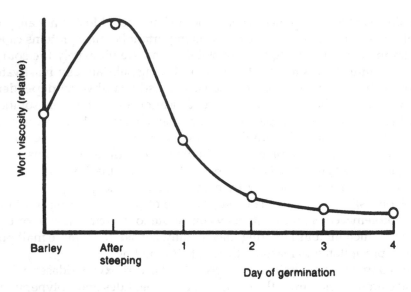

Figure 13–7 Viscosity of wort from a 70°C mash made with malt from several stages of germination. β-Glucans are first released from endosperm cell walls by 'solubilases' which cause an initial increase in viscosity. Viscosity then rapidly declines as β-glucanases are released by the aleurone.

Addition of exogenous β-glucanases (from microbial sources) during mashing is a feature of some mashing systems, especially those using barley as adjunct in which this problem is exacerbated. This added enzyme which is fairly heat stable can promote ease of extract recovery (run-off from the lauter), and promotes beer stability and ease of beer filtration during finishing.

13.8 ADDITION OF ENZYMES

β-Glucanases, proteases and amylases are available commercially from bacterial and fungal sources for addition during mashing, in the fermenter or in the finishing cellars. Where possible brewers prefer to add an enzyme to the mash because it is then inactivated in boiling. This is especially true of proteases and amyloglucosidases (which produce glucose from starch or dextrins in the manufacture of calorie-reduced beer), for example, because their extended action in finished beer may be particularly damaging, and they may cross-contaminate other products with devastating results.

Brewers add enzymes only when needed to solve an intractable practical problem. Such problems usually arise when brewers step outside of the normal bounds of brewing practice—(for example using barley as adjunct. In this case added β-glucanase helps solve problems associated with viscosity. See above). If excessive amounts of starchy adjuncts are used, added α-amylase, especially a heat-stable one, can supplement or replace the malt added in the cereal cooker for example. Enzymes also permit the manufacture of new products. Amyloglucosidase hydrolyses the residual dextrins remaining after the action of malt amylases to produce glucose. Such a wort is completely fermentable and the resulting beer contains no calories in the form of dextrins because they are broken down and converted to alcohol. The high-alcohol beer produced can then be diluted to the required alcohol (calorie) content. Amyloglucosidase is commonly added during fermentation. Proteases, such as papain are added to beer for haze stabilization, though this is rarely done these days because insoluble chill-proofing agents are available and preferred.

CHAPTER 14

Hops technology

14.1 HISTORICAL BACKGROUND

Hops were in use in brewing in central Europe in the 11th century AD. Why the practice of using them arose in the first place is a matter of conjecture. Fermented unhopped malt extract is an extremely heavy, cloying and satiating product that is difficult to consume in quantity! There are many records in the early brewing literature of various types of 'flavor enhancers' being used in brewing. These included chemical substances such as common table salt and salts of iron, even strychnine used to impart 'bite'! Various fruits, fruit infusions or extracts (e.g. raspberry) were also used. Other enhancers were various herbs such as bog myrtle, thyme, yarrow and rosemary. It is easy to imagine brewers experimenting with a variety of herbal products, and cold and hot water extracts of them, to enhance beer flavor and stability. From such experiments, the use of hops probably arose.

Hop constituents are partly extracted into an aqueous phase and chemically modified by heat during the brewing process (Chapter 10). There is no doubt that the unique bitterness and aroma that they impart to beer increase its palatability and beer would not be beer without hop bitterness and aroma.

14.2 BOTANY

The hop plant, *Humulus lupulus*, is a member of the family *Cannabinaceae*. Some members of this family are well-known for their content of pharmacologically active resins; however, the resinous constituents of the hop do not have such properties! The hop is a climbing, perennial plant native to Europe, Asia and North America. Although not native to the Southern hemisphere, it is successfully cultivated in Australia, New Zealand and South Africa, for example. The female plant is cultivated commercially because it bears the hop cones which contain the hop resins and oils. In some countries great care is taken to exclude the possibility of fertilization of flowers by removing wild male plants from the hop-growing regions. The presence of seed is claimed by some, especially lager brewers, to reduce the brewing quality by imparting undesirable flavors. However, although fertilized hops of many varieties produce lower yields of cones, increased yield of bittering substances as α-acid per hectare are obtained, and seeding could protect the vine from diseases especially in damp climates.

The hop cone comprises a central strig carrying several nodes. Each node has two leaf-like bracts and four smaller leaf-like bracteoles. The brewing value of the hop is found in the resins and oils contained in the lupulin glands found most extensively at the base of each bracteole (Figure 14–1).

14.3 BREWING VALUE

It is the mature cones which are harvested and processed. The gross chemical composition of commercial hop cones is shown in Table 14–1. The resins contribute to bittering value and the essential oils to aroma. The tannin constituents may assist protein precipitation during boiling, but the other constituents probably have negligible value to the brewer.

The claim that hop resins have anti-microbial character is often made, and this is true if the iso-α-acids are in sufficient concentration and at a low enough pH so that they are undissociated (below their pK_a). The levels at which the resins are effective are somewhat higher than those in most beers of today, though beers of former years tended to be more highly hopped. Only Gram-positive bacteria are susceptible and most bacteria indigenous to a brewing process are, in any event, resistant to hop compounds. The wort boil, necessary to extract the hops, clearly kills all vegetative microbes derived from raw materials. Therefore, by virtue of this boil, using hops will improve the overall microbiological stability of beer, providing that good hygienic practices are used in the remainder of the process.

14.4 CULTIVATION

Selection and breeding of hops is widely practiced. Breeding programs are directed at creating disease-resistant varieties, high yields of bittering substances (so-called high-alpha varieties because of their content of alpha-acids; see Chapter 15) and for aroma characteristics. Thus it is com-

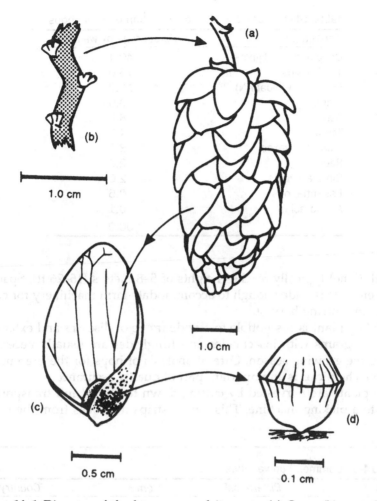

Figure 14–1 Diagram of the hop cone and its parts. (a) Cone; (b) central strig within cone; (c) bracteole (leaf of cone); (d) lupulin gland (seen as a greenish yellow 'dust' at the base of the bracteole).

mon practice to refer to a particular variety being for bitterness or aroma (Table 14–2).

The hop may be propagated from cuttings taken from the perennial rootstock, from underground rhizomes or from softwood cuttings.

The plant, growing from the perennial rootstock, is trained up string or thin wire attached to permanent wire supports (Figure 14–2). The twining

Table 14–1 Gross chemical composition of whole hops

Constituent	Percentage by weight
Cellulose and lignins	40.4
Total resins	15.0
Proteins (Kjeldahl N)	15.0
Water	10.0
Ash	8.0
Tannins	4.0
Fats and waxes	3.0
Pectin	2.0
Simple sugars	2.0
Essential oils	0.5
Amino acids	0.1
Total	100.0

growth (bine) typically reaches heights of 5–8 meters (16–26 ft). Spacing between rows is wide enough to accommodate farm machinery for cultivation and during harvest.

The hop plant is susceptible to a wide range of diseases and extensive spray programs with insecticides and fungicides are usually necessary during the growing season. Careful analysis of hops for the presence of residual chemicals is an important part of quality assurance.

The plants are harvested by cutting down the bines and transporting them to a picking machine. This device strips the cones from the bines.

Table 14–2 Common hop varieties

Variety	Bitterness[a]	Aroma	Country
Chinook	13	—	USA
Clusters	6	Moderate	USA
Eroica	13	—	USA
Goldings	5	Very good	UK
Hallertau	4	Very good	UK
Saaz	4	Very good	Czechoslovakia
Tettnang	5	Very good	Germany
Willamette	6	Good	USA
Wye Challenger	8	Very good	UK
Yeoman	11	—	UK

[a] % α-acids
— Indicates a variety not used as an aroma hop

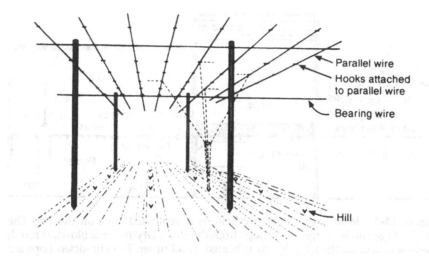

Parallel wire

Hooks attached
to parallel wire

Bearing wire

Hill

Figure 14–2 Diagram of supporting framework for cultivation of hops.

The cones are then separated from leaves and twigs, etc. mostly by siev-ing and air-jets, and transferred to a kiln (Figure 14–3). Kilning is neces-sary to reduce the moisture content of the cones from up to 80% at harvest to about 10%. Dried cones when compressed and cold can be safely stored. Traditionally the hop kiln was a simple single deck drier and the hop cones were spread on a perforated floor and dried by forced circulation of heated air (60–80°C, 140 to 176°F). More efficient use of energy is achieved (as in malt kilning; Chapter 10) using a multi-stage process or by re-cir-culating part of the air during the later stages of the drying process. In some countries, the practice of sulfuring the hops was used. Elemental sulfur was burned to generate sulfur dioxide which was mixed with the hot air stream. This process 'bleached' the cones, improving their appear-ance by making them a uniform color. This practice has largely been aban-doned because it contributed unnecessarily to the sulfur dioxide level in finished beers. In some countries, SO_2 is permitted as the sole preserva-tive in beer but levels must not exceed 50 mg/l (10 mg/l in the USA)

The dried cones, referred to as **whole hops**, are packaged whole in pockets (UK), ballots (Germany) or bales (USA). A pocket is a jute (or woven polypropylene) sack (approximately 2000 × 650 × 600 mm) con-taining about 80 kg of hops. A bale is approximately 1500 × 750 × 500 mm and contains 90 kg. Quantities of hops are often measured in Zent-ners (50 kg).

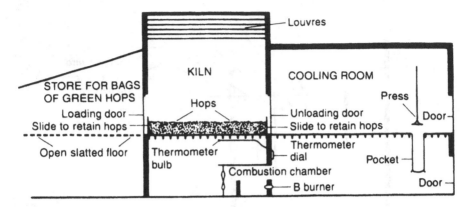

Figure 14–3 Clean, 'green' hops are held in a ventilated store on the left. The single-stage kiln is charged with hops from the store. Drying air is blown through the bed of hops in the kiln; this air is heated by a burner. The kiln-dried hops are moved into the cooling room and when cool are packaged (pressed) into pockets (shown) or bales. This modern type of oast-house has replaced the smaller, cone-shaped, brick ones which were a characteristic architectural feature of hop growing regions.

14.5 PROCESSED HOPS

The brewing value of hops deteriorates on storage. This deterioration is retarded by storing at low temperature. Whole hops are added at the kettle boil and may be used in finished beers (dry hopping). The whole hop is a bulky product, expensive to transport and store and extracting the brewing value during the kettle-boil is a relatively inefficient process. Whole hops can be processed into useful products which retain their brewing value, reduce their bulk and increase the efficiency of extraction in brewing (Figure 14–4).

The simplest procedure uses compression. Compressed pockets are about half the volume of the uncompressed ones yet contain about twice the weight of hops. Compressed whole hops may also be produced as cylindrical plugs each containing about 15 g (0.5 oz). These products are mainly used for dry hopping since they provide a ready means of controlling the amount of hops added to the cask.

Hop cones, screened to remove debris, may be reduced to powders by hammer milling. During the milling process the cones are kept cool. Powders are usually produced to specification by blending hop lots before or

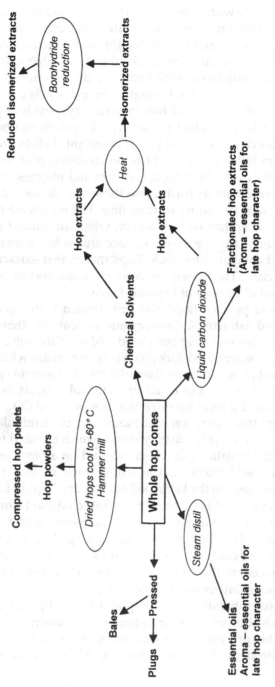

Figure 14-4 Flow diagram of hop processing.

after milling, possibly with some extracts to adjust and standardize brewing value, and are then processed into **pellets**. A uniform powder is produced in a rotating drum and then forced through a metal plate (die) containing holes about 6 mm in diameter. Solid CO_2 is added to dissipate the heat formed which otherwise would damage the hop material. Rotating knives cut the extruded material to length. Typical pellets are 10 mm long and contain 90% of the original hop material (type-90 hop pellets). By removal of about half the plant matter type-45 pellets can be produced which contain much more α-acid per unit weight. Pellets are packaged under vacuum in foil containers which, when stored cold, show greatly improved stability over whole hops. Calcium and magnesium salts may be added to increase stability further and these also aid the extraction and isomerization of α-acids during wort boiling. The pellets are firm and do not disintegrate when handled. However, when introduced to hot wort they readily disintegrate. The greater surface area of hop material exposed to the hot liquid compared to whole hops means that extraction is very much more efficient. The brewer recovers the extra cost of pelleting the hops in improved utilization of hop substance.

Hop powders or pelleted hops may be extracted with solvents to produce concentrated **extracts**. Various organic solvents (methanol, ethanol, trichloroethylene, hexane) have been used and once the solvent is evaporated off a highly concentrated dark green extract remains which is usually packaged in small cans. Concerns about the residual solvents passing into beer have resulted in the use of all but ethanol extracts being greatly reduced. The favored solvent for extraction is now liquid CO_2. This has the added advantage that extraction conditions may be manipulated to produce extracts containing either mainly resins, which are used for bittering, or mainly essential (volatile) oils which are used for aroma, and can separate the α- and β-acid fractions. The aroma fractions may be added to replace late aroma hops in the kettle boil and/or dry hops in finished beer.

Hop resin extracts still need heat in the kettle boil to isomerize the α-acid constituents and produce bitterness. This isomerization process can be conducted directly on the extracts by heating them with potassium hydroxide or adding calcium or magnesium salts and applying heat. The isomerized extracts are then direct sources of bitterness and e.g. can be used to finely adjust bitterness in finished beer.

Certain hop constituents are decomposed by sunlight and produce a marked unpleasant aroma in beer ('skunky' or 'sunstruck'; Chapter 15; Figure 15–4). This process can be eliminated if the hop constituents are chemically reduced. Such products, as reduced isomerized extracts, are widely used.

Hop chemistry and wort boiling

15.1 INTRODUCTION

Hops and hop-derived products are used primarily to impart bitter flavor to beer. Bitterness in the form of iso-α-acids is the primary flavor impact of hops though under some processing conditions hop aroma character can survive into beer. Iso-α-acids also support foam formation and, in some beers, have anti-microbial activity. The bright yellow, glistening lupulin glands of the hop cone, located at the base of the bracteoles, contain all of the material of interest to brewers. These materials comprise two components called the **total resin** and the **essential oil**.

15.2 THE RESINS

The total resin of hops is that portion of ground hops soluble in methanol or in diethyl ether. After the solvent has been evaporated the total resin can then be extracted with hexane which dissolves the soft resin only and the hard resin remains behind. The soft resin fraction contains the α-acids of hops (which are the source of the primary bittering components in beer and govern the brewing value of hops), a closely related group of compounds called the β-acids and some uncharacterized material. β-Acids have a less obvious function in bittering of beer than α-acids. Neither the α-acids nor β-acids survive to any significant extent into beer in normal processing, but change into different chemical entities that do.

In contrast to these components of the soft resins, the hard resins and the uncharacterized soft resins have no clear effect on beer flavor or the brewing value of hops, but simply dissolve in extracting solvents along with the soft resins. Hard resins probably arise from the oxidation and polymerization of the α- and β-acids because during prolonged storage the hard resins increase at the expense of the α- and β-acids which decrease in concentration.

259

15.2.1 Resin analysis

The solubility of hop resins in organic solvents provides a ready vehicle for their analysis. In laboratory methods the first extraction of the hops is made with benzene or toluene. This extract is then diluted with methanol for analysis. Analysis must always be conducted on an adequate sample of the whole, which, in the case of hops is a difficult challenge. Standard methods specify sampling methods, which vary depending on whether the sample is from unpressed hops, bales, pellets, hop powders or extracts. Generally analysts take 200 g samples from 10% of the units to be analyzed.

Although a determination of total and soft resins by fractionation with solvents as described above (Wollmer separation) may be a useful and sufficient analysis in some settings, determination of α-acids more specifically defines the brewing value of hops. Brewers therefore wish to have methods of measurement which quantify the α-acid component of the total resin and which can be used for purchase of hops. α-Acids have three chemical properties that can be used for analysis: they form lead salts which are insoluble in methanol, they absorb light in the ultra-violet (UV) range and they rotate the plane of polarized light. Precipitation with lead or UV light absorbance are the primary methods used.

Precipitation with lead

If a solution of lead acetate in methanol containing some acetic acid is added to a methanolic extract of hops a bright yellow precipitate forms which is the lead salt of the α-acids. In principle the precipitate could be separated from the methanol solution, dried and weighed (gravimetric analysis). This is clumsy in practice and (because the lead precipitate is somewhat soluble in excess lead acetate solution) rather inaccurate. The standard methods therefore require that the lead acetate solution be titrated slowly into a dilute methanol solution of the hop extract while monitoring the conductance or resistance of the solution with a suitable instrument. The result is a curve with two straight line components. Extrapolation of these linear parts of the curve gives an intercept which is the 'true' end-point of the titration (Figure 15–1b).

The end-point is called the conductometric value (CV) or the lead conductance value (LCV) of the hop sample. This value forms the primary basis for commercial transactions because the method yields a good (though not exact) estimate of the α-acids and, perhaps more importantly, estimates the bittering potential of the hop sample rather well. The mea-

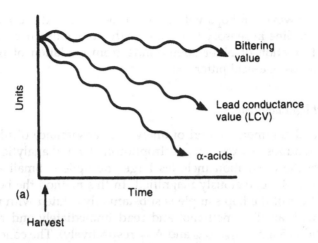

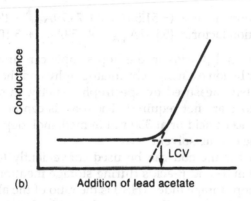

Figure 15–1 (a) Analysis of hops and the relation of the lead conductance value (LCV) over time to α-acids and the bittering value. (b) Measurement of LCV.

surement must be made on hops shortly after harvest because as hops age the CV or LCV changes in a curious way (Figure 15–1a). On older hops the method gives a higher value than the true value of α-acids present, but a lower value than the sensory bitterness of hops would predict. This means that during storage of hops α-acids decline but presumably new bitter compounds are formed, largely from β-acids. These replace the sensory bitterness lost with the α-acids and, if the α-acid to β-acid ratio is about unity as is commonly the case, sensory bitterness remains more or less constant with storage. Some but not all of these new compounds form lead salts. However, in hops with a high α-acid : β-acid ratio, there is a significant decline in sensory bitterness with storage because there is insufficient formation of bitter compounds from oxidation of β-acids to make good loss of α-acid bitterness.

Ultra-violet light absorbance

The second common method of analysis is absorbance of ultra-violet light. This requires an excellent spectrophotometer and analytical expertise because the calculation includes large multipliers. Small analytical errors are thereby egregiously magnified. In this method the benzene or toluene extract of the hop sample is substantially diluted with methanol and then with alkaline methanol and read immediately and rapidly at 355, 325 and 275 nm (A_{355}, A_{325} and A_{275} respectively). The concentration of α- and β-acids can then be calculated from the regression equations:

$$\alpha\text{-acids (\%)} = \text{dilution factor} \times (-51.56A_{355} + 73.79A_{325} - 198.07A_{275})$$
$$\beta\text{-acids (\%)} = \text{dilution factor} \times (55.57A_{355} - 47.59A_{325} + 5.10A_{275})$$

Alternatively the α- and β-acids in the hop sample can first be separated from each other by ion exchange chromatography and the α- and β-acid fractions separately measured by spectrophotometry. In such cases the regression equations are not required. The α-acids can be estimated directly at 280 nm in acetic acid or at 330 nm in methanol, depending on the resin used and the eluting solvent.

The spectophotometric method can be used conveniently to monitor the oxidative change in α- and β-acids during storage mentioned in the last paragraph. The hop storage index (HSI) is the ratio of the absorbance of a hop extract at A_{275} (where oxidized α- and β-acids absorb) and A_{325} (where the unoxidized acids absorb) or A_{275}/A_{325} (Figure 15–2). Fresh hops enjoy a HSI of 0.22 to 0.26, and aged hops a value of 1.5 to 3.0 times higher.

The α-acid components of the hop resin and their derivatives can also be analyzed by HPLC (high pressure liquid chromatography) methods.

15.2.2 Analogs of α- and β-acids

The terms α- and β-acids are generic names for two groups of similar compounds (Figure 15–3). There are several (perhaps many) analogs of the α-acids which differ from each other only in the acyl side chain (R) of the molecule (Figure 15–3). The β-acids comprise a similar range of analogs. The most important analogs of α-acids are humulone (R = isovaleryl),

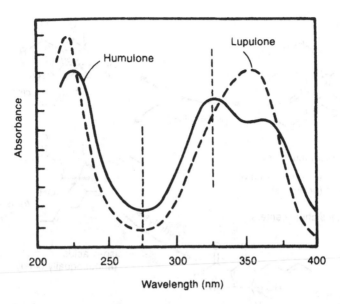

Figure 15–2 Absorption spectra of a humulone (solid line) and lupulone (broken line) complex in alkaline methanol. The ratio of absorbance at 275 nm to that at 325 nm is taken as the hop storage index (HSI).

cohumulone (R = isobutyryl) and adhumulone (R = 2-methylbutyryl) and of the β-acids lupulone (R = isovaleryl), colupulone (R = isobutyryl) and adlupulone (R = 2-methylbutyryl). The α- and β-acids each comprise roughly half of the soft resin fraction of the hop; that is, the α-acid : β-acid ratio is often between 0.8 and 1.2. In some European hops famed for their aroma this value may be 0.5 and in some American high-alpha varieties it may exceed 3.0. If hops contain a high level of soft resins, a high proportion of which is α-acids, the hops are considered **high-alpha hops**. These have special value for preparation of hop products, especially extracts. In most varieties, adhumulone forms a fairly constant part of the total α-acid content (10–15%) though the humulone and cohumulone vary widely. Generally cohumulone makes up 20–50% of the α-acid fraction and humulone the remainder (say 45–70%). The α-acids contain an asymmetric carbon atom (chiral center) at C–6 and so each analog can exist in two forms (enantiomers) called *cis-* and *trans-* which rotate the plane of polarized light in opposite directions. Natural humulone is levorotatory. The iso-α-acids derived from the three major α-acids therefore exist in beer in at least (other isomers of the iso-α-acids are possible) six forms. For-

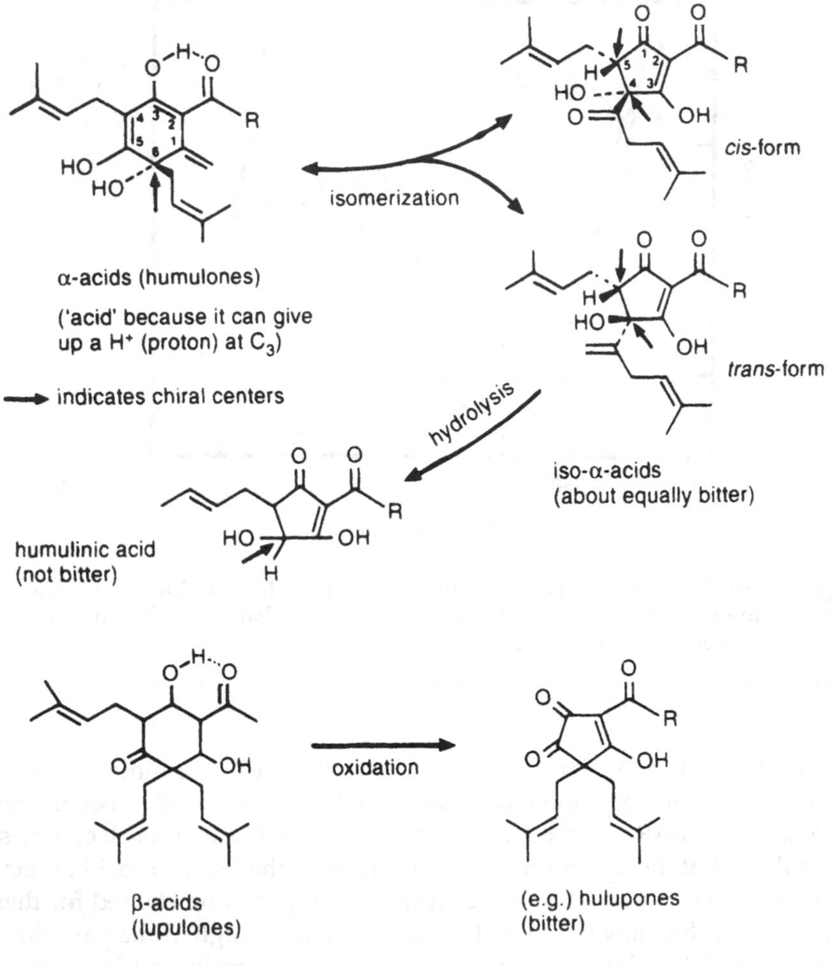

$R = CH_2CH(CH_3)_2$ isovaleryl = humulone (lupulone)
$R = CH(CH_3)_2$ isobutyryl = cohumulone (colupulone)
$R = CH(CH_3)CH_2CH_3$ 2-methyl-butyryl = adhumulone (adlupulone)

Small amounts of other isomers are present

Figure 15–3 Hop acids, structures and chemical reactions of brewing significance.

tunately there appears to be little difference in the bittering potential of the several humulones and their isomers. The LCV therefore reasonably reflects the brewing (bittering) value of a hop sample, regardless of the exact composition of the α-acid fraction. The lupulone, colupulone and adlupulone composition of the β-acid fraction is independent of the composition of the α-acid fraction, and it is common for the co- and adlupulones to be present in roughly equal amounts.

The amount of soft resin in the hop, the α-acid : β-acid ratio in the soft resin and the amount of each analog of α- and β-acid present, is a function of hop variety. The amount and composition of the soft resin is also affected by the region in which the hops are grown and the local weather each year, the presence of seeds (seeding tends to raise the α-acid content), agronomic practices including application of fertilizer, and illumination (in South Africa artificial light is used to extend day length), ripeness at harvest, the position of the cone on the plant and so on. Most of these factors also affect overall yield. Generally, however, all else being equal, each hop variety expresses part of its genetic character in the amount and composition of the soft resin it produces.

15.2.3 Storage effects

The soft resin fraction containing the bittering substances changes in composition during storage. α-Acids decrease linearly with time of storage. The rate of decrease is most accelerated by the access of air (which can relate to the way the hops are packed tightly in compressed bales and wrapped), lack of refrigeration, and high moisture content. This decline is general for all hops but some varieties (e.g. Bullion) lose α-acids faster than others. As the α-acid decreases so the hard resins (containing oxidized and polymerized forms of the α- and β-acids) increase and low-molecular weight acids derived from the acyl side chain (R) of the acids (isobutyric, isovaleric, 2-methyl butyric) accumulate, lending a 'cheesy' aroma to ill-stored, older hops. Formation of oxidation products of α-acids and especially β-acids that are soluble and bitter probably accounts for the fact that the bittering value of the hop declines less, or even much less, than the loss of α-acids would predict. Among these bitter compounds are the hulupones formed from β-acids and humulinones from α-acids; both contain the five-membered ring present in iso-α-acids (Figure 15–3).

10.3 WORT BOILING

When hops or hop products are boiled in wort the single most important reaction of hop components is the isomerization of α-acids to iso-α-acids. Though it is not clear whether iso-humulone for example is any more bitter than humulone itself, the point is moot since α-acids are insoluble in beer at its normal temperature and pH. Iso-α-acids account for the desirable bitter qualities of hops, whereas harsh and lingering bitterness is commonly ascribed to other bitter compounds derived from oxidation reactions of α- and especially β-acids and miscellaneous other reactions of α-acids and iso-α-acids under the influence of the kettle boil.

Success in the enterprise of kettle boiling is therefore best measured by the amount of iso-α-acids in beer. This can be accomplished for example by extracting beer to which a small amount of strong acid is added with iso-octane (2,2,4-trimethylpentane). The absorbance of ultra-violet light by the extract at 275 nm is directly related to beer bitterness:

$$\text{Bitterness units (BU)} = A_{275} \times 50$$

This method does not necessarily accurately measure iso-α-acids in beer made from all types of hop products including aged hops, and black beers can give curious results, but the value for BU is a good representation of the sensory bitterness of beer. A refinement of the method is available which permits the accurate measurement of iso-α-acids if needed, and HPLC (high pressure liquid chromatography) permits the analysis of individual iso-α-acids and their isomers in beer.

15.3.1 Hop utilization

The amount of iso-α-acids in beer never agrees with the amount of α-acids added to wort during boiling in the form of hop products. Losses occur which are always significant (more than 30%) and may be egregious (90%). This value is expressed as hop utilization (%)

$$\text{Hop utilization (\%)} = \frac{\text{iso-}\alpha\text{-acids in beer} \times 100}{\alpha\text{-acids added to the kettle}}$$

These losses depend on many factors including the kind and amount of hop product used, the intensity and length of the kettle boil, wort composition especially its gravity (low gravity increases utilization), and subsequent processing. The α-acids dissolve most easily from extracts, less easily from pellets (except stabilized pellets in which the α-acids may be partly converted to calcium or magnesium salts of iso-α-acids during production) and least with whole hops. The α-acids isomerize to a somewhat variable extent during boiling and can participate in many side reactions at the pH and temperature of boiling wort. For example, iso-α-acids can form humulinic acids (which are not bitter). High wort pH promotes α-acid dissolution and isomerization but at lower pH isomerization is slower. Iso-α-acids react with proteins of wort whence they are partially removed as trub or hot break which has an intensely bitter taste. High-gravity worts extract hops less efficiently than low-gravity ones, and a high hopping rate reduces extraction efficiency. Similarly, during fermentation iso-α-acids associate with the surface of the yeast cells present which renders brewer's yeast intensely bitter. Iso-α-acids, being surfactants, react with inert surfaces of all sorts and for example separate on gas bubbles to be deposited on the fermenter walls above the surface of the fermentation. This phenomenon is aggravated by top-cropping yeasts. These and other factors render the use of hops in the kettle boil an inefficient process at best.

Herein lies the advantage of products derived from hops, especially isomerized hop extracts which can be added directly to beer. Brewers can achieve virtually 100% hop utilization by this strategy as well as savings from more convenient storage and transportation (less space and no refrigeration needed), no losses of extract entrained in spent hops and no problems of spent hops disposal.

15.3.2 Isomerization

The chemistry of isomerization which is the most important reaction in hop chemistry is well understood (Figure 15–3). Given their choice of reaction conditions chemists could isolate and convert α-acids to iso-α-acids in very high yield. Conditions would include slightly alkaline pH established by dilute alkali, the presence of divalent metal ions especially Mg^{++} which catalyzes the conversion (and explains the use of such salts in the preparation of some hop pellets) and a short boil. Such specific choice of chemical reaction conditions can be used in the preparation of isomerized hop extract for example. However, if hops or unisomerized hop products are added directly to the kettle the reactions that ensue are far less predictable or quantitative and a complex range of compounds arise among which iso-α-acids are prominent but not alone. Some of the compounds formed arise from α-acids but are not iso-α-acids; other compounds are reaction products arising from iso-α-acids and others from β-acids. The oxidation products of α- and β-acids can also dissolve and react during boiling. Some of these compounds are more bitter than iso-α-acids (e.g. anti-isohumulones, alloisohumulones (which have a double bond shift in the isohexenoyl side chain) and humulinone arising by oxidation of α-acids); some are as bitter as iso-α-acids (e.g. hulupones from oxidation of β-acids) and some are not bitter at all (e.g. humulinic acids). A very small proportion of β-acids may undergo isomerization to form products with five-membered ring structures. Only these products of the α- and β-acids with the five-membered ring seem to be bitter. The small amount of isomerized β-acids may contribute a harsh and lingering bitterness.

Thus, although the chemistry of hops in model systems is well established, what happens in the kettle boil, and the relation of those events to the recovery and quality of beer bitterness is much less well known. Though many reaction products of the soft resins of hops could arise in beer, the list of those actually identified and present in all cases is short.

Beer exposed to light in green or clear bottles, or in drinking glasses, quickly acquires the light-struck, sun-struck or "skunky" aroma and flavor. This arises from the photolysis of the 4-methyl-pentenyl side chain of iso-α-acids to release an isopentenyl radical which reacts with a thiyl radical (.SH) derived from H_2S to form 3-methyl-2-butene-1-thiol (4-methyl-pentenyl-mercaptan, iso-pentenyl-mercaptan, Figure 15–4). This compound has the memorable and unpleasant aroma of skunks. Reduction of the carbonyl group in the side chain of the iso-α-acid with sodium borohydride to an alcohol (rho iso-α-acids) prevents this reaction (Figure 15–4).

Using chemical reduction as part of the strategy it is possible to produce tetrahydro (and hexahydro-) iso-α-acids from α-acids and, remarkably, from β-acids. These compounds are more bitter than the corresponding iso-α-

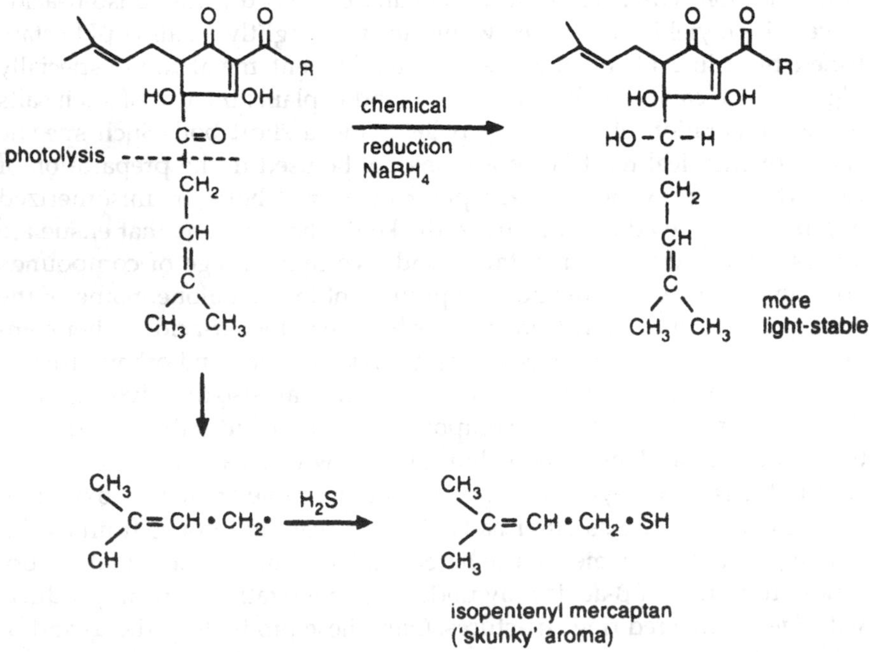

Figure 15–4 Photo-oxidation of hop constituents leading to sun-struck aroma.

acid, less soluble, considerably more bacteriostatic and remarkably enhance beer foam. Although they *do* break down under illumination because the side-chain carbonyl group is not reduced (as in rho-iso-α-acids) they do not form the skunky aroma because they do not form the iso-pentenyl radical.

15.4 THE ESSENTIAL OIL OF HOPS

Hops may contain 0.5% to as much as 3.0% by weight of essential oil which contributes intense aroma to hops and hop products. The quality of this aroma has long been used as a means of judging hops for purchase. Some hop varieties have been deemed **aroma hops** because of their particularly fine aroma, and are used in special ways (by adding hops late in the kettle or to the underback or dry-hopping) to transmit hop aroma to beers. In beers which are produced by dry-hopping (i.e. adding hops to finished beer) the aroma contribution of the hops to the beer results from those major components of the hop oil that are soluble in a dilute alcohol solution (beer). The hop aroma of beers in which hops are added to the kettle, on the other hand, is not well characterized, though it is clear that the major components of the essential oil fraction are *not* those responsible for the hop aroma of those beers. The desirable hop aroma may be due to a few intensely aromatic compounds present in low amounts or to many different compounds present in sub-threshold quantities (especially "late hop" or noble hop aroma).

The hop oil fraction can be isolated for analysis by steam distillation of whole hops or hop products. The essential oil comprises two major fractions: the hydrocarbons and the oxygenated fraction (which generally has more smell though is less volatile). The hydrocarbon fraction comprises 80 to 90% of the total oil fraction and contains myrcene (a monoterpene C_{10}) and caryophyllene and humulene (sesquiterpenes, C_{15}) as the major components (Figure 15–5), plus a whole range of other chemically related monoterpenes and sesquiterpenes. The range and concentration of these oil components is a varietal characteristic that has been used to establish identification schemes for hops and for use in hop breeding.

The hydrocarbons are extremely volatile and appear only in dry-hopped beers to which they impart a dramatic aroma quality. Desirable hop aroma in beer made by kettle-hopping practices, which is sometimes called late-hop or noble hop aroma, is probably derived from elements of the oxygenated fraction and the aroma imparted is quite unlike that of whole hops. Though the precise chemistry is unknown, it is possible to produce essences that imitate the late hop character from fractions of CO_2-extracts of hops.

Examples of hydrocarbons

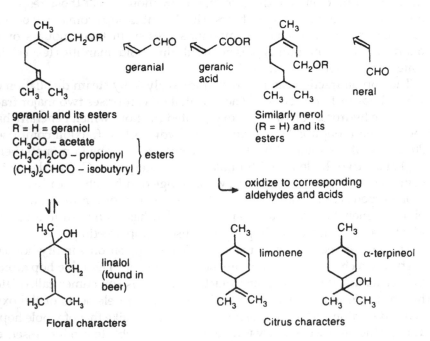

myrcene
(C-10)

caryophyllene
(C-15)

humulene
(C-15)

β farnesene   (not present in all
(C-15)   varieties)

Complex group of compounds
useful in varietal identification
and breeding

Examples of oxygenated compounds

geranial

geranic
acid

neral

geraniol and its esters
R = H = geraniol
CH₃CO – acetate
CH₃CH₂CO – propionyl   } esters
(CH₃)₂CHCO – isobutyryl

Similarly nerol
(R = H) and its
esters

oxidize to corresponding
aldehydes and acids

linalol
(found in
beer)

limonene

α-terpineol

Floral characters

Citrus characters

Figure 15–5 Constituents of hop essential oils.

The oxygenated fraction comprises oxygenated forms of the terpenes and sesquiterpenes. They are mostly acids, alcohols, esters and ethers (Figure 15–5).

Hop oil also contains a quantitatively minor fraction of sulfur-containing compounds. These are perhaps the most potently aromatic material present and may have an importance much greater than their concentration in hops would suggest. Their concentration in hops is partly influenced by treatment of the vines with sulfur and with SO_2 in the kiln. They have undesirable aromas at above-threshold concentrations.

In practical brewing, components of the hop oil, especially the hydrocarbon fraction but also largely the oxygenated fraction, do not substantially survive kettle boiling, even with late addition of hops, because they are volatile in steam. This explains why most hop oil is driven off in the kettle boil and why the typical aroma around a brewery is from the exhaust stack of the kettle. Not all materials are necessarily driven off however and some components of the oxygenated fraction can survive boiling into cooled wort. However, it is not known which hop oil component or mixture of components is responsible for the late hop aroma of beer. Such survival will be affected by the kind and amount of hops used and their composition, the intensity of wort boiling and the precise time of addition to the kettle. These materials are then subject to further modification during fermentation (esterification of acids and reduction of ketones for example) as well as evaporation in the stream of fermentation gas. All these processes are probably selective and particular to individual breweries so that the consistent survival of late hop aroma into beer depends, as so many other things do, on consistent processing

In micro-breweries hops can be added to the fermenter after the initial fast fermentation subsides. Some selective vaporization of hop oil and metabolic alteration occurs and a quite satisfactory even intriguing hop aroma can be achieved. When pellets are added to the fermenter the beer is more bitter.

15.5 KETTLE BOIL

Boiling wort in a kettle has indubitably been a part of brewing technology since early times, though when it was introduced is not clear. Until comparatively recent times boiling was done with external sources of heat. Copper was (and still is) the preferred metal of construction for such kettles because it has a high thermal conductivity compared with stainless steel. Heat from an external burner therefore is more efficiently transferred to the wort in a copper kettle than in a stainless steel one because the flux of heat through the metal is greater. Copper is also a more easily wetted metal than stainless steel. As a result steam bubbles leave a copper surface readily and therefore are less likely to act as an insulating layer than they could in a stainless steel kettle. Kettles for microbreweries are usually built of stainless steel and are heated by an external gas flame or external steam coils. The disadvantages of this material are plainly seen in most designs. In Britain, kettles are stilled called 'coppers' although new ones are made of stainless steel.

In most modern breweries the heat transfer surface (calandria) is usually contained within the kettle and the poor thermal conductivity of stainless steel acts advantageously as an insulation. So long as the heater surface itself is made of copper the poor thermal and wetting properties of stainless steel are not important. Its greater strength and ease of fabrication and, in latter days, cheapness, as well as its resistance to aggressive cleaning agents and inert reaction with wort render it a most suitable metal in this application. Nevertheless, some brewers incorporate copper in the heater and/or at some other point in the wort handling system of the brewhouse because they believe that traces of copper picked up from contact with a copper surface improves beer, being especially useful in the control of sulfury aromas and in yeast nutrition.

15.5.1 Kettle design and size

The length and intensity of the boil influences beer properties including its stability, color and flavor. A kettle is simply a device in which wort can be boiled efficiently. Efficient boiling requires a 'full rolling boil' meaning intense and rapid motion as well as evolution and removal of steam. Under these conditions brewers most easily achieve the purposes of wort boiling which are as follows: (a) Boiling evaporates water in the form of steam which concentrates the wort. (b) The steam carries with it unwanted volatile materials of the wort and hops, including for example the bulk of the essential oil of hops and sulfur-containing volatiles of wort with the potential to form sulfury aromas in beer. (c) Boiling stabilizes the wort in three senses: it sterilizes the wort rendering it bacteriologically stable; it inactivates any enzymes that survive mashing; and it precipitates a protein–carbohydrate–tannin complex called **hot break** or **hot trub**. This comprises high-molecular weight proteins which might otherwise persist into beer rendering it prone to chill haze. (d) Boiling also solubilizes and isomerizes the bittering materials of hops and otherwise adjusts flavor through the formation of new flavor compounds from the Maillard reaction, for example (Figure 15–6), which might also be responsible for some color pick-up (not always desirable).

Kettle designs assure these results primarily by providing adequate heat exchange surfaces (calandria). These usually provide intense heat at the center of the vessel causing a vigorous upward rush of boiling wort and steam that fountains in the middle of the vessel. This impacts an inverted cone that dashes the wort towards the sides of the vessel. Wort then recirculates to the bottom of the vessel to rise again through the calandria. This is the 'full rolling boil' that heat exchange surfaces are designed to achieve. Calandria may be placed eccentrically to promote rolling motion but two more radical approaches to kettle design provide either a sloped heating surface with impellers or external calandria (Figure 15–7). One external calandria may serve two kettles.

Heat exchange surfaces are easily fouled by organic deposits from the wort and brewers must aggressively clean kettles, especially the calandria, with in-place jets and spray balls to assure the efficiency of the heat exchange surfaces.

In modern breweries kettles accept one brew, and the kettle size is therefore matched to the capacity of the mashing/lautering system. This was

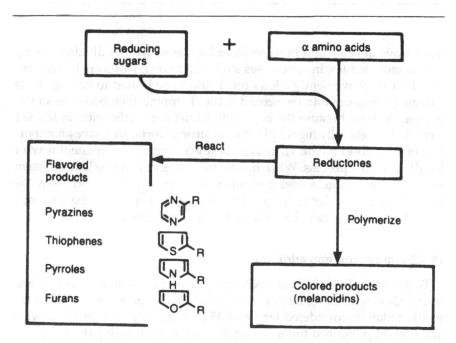

Figure 15–6 Formation of color and flavor by the Maillard reaction.

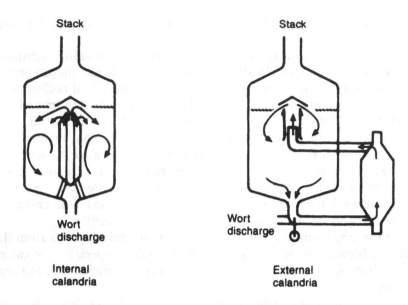

Figure 15–7 Schematic representations of brewery kettles.

not always the case and in some older breweries wort is divided among two or more kettles. In some cases strong and weak worts are boiled separately. This conveniently allows one kettle to be heated to boiling while sparging continues into the second kettle. Hopping then becomes an important decision because the hops will extract more efficiently in weaker wort with a generally higher pH than in strong worts, and have an important effect on beer flavor. There is no best way, merely a consistent way, of handling such a process. When one kettle serves one mash/lauter system a temporary storage vessel commonly receives the first wort from the lauter because the kettle is still occupied with wort from the previous brew. This vessel is called a **holding kettle** or underback.

15.5.2 Timing and hop addition

Timing of the kettle boil usually begins when all the wort is in and the boil starts. Although boiling formerly lasted for 3 hours or more, half that time would today be considered long and 45 to 60 minutes is typical. Brewers add hops at prescribed times during the boil. In ale brewing there is a tendency to add all the hops at the beginning of the boil, and this helps to control the initial high foaming as the wort comes to the boil, but lager brewers often add hops at different times during the boil. Aroma hops added shortly before the end of the boil can impart a unique late hop flavor character.

Brewers weigh out the hops required into large containers for manual additions at the right times, though automatic machinery exists for handling pellets in some modern breweries. The hops used are chosen from the spectrum of hops currently available at the brewery both in terms of variety and age. The objective is consistency. If the charge of bittering hops for example is made of a mixture of half-a-dozen varieties, the temporary absence of one variety will not unduly upset the hop balance. Similarly, new varieties or the new season's hops can be introduced stepwise with such a blending practice. Small breweries often suffer from inconsistency from this source because they cannot afford a sufficient inventory of hops, both from the point of view of their cost and because an excessive inventory will age and deteriorate.

Hop extracts for use in the kettle are packed in lightweight metal cans which can be punctured and suspended in the boiling wort. The thick extract soon melts and disperses.

15.5.3 Wort removal

At the end of the boil the wort must be clarified, cooled, aerated, pitched with yeast (yeasted) and then sent to the fermentation cellar. The suspended matter comprises bulky expanded hop cones, part cones and individual bracts if whole hops were used in the boil or quite dense and small hop particles if pellets were used. There are no residual hop particles with kettle extracts. In addition the wort contains the **hot break** which is present as a thin suspension of large and quite dense flocs. Different hop products cannot be removed by the same machinery. Whole hops must be removed by some kind of filter or strainer.

In a traditional hop back, with a perforated floor reminiscent of a mash tun, the hops form a deep filter bed through which the hot wort is circulated until it is clear (bright). The filter bed of hops removes the hot break and small hop particles. The bed is rinsed to recover entrained wort. In a hop strainer or hop jack the wort is discharged onto a sloping or scraped screen through which the wort passes and the hops that collect are sparged (rinsed) with hot water and squeezed with a screw device to recover entrained wort.

The residue from pellets cannot be recovered in these ways. It can be settled out in shallow vessels or, more usually today, the wort is discharged to a **whirlpool separator** (Figure 15–8a). By propelling the wort tangentially (or at least off-center) into this tank, which is typically roughly equal in depth and diameter, the wort swirls in a consistent motion to create currents in the wort akin to the motion in a stirred cup of tea, i.e. the wort moves faster at the outer perimeter of the tank than at the center. The particles move out to the periphery of the tank, sink down the sides and are then propelled to the center (Figure15–8b) depending on the characteristics of the wort and the particles. A whirlpool cannot settle particles that would fail to settle by gravity alone but, like a centrifuge, accelerates settlement. Under ideal conditions the suspended particles settle to a well-compacted and quite stable mass at the center bottom of the whirlpool and in 20 minutes or so clear wort can be withdrawn. In some breweries and microbreweries the kettle conveniently acts as a whirlpool upon completion of the boil.

A few breweries strip clarified hot wort by passing it in a thin film over corrugated plates or tubes and expose it to a stream of hot air. This removes a good deal of water as well as additional volatiles not completely removed by boiling. The wort is cooled somewhat. In all breweries hot wort is cooled to fermenter temperature as rapidly as possible, usually in an enclosed heat exchanger to protect the wort from contamination by aerial bacteria. Air or oxygen is injected as part of this process. A heat exchanger is effectively a place where the rate of wort flow slows by passing it over a broad heat-conducting surface that separates the hot wort from the coolant (Figure 15–9). Heat exchangers usually have at least two stages. The first cools the hot wort with water to yield a flow of useful recovered hot water, and the second stage reduces the wort temperature to that required for fermentation about 8–12°C (46 to 53°F) for lagers and 14–17°C (57 to 62°F) for ales using a secondary coolant.

Cooling, even of perfectly bright wort, causes the precipitation of **cold break** which is chemically similar to hot break. This material may be sent on to the fermenters to be removed later with the spent yeast. Alternatively the cold break may be removed by wort settlement in shallow vessels called **brinks** or clarified by centrifugation. Following these steps the yeast is added to wort (pitching the yeast), usually by metering in the yeast in known amount from the yeast storage tanks.

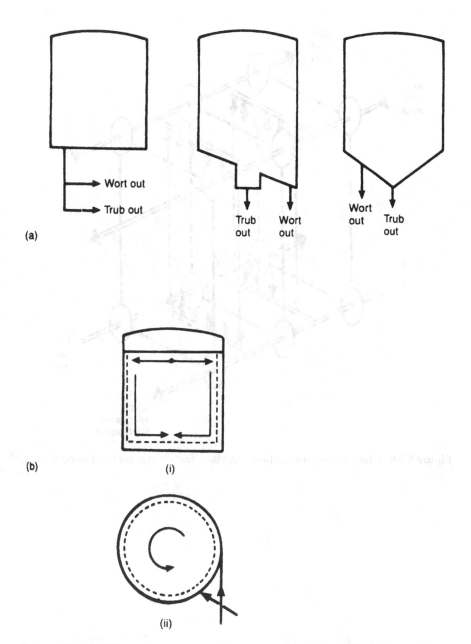

Figure 15–8 (a) Examples of whirlpool designs. (b) (i) Centrifugal force throws particles towards the walls of the vessel where friction with the walls and bottom causes liquid to slow. Wort circulates and particles travel to the center of the base. (ii) Entrance of the wort may be tangential or at up to a 30° angle.

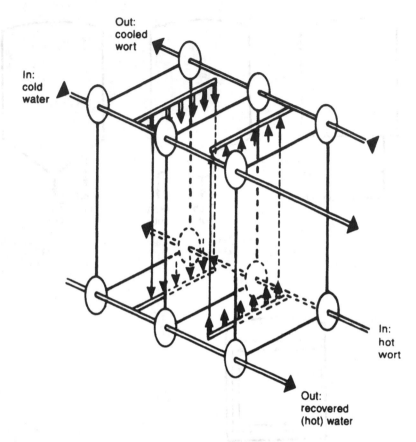

Figure 15–9 Schematic representation of a plate heat exchanger used to cool wort.

=== CHAPTER 16 ===

Brewer's yeast

16.1 INTRODUCTION

Yeast is the most important microorganism for producing fermented beverages. The growth and multiplication of this living organism are inseparable from the metabolic processes that produce ethanol, carbon dioxide, and the whole range of metabolic products that contribute to the flavor of the finished product. The brewer rightly attributes great significance to the strain of yeast used and often will not make it freely available to competitors.

The yeasts used to brew beer have evolved and been selected along with the brewing process. In this evolved state, they have no direct counterparts in nature.

The brewer uses yeast taken from one fermentation to start the next. This operation is unique to beer brewing and without doubt has contributed to the selection of brewing strains. This selection is perhaps most evident in the flocculent nature of brewer's yeast. This property depends on the composition of the cell wall and the availability of calcium ions. Brewer's yeast typically flocculates well enough to provide a crop for removal and leave sufficient yeast in suspension for so-called **secondary fermentation**. It is also notable that many yeasts that crop at the top, as opposed to the bottom of the fermenter, form small chains of unseparated cells. The pragmatic approach of the brewer, relying on experience and the sensory evaluation of the yeast and beer, would also ensure the selection of yeast giving "good" beers.

For brewing purposes, the brewer needs to know that the yeast he or she proposes to use is: the chosen strain, free from other microorganisms, "healthy," able to ferment actively, and in the right concentration to start another fermentation.

16.2 THE CHOSEN STRAIN

The development of microbiological techniques from the nineteenth century onward has enabled the brewer to isolate and preserve selected

279

individual strains. It is now common practice to maintain stocks of such strains (pure cultures) (Figure 16–1). Pure cultures are either maintained in a growing state by sub-culturing or preserved in a non-growing form. Maintaining strains is done using liquid or solidified media. Once grown, stocks are kept at 2–4°C in order to slow down the process of cell death. Preservation is by freeze-drying or, alternatively, storage in the presence of a cryoprotectant (glycerol) under liquid nitrogen (−192°C). It is argued that preservation is the superior process because the sub-culturing and growth required by maintenance techniques increase the opportunity for contamination and selection of mutant strains. Yeast may also be preserved in cryoprotectant (15–25% v/v glycerol) at higher temperatures (−20°C better −70°C); however, these temperatures are not as effective at maintaining viability as that obtained with liquid nitrogen. Bulk yeast may be transported over long distances provided it is kept cold, preferably in vacuum (dewar) flasks. Small samples on maintenance media or in freeze-dried form may be mailed. However, special arrangements apply for mailing live cultures of microorganisms. Yeast cultures that are air-dried on small squares of sterilized blotting or filter paper and sealed in sterile foil remain viable for several weeks and are especially suitable for mailing. These cultures can be revived in liquid maintenance media or agar containing maintenance media in Petri dishes. It is clearly most important to take extreme care to avoid contaminating the desired culture during preparation for storage whatever the method used.

16.2.1 Preparing yeast for fermentation

To obtain the large amount of yeast needed for fermentation, cultures on maintenance media or in their preserved state are first grown in the laboratory before transfer to specialized equipment (yeast propagation or culture plant; see below and Figure 16–5). This equipment is designed to hygienically produce the large amount of yeast needed for fermentation. This guarantees a regular supply of high-quality, healthy yeast, free from other microbes and, by reducing variation, assists in assuring both consistent fermentation and beer quality. Recourse to the laboratory stocks occurs at regular intervals (perhaps twice yearly), but fresh yeast is drawn from the propagator at much more frequent intervals. Typically, yeast will be replaced after five to ten successive fermentations. Most fermentations are, therefore, conducted with yeast drawn from a previous one and not from a propagator. Brewers involved in small-scale operations often use mixed cultures (more than one strain) and have very limited (if any) laboratory facilities. Provided that the brewer is scrupulously hygienic in

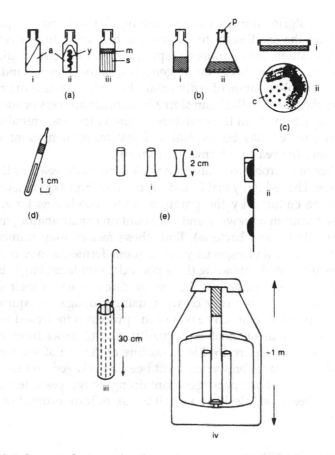

Figure 16–1 Some techniques for the maintenance or preservation of pure yeast cultures. Maintenance techniques relying on growth and sub-culture: (a) (i), (ii), and (iii) Medium containing agar. (i) and (ii), Slant cultures (slopes); a, nutrient medium solidified with agar; y, growth of yeast. (iii) Stab culture; s, vertical stabs of yeast growth in soft agar nutrient medium. The stabs are made by aseptic transfer of culture using a sterile needle; m, mineral oil overlay; this maintains anaerobic conditions and the culture remains viable for longer. (b) Liquid medium. (i) Bottle culture; (ii) flask culture. p, Sterile air-permeable plug. (c) Medium containing agar. Streak culture on medium with agar in a Petri dish; (i) viewed from the side; (ii) viewed from above. c, isolated colonies of yeast. Preservation techniques (suspended animation): (d) Freeze-dried culture. The culture is on a strip of filter paper in the vial, which is sealed under vacuum. (3) Storage under liquid nitrogen. (i) Short sections of sterilized plastic drinking straw are heat-sealed at one end, filled with cryoprotected yeast suspension and then closed by heat-sealing the open end. The straws are placed in small plastic vials and their contents cooled to −30°C. (ii) The vials are clipped onto a metal strip (cane). (iii) The canes are placed in a holder, which is then immersed in liquid nitrogen in a cryovessel. (iv) The well-insulated vessel restricts the evaporation of the liquid nitrogen; however it needs to be topped up on a regular basis.

handling and storing the yeast and sanitizing all plant and equipment, it is feasible on the small-scale to maintain a healthy culture over many years of operation. This requires keeping records of fermentation performance and regular checks on its general quality (sensory and microscopic analysis, observation of sedimentary behavior). In case of problems with yeast, the brewer will obtain slurry from other brewers or sometimes re-isolate the strains from their mixture. In the latter case, initial fermentation performance may be atypical and several fermentations may be needed before the yeast performs satisfactorily.

A number of microbrewers are using dried brewer's yeast as their primary source. This is highly viable and supplied in large (kg) amounts. The yeast may be cultured by the producer under conditions far removed from those found in a brewery and may contain contaminating microbes (particularly lactic and bacteria). Both these factors may influence the quality of the first and especially subsequent fermentations conducted with the yeast. Notwithstanding these potential problems, larger brewery companies are also examining the use of dried yeast in their process. Culturing yeast in the brewery, on wort, using propagation equipment is clearly an expensive option. The cost of the process is increased when the necessary yeast management and quality control procedures are also taken into account. Figure 16–2 is an outline of the use of yeast in brewery operations. In large breweries, most beer is packaged and sold free of yeast. Live beers, those packaged containing active yeast (e.g., bottle-conditioned beers; real ales), are a small but significant proportion of beer produced.

16.2.2 Yeast morphology

Brewer's yeast is a eukaryote and belongs to the Fungi. For scientific, as opposed to brewing purposes, all beer-brewing strains of yeast are placed in the genus *Saccharomyces* (= sugar fungus) and species *cerevisiae*. The brewing industry, however, retains an earlier classification into two types, ale-yeasts (*S. cerevisiae*) and lager-yeasts (*S. carlsbergensis*). The distinction is kept so as to separate yeasts used to make "ales" from those used to make "lagers." Since the major characteristics of the two beer types are more dependent on both the raw materials used and the manufacturing processes than on the nature of the yeast, the distinction, although useful, is artificial. Such a classification of yeast is of little value when comparing one brewing strain with another. In these circumstances, tests of performance under simulated brewing conditions, and examination of the growth

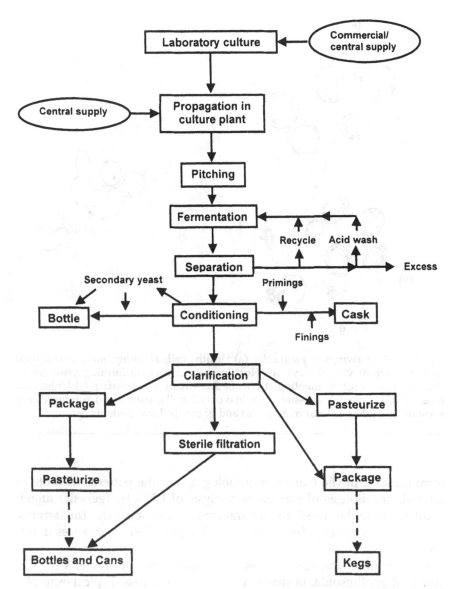

Figure 16–2 Flow diagram for the use of yeast in a brewery. The bulk of yeast is removed as excess at the end of the fermentation stage. Residual yeast or additional yeast added remains in package (live beers) or is largely removed by clarification (centrifugation but more usually filtration). In sterile filtration processes any yeast remaining in the beer is removed on the filter. Effective pasteurization kills the yeast remaining in other products.

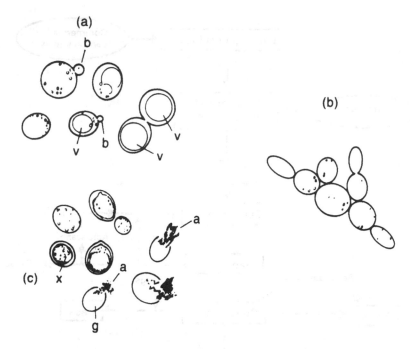

Figure 16–3 Drawings of yeast cells. (a) Healthy cells showing bud formation (b) and presence of vacuoles (v). (b) Healthy cells of a chain-forming yeast (often characteristic of top fermenting strains of *Saccharomyces cerevisiae*). (c) Unhealthy cells showing granular appearance, x; loss of cell wall integrity and autolysed cells showing extruded cellular material (a) and ghosts lacking contents (g).

form on solid media (colony morphology), and the patterns of proteins after electrophoresis of extracts or analysis of DNA by (genetic) finger-printing may be used to "characterize" and classify the strains. Microarray technology (see 18.9) is a technique offering advances in this area.

Viewed under the light microscope, a single cell of brewer's yeast is spherical or ellipsoidal in shape and is 5–13 μm across. Typical examples are shown in Figure 16–3. Each cell is bounded by a rigid cell wall enclosing the cell membrane (plasmalemma). Inside the plasmalemma lies the cytoplasm, which is permeated by membranes (endoplasmic reticulum and Golgi body). Within the cytoplasm are the nucleus, vacuole (the most prominent structure), and various inclusions.

The nucleus, membranous structures, and other inclusions are only clearly revealed by staining or by using the higher resolving power of the electron microscope.

16.3 FREEDOM FROM CONTAMINATING MICROBES

The presence of contaminating microorganisms in a yeast culture may be seen using the light microscope. Typically, a magnification of 400× is needed; at this magnification, 10 000 to 100 000 microorganisms/ml of sample can be readily detected. The higher the magnification, the less sensitive (or more tedious) the procedure because smaller volumes of sample are viewed in each field. At lower magnifications, bacteria are difficult to distinguish from the particulate matter and cell debris invariably present in brewery samples. Best results are obtained using contrast-enhancing optics such as phase or differential interference methods. Staining methods can also be used, but the fixation procedure often reduces the size of the cells. Contaminating yeasts are very difficult to detect because many are morphologically similar to brewing cells.

A trained and experienced operator can gain much useful information from using the light microscope. It is said that Pasteur was invited by a major London brewer to predict accurately which batches of beer were likely to spoil. Using his microscope to great effect, he did make accurate predictions. The brewer concerned then purchased a microscope! However, the microscope is not a particularly sensitive means of detecting contamination. Accordingly, more sensitive techniques employing plating samples on selective media are used (Chapter 5), though these do not, however, give results quickly and much research is directed at developing rapid methods of analysis (Chapter 5).

16.4 THE HEALTH OF YEAST

The light microscope can be used to monitor the general condition of yeast (Figure 16-3). Healthy yeast cells show clearly defined single vacuoles. Early in fermentation, many cells can be seen to be reproducing by budding, but toward the end most cells will have ceased to divide. The cell giving rise to a bud is referred to as the mother, and the bud the daughter. Early in fermentation, the daughters are small but as they age they increase in size and produce buds. At the end of a fermentation all cells tend to be large. The typical life-span of a cell (in laboratory culture) is about 25 generations. Dying cells show many smaller vacuoles, become granular in appearance, and develop a visually empty vacuole, which fills the cell. Ultimately the cells lyse. The process of lysis (autolysis = self-digestion) results in the release of cell contents leaving empty shells (ghosts) of predominantly cell walls (Figure 16–3).

16.4.1 Viability

The microscope may also be used to assess the **viability** of yeast. Viability (the proportion of living cells in a sample) is a measure of ability to ferment—a property not possessed by dead cells! It can be estimated using vital stains; these are absorbed and bound by dead but not living cells. The most common are methylene blue and eosin, which stain dead cells blue or pink respectively. Typically, a healthy culture would contain ≥95% viable cells and it would be inadvisable to use cultures with viabilities of <85%. It is generally accepted that measurement of viability alone is insufficient to give a true indication of fermentative ability. The ability of a yeast cell to absorb dye is not directly related to its vigor. This is certainly true where yeast has been subjected to physical or chemical shock such as exposure to high temperature, pressure, or extremes of pH. Such treatments lower the fermentation ability of many yeast strains without affecting the ability of cells to take up vital stains. Similarly, yeast showing poor (but acceptable) viability by staining tests often are unable to ferment efficiently. Measurements of fermentative ability can be made directly on a small scale using a high concentration of yeast to produce a rapid fermentation of brewer's wort. Fermentation can be detected in a few hours. Alternatively, the rate of oxygen uptake may be followed using an oxygen electrode. Healthy yeast takes up oxygen rapidly; this test can be semi-automated and takes a few minutes to complete. An alternative technique uses a conductivity cell to measure the capacitance of yeast. This parameter correlates with yeast viability.

16.4.2 Amount of yeast

The number of yeast cells in a suspension may be determined using a microscope and a counting chamber. With a chamber of the Thoma type (depth 0.1 mm), then 10 cells in the smallest ruled area ($1/400$ mm^2) corresponds to an original suspension of 40 million cells/ml (Figure 16–4). This approach is deceptively simple, but for reliable measurements several counts each of at least 600 cells should be made. Special care is needed when forming the chamber. The top is formed by a coverslip and this must be positioned accurately, otherwise the volume of the chamber

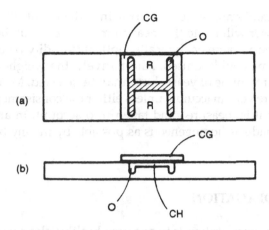

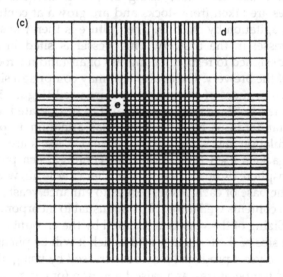

Figure 16-4 Counting chamber for measuring yeast cell concentration. (a) Plan view. (b) Side view. (c) Enlargement of one ruled area. CG, cover glass, O, overflow troughs; CH, chamber; R, ruled area. The depth of the chamber is typically 0.1 mm. The complete ruled area shown has the dimensions 26 × 26 mm; each square (d) in the ruled area has dimensions 0.2 × 0.2 mm. In the central area, each individual small square in ruled area (e) has the dimensions 0.05 × 0.05 mm. So yeast cells in the smallest square occupy a volume of 0.1 × 0.05 × 0.05 mm³. If the average number counted in each small square is n, then the number in 1 mm³ is $n/0.1 × 0.05 × 0.05 × n = 4000$ and the number of cells in 1 cm³ would be $n ×$ 4 000 000.

will vary significantly and introduce error into the count. It is often the case that the brewer will relate the measurement of cell number to a less tediously obtained parameter. Typically, optical turbidity or centrifuged packed cell volume could be used. Alternatively, the weight of pressed yeast or weight or volume of yeast slurry may be selected. Measurements based on slurry are very inaccurate unless either the consistency is known to be constant or it is measured and taken into account. In any event, a slurry must be made as homogeneous as possible by mixing before samples are taken.

16.5 YEAST PROPAGATION

The objective of propagation is to produce healthy, clean yeast in sufficient amounts for fermentation. The process initially begins in the laboratory under controlled conditions using aseptic techniques and sterile media. Cultures are taken from stocks and are grown at controlled temperature, with agitation by shaking. The culture is then transferred to a propagation vessel in the brewery. This vessel is sited in a separate enclosed area designed to restrict the ingress of air (and microorganisms) from the rest of the brewery. The equipment may comprise a single vessel but usually two or more of increasing size are used (Figure 16–5). Each vessel is designed to be cleaned in place and to be operated to the highest possible standards of hygiene. Particular attention is paid to the design of inoculation and sample ports (if any). Vessels are enclosed in dual-purpose jackets used for heating and cooling. Each propagator is vented to atmosphere through a sterilizing filter. Sterile air, which may be sparged into the base of each vessel, serves to enhance yeast growth and keep the vessel contents well mixed. It is not usual to incorporate mechanical agitators. Sizing of the vessels is dictated by the amount of yeast produced and the size of the fermenter into which it will be pitched.

The amount of yeast produced may be increased by using high-gravity worts, elevated temperatures, and raised aeration (or oxygenation) rates. For an ale yeast, propagation for 2 days, using a wort of specific gravity 1.045, and a temperature of 25°C may be used. The aeration rate would be 0.2 volumes air/volume wort for 1 minute in every 5 minutes. The yield of yeast would be about 10^8 yeast cells/ml.

A typical concentration of yeast in a fermenter at **pitching** is 10^7 cells/ml. A propagator with a working capacity of 100 hl used in this way would provide sufficient yeast to pitch, at a normal rate, a 1000-hl fermenter.

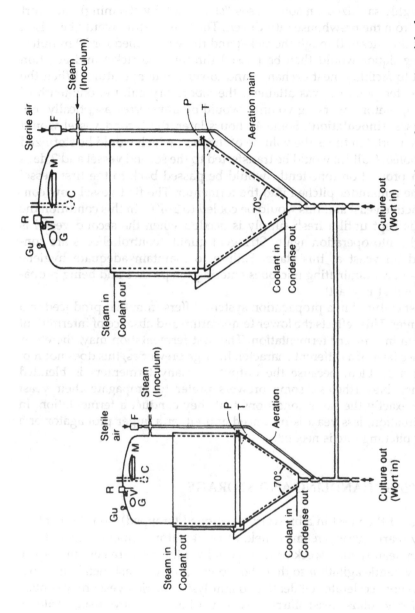

Figure 16–5 Two-stage yeast propagation system. The smaller vessel is about one-tenth the working capacity of the larger. The details of operations are given in the text. Gu, pressure gauge; R, pressure valve; G, sight glass; V, vent; C, cleaning spray ball; M, manway; F, sterilizing filter; P, sample point; T, temperature probe.

In a two-vessel propagation system, the first and smallest, would be thoroughly sanitized (in some cases "sterilized" by steaming) and wort direct from the brewhouse introduced. The temperature would be raised by passing steam through the jacket and the wort boiled for 30 minutes. Cooling liquor would then be passed through the jacket and aeration begun to facilitate heat exchange and lower the temperature. When the desired temperature was attained, the laboratory culture, one-tenth of the propagator's working volume, would be transferred aseptically into the vessel (**inoculation**). Some 48 hours later, the second vessel would receive wort, ten times the volume of the first, which would be sterilized and cooled. Culture would be transferred to the second vessel and 2 days later a proportion (one-tenth) would be passed back to the first vessel and the remainder pitched into the fermenter. The first vessel now contains active yeast and this would be cooled to 2–4°C. In this condition the yeast is kept until a fresh supply is needed when the second vessel is brought into operation again. Stringent quality control checks are conducted on yeast at this stage. Failure to maintain adequate hygiene results in contaminating microbes entering the process and being propagated with the yeast!

Yeast cultured in a propagation system differs from that produced in a fermenter. This reflects the lower temperature and absence of intermittent aeration in brewery fermentation. The first fermentation may, therefore, produce beer of a different character. In large breweries, this does not represent a problem because the output of many fermenters is blended together. Nevertheless, some brewers prefer to propagate their yeast under exactly the same conditions that they conduct a fermentation. In this situation, less yeast is produced and either a larger propagator or a lower pitching rate is needed.

16.6 YEAST HANDLING AND STORAGE

Most of the yeast in a brewery is removed (cropped) from fermenters. Slurry (barm, yeast cream) is held at refrigerator temperatures and will remain healthy for a week or so, especially if the cells are kept in suspension by gentle agitation so that "hot spots" (due to yeast metabolic activity) do not accelerate cell death and autolysis. Slurried yeast may contain 5–15% dry solids. Yeast slurry may be washed on a sieve using water at 1°C to remove impurities such as wort solids, dead cells, and possibly bacteria. The flocculent nature of the yeast makes yeast flocs denser than the contaminants.

Yeast that contains bacteria may be washed at acid pH to effect a reduction in the level of contamination. Various acids—tartaric, sulfuric (with and without ammonium persulfate) and phosphoric—may be used. Typical pH values would be in the range 2.1–2.4 and contact at 2–4°C for 30–60 minutes. Some yeasts, although remaining viable by staining, show reduced fermentative ability after such treatment.

Some breweries press yeast, using filter presses, which produces "cake yeast" (20–30% dry solids). The design and operation of modern "membrane presses" are similar to those described for mash filters in Chapter 12. Pressed yeast may be used to pitch fermenters and before use it is also stored at refrigerator temperatures.

After assessing the quality and quantity of yeast, it is pitched into wort. Typical rates are 0.3 kg/hl using pressed yeast (25% dry solids) and proportionally more for slurry.

16.7 IMPROVED BREWING STRAINS

The general techniques of strain improvement by breeding are in theory available for use with brewing yeast. The fungal equivalent of the gametes of animals are the **ascospores**. Cell lines derived from ascospores may be "crossed" to give new combinations of genetic material. Brewer's yeast, however, tends to sporulate poorly and show poor spore viability, so this route to strain development has not proved popular. Other techniques of forming asexual hybrids (protoplast fusion and rare mating) have also been investigated as a means to improve strains. The difficulty here is not a technical one but more a question of defining the desirable traits to be pooled together and more importantly controlling the outcome or selecting for the desired products.

The techniques of **gene cloning** and **genetic engineering** are readily applied to selected strains of *S. cerevisiae*. However, commercial brewer's yeast is polyploid and this can create difficulties with a technology devised primarily for use with haploid strains. With brewing strains, research has focused on changes that improve the fermentation process, modify beer flavor, or increase the value of waste yeast. Strains capable of hydrolysing dextrins and β-glucans have been reported. These increase the fermentability of wort where the enzymes hydrolyze polymeric carbohydrates to fermentable sugars. Removal of β-glucan and the resulting decrease in viscosity, may also increase efficiency of beer filtration by extending the length of time a beer filter can be used. Strains that prevent the growth of wild yeasts (show anti-contaminant properties) or are engineered to produce lower levels of hydrogen sulfide and diacetyl have been developed. The flocculent nature of brewing strains has also been changed by genetic modification. To our knowledge, none of these is in use in beer production.

A strain manipulated to synthesize human serum albumin and thus greatly increase the value to the brewer of waste yeast is a key development. Brewer's yeast can be manipulated to produce non-yeast (heterologous) proteins and, of course, the yeast is available in large quantities. The aim would be to remove yeast from fermentation and then cause it to synthesize the heterologous product; this, of course, would be of high value. The extent of use of these and other strains, both now and in the future, is difficult to assess. Considerable consumer resistance is likely to be expressed towards such developments and the costs of meeting current and possible future legislation pertaining to food and beverage standards may make developments uneconomic.

16.8 WASTE YEAST

The brewer can expect to produce about five times the amount of yeast he or she uses. Discharging this material to the drains is an increasingly expensive operation. Some breweries have installed effluent treatment equipment to handle the majority of brewery wastes. Excess yeast has a value as a by-product for use by distilleries, food processors (including some brewery companies), and animal feed producers. The distilling industry can provide up to 50% of its requirement for yeast from waste brewer's yeast. It is less expensive than pure distiller's yeast and adds additional flavor components (congeners). The high content of nucleic acids restricts the use of yeasts in human foods to the status of a minor component, a dietary supplement. It is a valuable source of vitamins

Table 16-1 Partial composition of dried, unfortified brewer's yeast

Major component (g/100g)	Concentration (mg/100g)	Vitamin	Concentration
Protein	50	Niacin	50
Carbohydrate	42	Thiamin	15
Ash	7	Pantothenate	10
Fat	6	Riboflavin	7
Moisture	5	Folic acid	4
		Pyridoxine	3
		Biotin	0.2

(Table 16–1) and is used, after processing, in the form of powder, tablet, flake, or liquid concentrate. It is usually necessary to debitter the yeast (remove adsorbed hop bitter substances) before using it this way. A novel approach uses yeast in the manufacture of pretzels. Yeast extract is also used as a carrier of flavors in potato chips (crisps), packet soups, and in its own right as a source of "meatiness." Yeast mixed in small proportion with spent grains is used as cattle feed. Yeast slurry treated to kill the cells (using heat, propionic, or formic acids) may be fed directly to pigs.

======== CHAPTER 17 ========

Fermentation— overview, process, and technology

17.1 OVERVIEW

The cooled aerated wort made in the brewhouse is fermented by yeast to make an immature beer (green beer). During this process, yeast reproduces. The main fermentation (primary) in which green beer and yeast are produced is often followed by a slower process at lower temperature in the presence of lesser amounts of yeast. This is referred to as **secondary fermentation**, conditioning, or even maturation. Finally, beer may be freed from yeast and aged (matured) at low temperature before packaging. The quality of the green beer depends upon the composition of the wort, the yeast strain and amount used, and the process and technology of primary fermentation. This product is the raw material for later processes in the brewery.

It is an axiom in brewing that if the wort is of uniform composition, the yeast is healthy and free from contaminating microbes, and the fermentation conditions are accurately controlled, then the consistent quality of the green beer is assured. The whole process of primary fermentation is an exercise in nurturing and cajoling yeast to do the work of a living cell, in an environment specified by the brewer. This work, from the stand-point of the yeast, is to grow and multiply. It is the by-products of this labor that both make a major contribution to the chemical composition of beer (alcohol, carbon dioxide and flavor compounds) and generate new yeast for subsequent fermentations.

For yeast to live and grow, wort must contain a sufficient supply of nutrients. Yeast needs fermentable carbohydrate, assimilable nitrogen, molecular oxygen, the vitamin biotin, sources of phosphorus and sulfur, calcium and magnesium ions, and trace elements such as copper and zinc ions. In all-malt worts made using brewing water containing salts, these

295

are in adequate supply. Some worts may be supplemented with copper, zinc, and yeast food. The latter is a source of biotin (yeast extract) and assimilable nitrogen (ammonium salts). Wort is routinely analyzed for content of fermentable sugars (fermentability tests) and assimilable nitrogen (free amino nitrogen—FAN—test). Typical values would be 70% fermentable and 140 mg/l FAN for a wort of 1.040 specific gravity. Specific measurements of ammonium ions and amino acids in wort have shown that the FAN test greatly underestimates the assimilable nitrogen content. It also takes no account of the fact that some amino acids, notably glutamic and aspartic, are more readily used by yeast than others.

The dissolved oxygen content of wort is an important parameter. Research has shown that different yeasts have different requirements for molecular oxygen. Sufficient must be supplied at the beginning of the fermentation to satisfy the yeasts' requirements. Once this is done, no more is needed and, indeed, anaerobic conditions are established and necessary to prevent undesirable oxidation reactions, which spoil finished beer. Given traditional brewing operations, most brewer's yeast will have been exposed to air-saturated worts. Worts of specific gravity 1.040 contain about 6 ppm dissolved oxygen at 20°C when air-saturated. The opportunities for yeast to be exposed to air were certainly greater in traditional practices (see below) than in modern ones. It is not surprising, therefore, that in modern practice, 8 ppm dissolved oxygen is often used and in some cases more. The quantity of oxygen in solution is inversely proportional to both specific gravity and temperature. Therefore, air saturation gives different dissolved oxygen concentrations with worts of different gravities at the same temperature and *vice versa*. Brewers, therefore, tend to use molecular oxygen as source and carefully monitor the level in wort using in-line oxygen electrodes. Too much oxygen results in too vigorous fermentation (with consequent changes in beer flavor) and excessive yeast growth at the expense of alcohol production. Too little causes accelerated loss in yeast vitality and viability.

In nurturing yeast, the brewer will provide a controlled temperature and remove the excess heat produced by yeast metabolism. Even on wort, yeast must be cajoled into producing the maximum amount of alcohol and minimum amount of growth by restricting the availability of oxygen and using lower fermentation temperatures (9–20°C) than those preferred by the organism (25–28°C).

17.2 PROCESS AND TECHNOLOGY

The main consideration in primary fermentation is to ferment wort to the desired gravity (degree of attenuation) in the required period of time.

This is largely achieved first by accurate pitching of wort and secondly, by controlling the temperature of the process. The progress of fermentation is usually monitored by measuring **specific gravity**. Intermittent sampling or on-line analysis may be used and measurement of CO_2 or ethanol in the exhaust gases are alternative procedures. Typical fermentation profiles for ale and lager production are shown in Figure 17–1a,b). The decline in specific gravity is matched by the growth of yeast as sugars are used and ethanol is produced. The pH falls as ammonium ions and amino acids are consumed and the yeast secretes organic acids. Most flavor compounds are also secreted in line with yeast growth. The concentration of flavor compounds may decline towards the end of fermentation and dur-

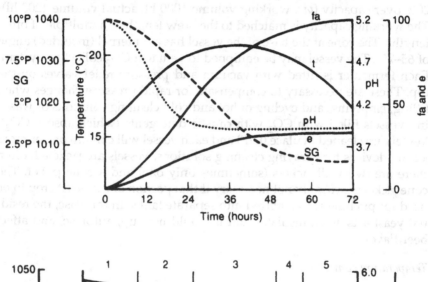

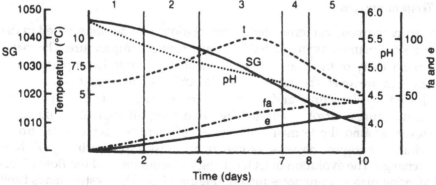

Figure 17–1 Time-course of fermentation for (a) ale and (b) lager. °P, extract in degrees Plato; SG, specific gravity; fa and e, levels of fusel alcohols and esters (mg/l); t, temperature (°C).

ing storage. This decline arises as some volatiles are carried out with the evolving carbon dioxide or absorbed and metabolized by the yeast. Primary fermentation in traditional vessels (rectangular boxes or horizontal cylinders) takes about 4 days at 20°C and 10 days at 12°C. Lower temperatures mean longer fermentation because of the lower rate of yeast metabolism at lower temperatures.

17.2.1 Cylindroconical fermenters

There is a world-wide trend to replace old fermentation vessels with cylindroconical fermenters (Figure 17–2). Newly designed fermentation plant almost always incorporates cylindroconical fermenters. This type of vessel is usually fabricated in stainless steel (304 or 309 grade) and sized 20% over capacity (*viz.* working volume 1000 hl, actual volume 1200 hl). The working capacity is matched to the brew length or a multiple of brew lengths. The cone at the base of the vessel has an internal (included) angle of 65–75°. The vessel may be equipped to vent to a CO_2 collection system. Each fermenter is fitted with vacuum and pressure relief valves on the top. These are necessary to compensate for rapid pressure changes when filling, emptying, and cycling of hot and cold cleaning solutions or washing vessels filled with CO_2 with caustic detergents (which absorb CO_2). Vessels are cleaned in place (CIP) and each vessel will contain one or more jetting devices for spraying cleaning solutions. Vessels are jacketed; often there are two wall jackets (sometimes only one) and a cone jacket. The cone jacket is probably not necessary if that portion of the yeast crop to be used for pitching is transferred into separate tanks. In this case, the residual yeast acts as an insulator, but it would heat up, autolyse, and affect beer flavor.

Temperature control

In cylindroconical fermenters, fermentations that do not proceed in the expected manner are regulated by adjustment of **temperature**. The metabolic activity of the yeast and thus fermentation rate is proportional to temperature over the range about 2–30°C. So, fermentations may be slowed by lowering and speeded up by raising the temperature. Temperature control is effected by circulating attemperating liquid in jackets around the fermentation vessel. Efficient temperature control is achieved because the vessel contents are mixed, promoting good heat exchange. The evolution of CO_2 during fermentation and the flow of fermenting material promotes mixing (Figure 17–3). This system lends itself readily to temperature programming. The ease of fermentation control is

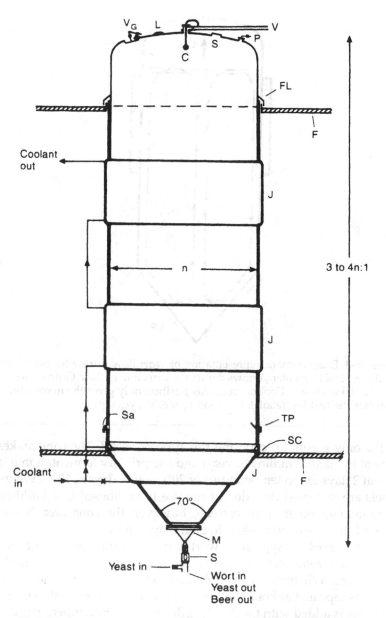

Figure 17-2 Diagrammatic representation of a cylindroconical fermenter. n, Diameter of cylinder; V_G, vacuum relief valve; L, inspection light; C, cleaning-inplace jets; S, sight glass; P, pressure relief valve; V, vent and CO_2 collection main; FL, flange; F, floor; J, cooling jacket; Sa, sample point; TP, temperature probe; SC, supporting collar; M, manway.

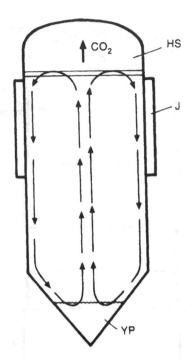

Figure 17–3 Diagrammatic representation of wort flow during fermentation in a cylindroconical fermenter. Arrows depict the direction of flow. Only a single cooling jacket (J) is shown. Cooling from this position only gives the maximum circulation (see the text for details). HS, head space; YP, yeast plug.

not the only reason for choosing this type of vessel. The cone makes for efficient hygienic handling of yeast and the primary fermentation is complete in 2 days at top temperatures of 20°C and 5 days at 12°C. Where the vessels are insulated they do not require to be housed in a building (in common with other large vessels); however, the cone area is usually enclosed for ease of access and hygienic operation.

Typically, cooled oxygenated wort is pumped into the base of the vessel and yeast from a yeast tank or the cone of another vessel is simultaneously added. Where the vessel volume is a multiple of brew lengths, this process is repeated although in some operations all the yeast to ferment a full vessel is added with the first length of wort. The temperature of wort at pitching is pre-determined (typically 15–17°C for ales and 10–13°C for lagers) and as yeast begins to ferment, the temperature rises. Cooling is automatically applied to control both the rate at which the temperature rises and the maximum temperature reached (20–23°C ales; 12–17°C lagers). As fermentation progresses, yeast will flocculate and tend to settle in the cone. At the base of the cone the hydrostatic pressure ensures that the concentration of CO_2 in solution is greater than at the top of the vessel. This concentration gradient encourages the evolution of a stream of gas up the center of the vessel. This gives the impression that the center of the wort surface is boiling. Although the gas escapes, the liquid cannot, so its momentum carries it back down the walls of the vessel. This mechanism ensures yeast and wort are well-mixed and explains why fermentation is so rapid. As the yeast nutrients are depleted, fermentation slows, less heat is evolved, and decreasing amounts of CO_2 are made. The vessel cools and increased cooling may be used to bring on a quiescent phase and assist the bulk of the yeast to flocculate and settle in the cone.

Cooling may be to a few degrees below original pitching temperature and the vessel held for "warm conditioning" (see Chapter 19). Eventually the contents will be cooled to 4°C. It is not usual to attempt to cool below 4°C. At this temperature beer is at its most dense and in the absence of mixing, heat exchange is poor so the fermenter would be unavailable for some time. Green beer is transferred to further tanks (often horizontal cylinders) for maturation and either chilled to 0 to −1°C through a heat exchanger or *in situ*. This practice allows both more efficient cooling and the freeing of cylindroconicals for fermenting.

Yeast may be removed at any time from the base of the cone. Common practice is to transfer it in part or in whole to a chilled yeast slurry tank at the end of fermentation. This may then be used as the source of pitching yeast. Alternatively, the yeast may be used directly from the cone to pitch another vessel, the remainder being transferred to a yeast tank. The excess yeast is processed (see Chapter 16).

17.2.2 Fermenter design

Many breweries use large fermentation tanks that do not have steep conical bottoms (Rainier, Asahi tanks). In these systems, yeast handling is not as efficient and more beer losses may be encountered because of the less-compact nature of the yeast sediment. Similar difficulties are encountered with any large rectangular fermenter.

Historically, the trend has been for breweries to effect economies of scale and automated operation and introduce larger and larger fermenters culminating in the large tanks described above. Nonetheless, many breweries still employ more traditional methods of fermentation. These are as important in defining the brewing operation as the different methods of mashing described in Chapter 12.

Early fermenters were small and constructed from readily available materials. Examples would be the earthenware jars of the Babylonians, the wooden rounds (fabricated from staves of wood sealed with pitch and bound with metal hoops), the slate (stone) square of Britain, and even the hollowed out gourds used by African cultures to produce local brews.

As brewing operations increased in size, these vessels gave way to bigger square or rectangular ones made from copper and eventually stainless steel. However, these larger vessel are still relatively shallow (2–4 m). Wooden rounds and copper squares are still in use in many breweries today.

Yeast handling in the early vessels was simple; it was either scraped off the top or allowed to flow out through a lip made in the side of the fermenter just above the level of wort. Yeast is, therefore, selected to crop at the top and the system is described as **top-fermentation**. As vessel size increased, the brewer devised improved methods of top-cropping. So the hygienically questionable parachute system was born (Figure 17–4a). The "Parachute" (a large funnel) protruded through a gland (sealed hole) in the bottom of the vessel and could be lowered into the fermentation to allow the yeast head to flow out and into a collecting vessel on the floor below. An improved system much in evidence today uses a vacuum line to suck the yeast into a holding tank. Fermentation naturally led to collecting yeast that, at the end of fermentation, had settled to the bottom. This system of **bottom-fermentation** adapts easily to the larger scale.

Specialized systems

The fermentation process, once developed as top or bottom, becomes enshrined in brewing practice. Associated with the system is the axiom that final beer quality depends on the process and there is therefore great resistance to change. This is most clearly seen in the development of two specialized systems of fermentation in Britain. The Yorkshire square and the Burton Union (Figure 17–4b,c); both systems are labor intensive in operation (especially cleaning) and effectively impossible to scale up to modern volumes.

The **Yorkshire square** (Figure 17–4b) separates yeast and fermenting wort on a false ceiling. Actively fermenting yeast rises through the central hole (0.6 m diameter, with a 5-cm high rim) and collects; fermenting wort flows back into the lower chamber (fermenting compartment) through narrow bore pipes (organ pipes). Wort can be pumped from the fermenting chamber and sprayed over the yeast to assist attenuation.

The **Burton Union system** (Figure 17–4c) is based on a series of 7-hl (153 imperial gallon) casks. Each cask is fitted with a swan neck that overhangs a copper trough mounted above a cask. A group of two rows, each of 12 casks, discharging to the same central trough is referred to as a set.

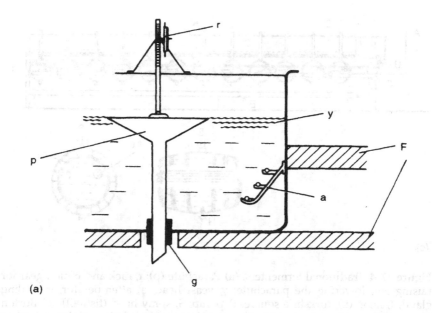

(a)

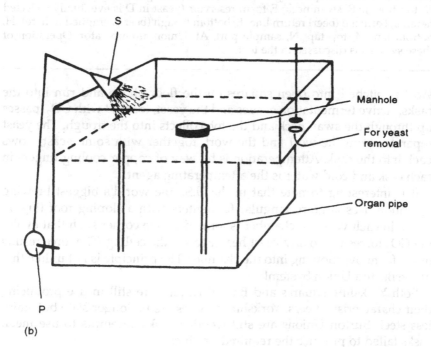

(b)

continues

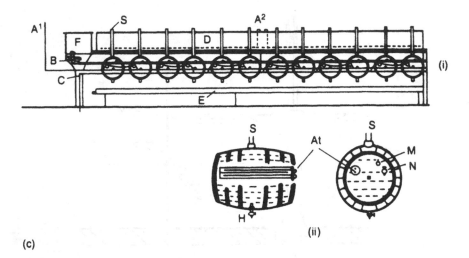

Figure 17–4 Traditional fermenters. (a) Parachute (p); r, rack and pinion gear for raising and lowering the parachute; y, yeast head; a, attemperator; g, sealing gland; f, floor. (b) Yorkshire square. P, pump; S, spray head (fishtail). (c) Burton Union. (i) One side of a set of 24; (ii) single Union cask and view (right); side view (left). A^1, Union attemperating water line; A^2, top trough attemperating water line; D, top trough; S, swan neck; F, foam reservoir (yeast in D is eventually collected here); B, barm ale (beer) return line; E, bottom trough (beer is emptied to here); H, bottom tap; M, top tap; N, sample port; At, Union attemperator. Operation of these systems is discussed in the text.

Wort is pitched into open squares on the floor above and run into the casks. Active fermentation ensues and the yeast, foamed with CO_2, passes up through the swan neck and the fob collects into the trough. The yeast separates from the wort and the wort, together with some yeast, flows back into the cask. Attemperation is by way of copper cooling fingers in each cask and cold water is the attemperating agent.

It is interesting to note that in the U.S., the world's biggest brewing operation uses large rectangular fermenters with a sloping roof (Figure 17–5). In each vessel, a chamber is formed at one corner such that evolving CO_2 forces the foamy yeast head against the ceiling. The force results in the foam overflowing into the chamber. The principle is not unlike that of the Burton Union System!

Both Yorkshire squares and Burton Unions are still in use producing their characteristic beers. Yorkshire squares are no longer slate but stainless steel. Burton Unions are still wooden casks; attempts to use metal casks failed to produce the required product.

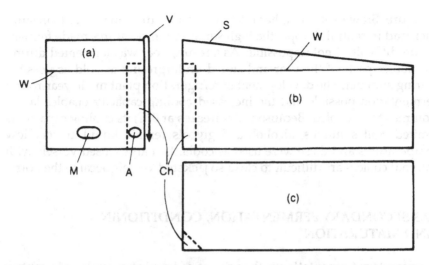

Figure 17–5 Large fermenter with overflow. (a) Front end; (b) side view; (c) view from overhead. W, wort level; S, sloping ceiling; Ch, foam chamber; M, manway; A, access port to foam chamber; V, vent.

The beers of Britain historically produced by top-fermentation were labeled as **ales** to distinguish them from **beers** produced in the rest of Europe. These latter were largely made by bottom-fermentation and called **lagers**. This nomenclature persists, even though the greater part of the volume of British ale is made by bottom-fermentation in cylindroconical fermenters. Top-fermenting, *S. cerevisiae*, ale yeasts sediment in these vessels. Should we now describe them as bottom-fermenters and the beers as lagers? Perhaps it is more sensible to abolish the ale and lager distinction, call them beer and use more specific descriptors based directly upon process.

Cooling system design

Increasing the scale of fermentation and changing vessel design, has necessitated changes to the cooling systems. In the smallest vessels, heat loss alone was sufficient for the temperature not to exceed levels where yeast activity is impaired. As the vessel size increases, the surface area to volume ratio falls, therefore there is less surface to lose heat from. Covering vessels better insulates them. Accordingly, cooling coils or panels were placed within the vessel and well-water used as coolant. No precise control of temperature was made and the top temperature depended on the temperature at which the wort was pitched and the ambient temperature. Brewers used higher pitching temperatures in winter than summer, and in central Europe, the high summer temperatures made fermentation difficult, if not impossible. As a result, beer was fermented during the cooler parts of the year and stored underground as cold as possible during summer. The development of refrigeration plant made year-round fermentation possible and the increased cooling capacity enables larger volumes to be cooled. Because water freezes at 0°C, its cooling capacity is limited. Salt solutions, alcohol, and glycols remain liquid at very low temperatures and these were used as coolants. Large closed vessels with internal coolers are difficult to clean so jacketed vessels became the norm.

17.3 SECONDARY FERMENTATION, CONDITIONING AND MATURATION

Secondary fermentation is the process of keeping green beer in contact with yeast after the primary fermentation is ended. Additional sugar or primings may be added. The practice is most widely used to produce live beers (cask beer, traditional draught, real ale), where it is called **conditioning**. Using the yeast still present or added (secondary) yeast at about 0.5–2 million cells/ml, live beers are made and sold in casks or bottles. The cask beer is fined so that the yeast is sedimented out and the beer bright before dispense. The fining agent, isinglass, is produced from the swim bladders of certain marine fish. Partial digestion of the bladders, which are mainly the structural protein collagen, gives a product that adsorbs and precipitates yeast. The adsorption process is temperature- and pH-dependent because it involves the interaction of charges on the surfaces of protein and yeast cell. In bottle-conditioned beers, primings, but not finings, are added. The yeast forms a sediment at the bottom the bottle. With some products (e.g., Cooper's ale from Australia), the yeast is considered an integral part of the beverage. With others (Worthington White Shield from Britain), the beer must be carefully decanted, leaving the yeast and as little beer as possible in the bottle. Consumption of live beers has become an almost "cult-activity" in recent times. Pressure from consumer organizations has resulted in an increased volume of production of these products and the opening of small pub- and microbreweries. Nevertheless, live beer forms a very small proportion of the total beer market.

In Germany, one traditional form of secondary fermentation is known as **kraüsening**. Kraüsen (= ruffles, frills) is the active stage of bottom fermentation when a yeast head develops and has the appearance of lace material used, in earlier times, to make the frilly cuffs of the gentlemen's shirts. The addition of kraüsen produces changes both in beer composition, especially removal of diacetyl and aldehydes, and in a closed vessel, provides "natural" carbonation to the beer.

In modern brewing practice, some of the benefits of secondary fermentation and conditioning may be obtained in a shorter time by **warm conditioning** the green beer. This is also called the "diacetyl rest." This practice holds the beer, after most of the yeast is removed, at near fermentation temperatures allowing the higher temperature to cause the yeast to do its work more rapidly than at lagering temperature. The beer is then chilled and filtered.

An alternative technology employs yeast cells immobilized (attached to) an inert support. Beer, after primary fermentation, is passed through the immobilized yeast in a continuous stream. The active yeast cells metabolize constituents of the green beer, thus achieving a novel form of "secondary fermentation."

One significant advantage of keeping yeast in contact with beer is that any oxygen that gains access will be absorbed instantaneously by the microbe. Once yeast is removed, the brewer must take stringent measures to prevent oxygen getting into the finished beer.

Maturation processes usually involve gradually lowering the temperature of the product and giving time to permit the settling out of yeast and chill haze particles (that form at low temperatures). Again, this practice (traditionally of 3 months' duration) is derived from mainland Europe and referred to as **lagering** (lager = store). This process is expensive, especially in terms of energy use, and many brewers accelerate it. In the case of chill haze, which comprises proteins and polyphenols, brewers can accelerate its formation, prevent its formation or accelerate its removal. The formation of haze is catalyzed by molecular oxygen. Formation may be accelerated by including additional polyphenols (tannin). Proteins may be removed by digestion with enzymes (papain) or with adsorption (silica hydrogels; nylon powders, or polyvinyl polypyrrolidone will adsorb polyphenols) and the activity of molecular oxygen may be guarded against by including an anti-oxidant (ascorbic acid). Chill haze is removed more quickly using finings, filtration, or centrifugation. These processes are carried out at temperatures of 0 to $-1°C$.

There can be no doubt that changes in the fermentation, conditioning, and maturation processes produce changes in the nature of the final beer. Scientific analysis of beer composition and flavor shows this to be true. It is also true to say, however, that the changes produced are probably no greater than those brought about by changes in raw materials or brewhouse practice that have occurred over the same period of time.

17.4 MOVEMENT OF FLUIDS

In large breweries, the fermentation and finishing cellars are a vast and complex network of tanks and pipes around which wort, water, beer, gases, and cleaning solutions are moved by many pumps and directed by many valves. The behavior of fluids in these circumstances is of intimate interest to brewers, and some broad understanding of why and how fluids do (and do not) flow should be useful. Breweries work well when fluids flow properly in well designed systems, but many factors can conspire to prevent this.

17.4.1. Fluid flow

Gases and liquids are called fluids because they distort to conform to the shape of the vessel that contains them. Gases are compressible fluids. Liquids are fluids that cannot be compressed. Water, wort, and beer are Newtonian fluids, which means that as we apply more shear stress, they move correspondingly faster. For all practical engineering purposes, the properties of water, wort, and beer can be assumed to be the same.

Brewers generally think of flow in a pipe in terms of volume per time (such as hectoliters per hour or gallons per minute). But this is a function of two separable items: (1) average liquid speed, u, as meters per second (m/s), and (2) the cross sectional area of the pipe, A (m^2). Volumetric flow rate is then V = Au (m^3/s) and mass flow rate, $\dot{m}$ = ρAu (kg/s) where ρ is the specific gravity of the liquid. Under the same applied force, liquids with a high viscosity move more slowly than those that are less viscous.

A **short** hose attached to a faucet or tap allows rapid movement of fluid through the hose. The **speed** and **weight** of flow generates a momentum and, at the hose-end, the water continues to travel through the air before falling to the ground. This is the **momentum** of the liquid and such jetting is how cleaning machines work, for example. This momentum also accounts for "water hammer," the noisy and potentially damaging blow delivered to ill-designed pipework when the liquid flow is suddenly stopped by closing a valve. Momentum, or inertial forces, therefore drive liquid forward.

In contrast, a **long** hose, especially if it be a narrow one, attached to the *same* faucet or tap yields much less flow at the hose-end, because of back pressure. There is a pressure drop along the length of the hose associated with **friction**, which results in reduced flow volume. Loss of energy to friction rises dramatically with increased fluid velocity and with viscous liquids. These friction or viscous forces are a function of liquid viscosity and (since they resist flow) they have the opposite effect of the momentum forces that drive the liquid forward. The Reynolds number is the ratio of these two kinds of factors affecting flow, that is, momentum forces and viscous forces:

$N_{Re} = \rho Du/\mu$ (a dimensionless number)

comprising fluid density (ρ), pipe diameter (D), and flow velocity (u), divided by viscosity (μ). A Reynolds number below 2100 implies laminar flow in which viscous forces dominate and one over 10,000 implies fully developed turbulent (mixing/eddying) flow.

Thus, flow in a pipe depends on the nature of the fluid, especially its viscosity (which will be affected by temperature), the amount of pipe wall (there is more pipe wall, relative to volume, in a narrow pipe than in a wide one) and, importantly, the friction between the liquid and the pipe wall, which is affected by Reynolds number and surface roughness. All these factors can be quantified and the data are available to engineers in the form of tables and charts.

In a piping **system**, additional factors come into play. In an arrangement such as that in Figure 17–6 we can suppose that flow will depend on the following factors:

- energy to overcome friction in the pipes and hoses (already mentioned) and friction due to valves and pipe fittings, such as elbows and Tees, and changes in pipe diameter, and friction in heat exchangers or filters, for example, which might impede flow; these are combined into factor E_f (energy loss due to friction) in the equation below.
- energy required to lift the liquid from level 1 to level 2: stated as distance Z_1 and Z_2 in the equation below. g is acceleration due to gravity.
- energy required to change the flow velocity of the liquid between points 1 and 2 (as might occur when the liquid at point 1 is at rest [velocity zero] or there is a difference in diameter between the pipes entering and leaving the pump): stated as u^2_1 and u^2_2 below. The value $\alpha = 1$ for turbulent flow (high Reynolds numbers).
- energy required to overcome any change in pressure in the system: shown as P_1/ρ and P_2/ρ in the equation below. In brewing, this term often disappears because $P_1 = P_2$. (P = pressure and ρ = liquid density).
- energy associated with the pump: this is the factor E_p (energy input by the pump) in the equation because the energy to move the liquid against these resistances comes from the pump.

Engineers use quantification of the energy factors listed to calculate the pumping power required in any given circumstance for transfer of liquid.

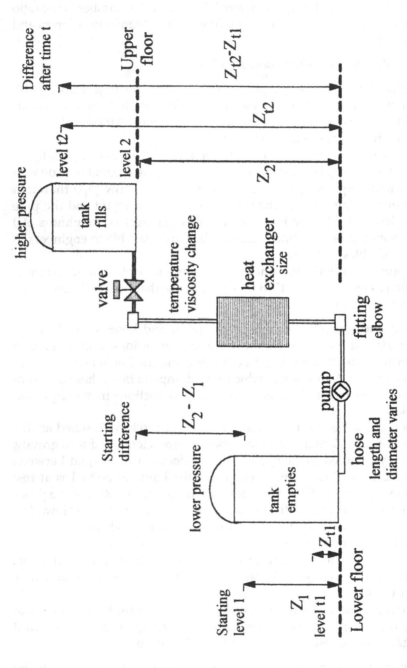

Figure 17–6 Diagrammatic representation of the factors that potentially affect the flow of a liquid in a piping system (see text). Point 1 is chosen as the liquid level in the feed tank, on the left, and point 2 is initially the bottom of the empty tank on the upper floor. As transfer of the liquid progresses the values for Z change dramatically.

All terms in the equation have units of energy per mass of flowing liquid (e.g., J/kg).

The diagram in Figure 17–6 might be divided at the pump and the factors that apply before the pump (suction side) can be compared to those that apply after it (pressure side). This mathematical arrangement is called the **Bernoulli equation**. It is, in effect, a total mechanical energy balance (potential energy plus kinetic energy plus pressure energy) of one side of the pump against the other:

$$gZ_1 + u^2{}_1/2\alpha + P_1/\rho + E_p = gZ_2 + u^2{}_2/2\alpha + P_2/\rho + E_f \text{ (J/kg)}$$

To solve for E_p (the pumping energy) the equation becomes:

$$g(Z_1 - Z_2) + (u^2{}_1 - u^2{}_2)/2\alpha + (P_1 - P_2)/\rho + E_f = E_p \text{ (J/kg)}$$

This calculated value for E_p (joules per kilogram, J/kg), the energy of pumping, must be multiplied by the mass flow rate, ṁ (kilograms per second, kg/s), and adjusted for the efficiency with which power is delivered to the pump, to yield the total pumping power requirement as joules per second, J/s (= Watts, W).

The Bernoulli equation nicely shows the factors involved in transfer of a liquid and might be consulted when flow rate is less than required or otherwise problematical. When using centrifugal pumps (as is typical in many breweries), pumping often becomes less efficient and the rate of beer transfer slows down as transfer goes on. This might be illustrated in Fig 17–6 in which it can be seen the factors Z_1 and Z_2 change quite dramatically (to Z_{t1} and Z_{t2}) as liquid is transferred from one tank to another over time. Or if this were a filtration in which the friction of (say) a filter increased during transfer, the flow rate would decline (and the pressure would change). The equation also warns of inserting devices into a stream that have significant frictional effects (e.g., a heat exchanger or long or narrow pipe runs, or hoses, or bends and valves), or elevating tanks or pressurizing them, or seeking to speed up flow, without reference to the reserve pumping power of the pump.

17.4.2 Pumps

Flow by gravity is free of charge and is used in breweries whenever it is convenient and effective. But for the most part, liquids in breweries move in pipes and hoses driven by pumps. There are generally two types of pumps—centrifugal pumps (most usual in breweries) and positive displacement pumps.

Centrifugal pumps (Figure 17–7a) are quite cheap, and work well in breweries because wort and beer are low viscosity fluids that can be moved at a high rate with generally small changes in pressure. Flow is easily regulated by a downstream valve, which can even be closed without damaging the pump (although the small volume of product in the pump casing might suffer damage!), and the pump can be easily cleaned as part of a CIP (cleaning-in-place) system. Liquid enters the pump at the center of an impeller or vane enclosed in a casing and driven by a motor. The rapidly revolving vane dashes the liquid to the periphery of the casing by centrifugal force from whence the liquid exits the pump at high velocity. The pressure so developed is called the **head**.

Centrifugal pumps are **not self-priming**. That is they must be primed, i.e., full of liquid, to work at all. Large gas bubbles in a liquid stream will cause such pumps to lose their prime and cease moving liquid although they continue turning. Consequently, these pumps are most conveniently installed low in a pipe/tank system so that they tend to prime automatically by gravity flow.

Positive displacement pumps (Figure 17–7b) are self-priming because they fill and empty an isolated and defined volume at each cycle of the pump. This force moves the product and so the speed or frequency of the cycle accurately controls flow rate; they are, therefore, useful for dosing additives into a flowing stream. They are necessary for moving liquids of high viscosity and so might find application, for example, for moving dense sugar syrups in a brewery. A piston in a cylinder is this type of pump; it is called a reciprocating pump. Rotary pumps of several kinds achieve the same objective by enclosing a pocket of liquid between a rotating vane and the pump casing. A lobe pump, for example, might be useful for moving yeast and trub. Axial flow pumps depend on an helical design and might be especially useful for slurries such as spent grain. Positive displacement pumps can cause damage if the delivery side is inadvertently blocked.

It requires power to drive a centrifugal pump. This power generates pumping head (pressure, expressed as meters of liquid pumped) and capacity of flow (volume). The greatest head is against a closed valve (no flow), accounting for the shape of the curve (Figure 17–8). As the power input increases, the flow rate rises; as a result, more pressure is lost to friction, and so the head declines and the efficiency of power utilization eventually passes through a maximum. Pumps are usually characterized by their **capacity** and **head** at maximum power efficiency.

A common occurrence when pumping is **cavitation**. This is the vaporization of a liquid inside the pump and commonly occurs, for example, with hot liquids such as wort in the brewhouse. The wort flashes to steam in the pump as a result of reduced pressure at the pump intake; the pump

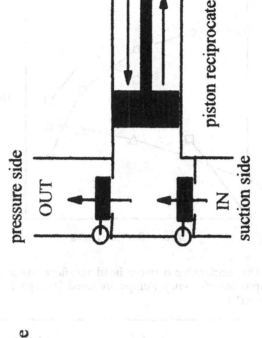

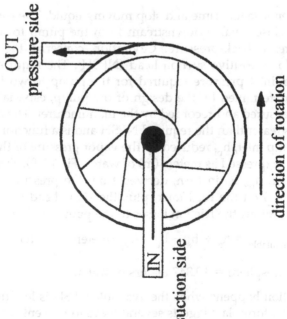

Figure 17-7 Diagram of two types of pump commonly used in breweries.

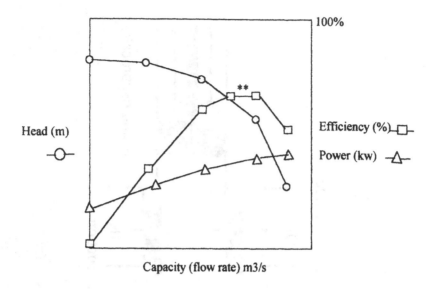

100%

Head (m)

—O—

Efficiency (%) —□—

Power (kw) —△—

Capacity (flow rate) m3/s

Figure 17–8 The relationship between head and flow rate of a pump, and showing power input and efficiency. Pumps are rated for capacity and head at maximum efficiency (**).

might then lose its prime and stop moving liquid. The easy solution is to partially close a valve downstream from the pump to reduce flow rate and increase back-pressure. Cavitation concerns the **available** and **required** net positive suction head (NPSH). The **required** NPSH is the minimum inlet pressure required for the pump to work properly. This value is determined by the design of the pump, especially the impeller, and is measured and recorded by the manufacturer. The **available** NPSH must be greater than the required NPSH and is a function of the pressure at the pump inlet, h_{pi} reduced by the vapor pressure of the liquid, h_{vp}, or $h_{pi} - h_{vp}$, expressed as meters (m) of water (Fig 17–9). The pressure at the pump inlet, h_{pi}, is, in turn, derived from the pressure (as the absolute pressure, h_a) on the feed tank, plus the static head of liquid above the pump, h_s, minus the friction losses to the pump, h_f (Fig 17–9). Therefore:

$$NPSH_{available} = h_a + h_s - h_f - h_{vp} \text{ (meters of water)}$$

One atmosphere = 10.347 meters of water.

Cavitation happens when the available NPSH is less than the required NPSH. The formula suggests several ways to prevent cavitation: increase available NPSH by increasing the static head before the pump (h_s, lower the pump or raise the tank), increase pressure in the delivery tank (h_a),

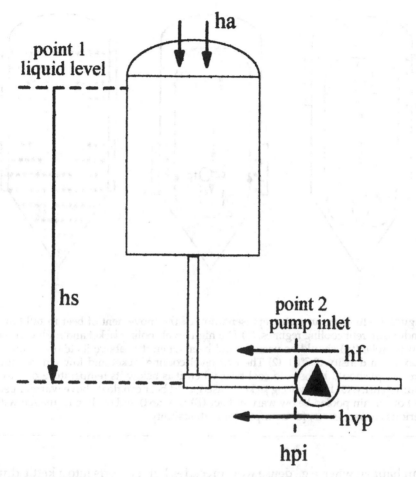

Figure 17-9 The factors involved in the available net positive suction head (NPSH) and the causes of pump cavitation. The NPSH is derived from the pressure at the inlet of the pump (h_{pi}), which is in turn a function of the positive factors h_a and h_s and the negative factors h_f and h_{vp} (see text).

lower the vapor pressure of the liquid (h_{vp}, make it cooler), simplify, shorten, and widen the delivery pipes to the pump (h_f). Also reduce pump speed and decrease flow rate (throttle output to increase back pressure).

17.4.3 Cooling beer in tanks

Liquids move by natural convection in response to their density. Dense liquids cannot easily mix with less dense liquids layered over them (as

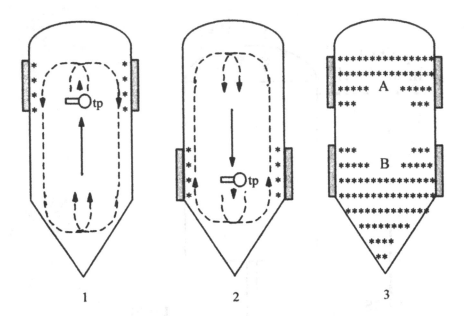

Figure 17–10 Diagrammatic representation of the movement of beer in tall tanks under different cooling regimes. (1) The high level cooling jacket and high temperature probe (tp) should be used to cool beer when it is above its temperature of maximum density (3.7°C). (2) The low level cooling jacket and low temperature probe (tp) should be used to cool beer when it is below its temperature of maximum density. (3) If the wrong jacket is used, cold beer can float above warmer beer (A) or remain pooled below warmer beer (B) (see text) and cooling is uneven with serious consequences, for example, for chill-stability.

can happen when e.g., dense wort precedes lighter worts into a kettle during the lauter run-off), but dense liquids easily sink through less dense liquids beneath them. These factors need to be taken into account when cooling beer, especially in large tall vessels.

When beer is cooled from, say, 15 or 20°C, it *increases* in density to about 3.7°C, which is the temperature of maximum density for beer. As cooling proceeds below this temperature, the density *decreases*. Cooling beer down to about 4°C and below 4°C, therefore, require different cooling strategies.

To cool beer from, say, 15 to 5°C in a tall tank (Figure 17–10), brewers should use the high cooling jackets and a temperature probe placed high in the tank. As the beer cools at the wall of the vessel, it increases in density and sinks. Eventually, the tank fills with cold beer, as detected by the probe, which will shut off the cooling. If the temperature probe is placed too low in the tank, it will detect the cold beer that settles first and will shut off the cooling prematurely. If the low level cooling jackets were used for this cooling duty, the cold beer produced cannot rise. If the temperature probe is placed at a high level in the tank, it will continue to call for cooling even when beer at the bottom of the tank is very cold and could conceivably freeze.

To cool beer from about 4°C to say minus 2°C brewers should use the low level cooling jackets and a low level temperature probe (Figure 17–10). The cold beer will rise as its density decreases. If the high level jackets were used for this duty, the less dense beer produced would simply float on the more dense beer below it and, because the temperature probe continues to call for cooling, the beer can easily freeze.

Ineffective cooling strategies (Figure 17–10) are more than a mere nuisance or waste of energy: the bulk of the beer is not being sufficiently cooled to precipitate the chill-haze tannin-protein complex that is necessary to achieve chill stability before filtration.

For these reasons, a common strategy for cooling beer below 4 to 5°C (to, say, minus 2°C) might include passing the beer through a heat exchanger (green beer chiller or trim chiller). This avoids the need to use the low cooling jackets and, of course, cooling is rather slow at these temperatures using natural convection only.

CHAPTER 18

Fermentation biochemistry

18.1 METABOLIC DIVERSITY OF YEAST

Yeast can grow aerobically or anaerobically. In the former case, the cells are said to be respiring and in the latter fermenting. Yeast has the ability to use a range of different sugars and even ethanol for growth. It can assimilate many nutrients. The composition of the environment influences the metabolic processes that the yeast uses. The most extreme example is the ability of yeast to ferment in the presence of high concentrations of sugar (producing alcohol and carbon dioxide) and then, in the presence of oxygen, respire the alcohol to give carbon dioxide and water. This is referred to as diauxie. Another is the fact that significant levels of glucose can inhibit the synthesis of enzymes responsible for the uptake and metabolism of maltose. Only the fermentative mode of growth applies in beer production. Pasteur defined fermentation as "life without air" and for yeast growing in a brewery fermenter, this is a good description.

Fermentation *in toto* cannot be separated from the process of yeast growth, which is the increase in cell size and number (by the process of budding). For growth to occur, brewer's yeast requires a nutritious environment. A brewer's wort, correctly made, contains all the nutrients required by yeast. Nutritional deprivation leads to incomplete and inadequate fermentation.

18.2 YEAST NUTRITION

Brewer's yeast needs readily used (assimilable) sources of carbon, nitrogen, certain vitamins, trace elements, and, under normal circumstances, a small amount of molecular oxygen. These are supplied in the form of fermentable sugars (mainly maltose), amino acids, B vitamins from malt, trace elements from malt, and brewing water (principal among these are the ions of calcium, magnesium, zinc, phosphate, and sulfate).

Molecular oxygen is supplied directly by the brewer at the beginning of the fermentation. Yeast foods used to supplement worts in some breweries contain yeast extract (for vitamins), ammonium hydrogen phosphate and zinc sulfate. Figure 18–1 gives a simplified overview of fermentation. Once pitched into wort, the yeast uses the nutrients to provide energy (ATP) and, in doing so, forms alcohol and carbon dioxide (CO_2). It produces reducing power (nicotinamide adenine dinucleotide phosphate, NADPH) for the synthesis of new yeast substance. The nutrients are also either directly assimilated into new cell components or used to generate intermediates for this process. Energy generation and energy utilization are intimately linked. Furthermore, they lead to the synthesis and secretion into the wort of a large number of minor metabolic products, many of which generate characteristic flavors and aromas.

18.3 METABOLIC PROCESSES—GENERAL CHARACTERISTICS

It is a convenient approach to consider metabolism as a series of biochemical pathways each complete in itself. However, the connections

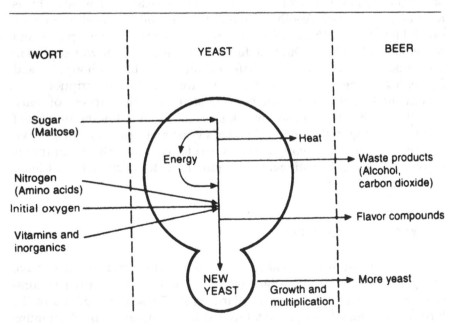

Figure 18–1 Overview of the main yeast-mediated biochemical events in a brewery fermentation.

between pathways are also very important. Few, if any, pathways are totally separated from others. A **biochemical pathway** is recognized as a series of reactions, each mediated by an enzyme in which the product of one reaction becomes the substrate of the next. Those pathways whose main purpose is to generate energy do so by chemical oxidation of substrates. The oxidation is usually achieved by transferring a hydride ion to an enzyme cofactor (nicotinamide adenine dinucleotide NAD^+). So NAD^+, the oxidized form of the cofactor, is converted to NADH, the reduced form. These oxidation–reduction reactions are conducted by dehydrogenase enzymes. Oxidation releases energy (a commonly observed non-enzymatic example is the heat obtained by burning fossil fuels). In metabolism, the oxidation process is carefully controlled so that some of the energy released is retained by the cell in the form of adenosine triphosphate (ATP). The reaction:

$$ADP + \text{inorganic phosphate } (P_i) \longrightarrow ATP$$

consumes energy and is driven by high-energy compounds formed during metabolism.

The reverse reaction (hydrolysis of ATP) releases energy that can be used to synthesize new products.

The energy transfer reactions are never 100% efficient, waste energy (in the form of heat) is always dissipated. This is the source of the heat arising during fermentation of wort. On a typical wort, enough waste heat is produced to raise the temperature of a perfectly insulated vessel some 17°C. A brewery fermenter is not, of course, perfectly lagged and some heat is lost to the environment; however, the remaining waste metabolic heat must be removed by cooling if the fermentation is to be controlled. The larger the vessel, the smaller its surface compared with the volume it holds. Thus, larger vessels need more efficient cooling systems than smaller ones.

Catabolic pathways are used to produce energy and generate intermediates for biosynthesis. Anabolic pathways consume reducing power, energy, and intermediates and produce end products (e.g., glycogen). Linking the two are so-called **anaplerotic reactions**, which top-up key biosynthetic intermediates. Some pathways (e.g., glycolysis—see below) have central roles and provide both functions and are said to be **amphibolic**.

The following section will be restricted to an outline of the main metabolic processes used by brewer's yeast in a brewery fermentation. A simplified approach is adopted in describing the major biochemical pathways. Those seeking more detailed information about these and general aspects of metabolism should refer to any modern textbook of biochemistry.

18.4 METABOLISM OF WORT SUGARS

18.4.1 Production of ATP and the formation of alcohol and carbon dioxide

The fermentable sugars in brewer's wort are maltose, maltotriose, and lesser amounts of sucrose, glucose, and fructose. The trisaccharide maltotriose and disaccharide maltose are transported into the cell and hydrolysed to glucose. Sucrose is hydrolyzed outside the cell to its constituents glucose and fructose. Glucose and fructose are transported into the cell (Figure 18–2). The enzyme invertase, responsible for the hydrolysis of sucrose, is located in the outer layers of the cell wall, however some escapes into beer. In common with any enzyme, it is readily inactivated by heat. This is exploited in brewing as a measure of pasteurization efficiency, with any invertase activity present after pasteurization showing that insufficient heat has been given. The test of enzyme activity takes a few minutes to perform.

Intracellular glucose is phosphorylated using ATP by the enzyme hexokinase to yield glucose 6-phosphate. Fructose is also phosphorylated to yield fructose 6-phosphate. This phosphorylation can be seen as a labeling reaction; only phosphorylated compounds are recognized by subsequent enzymes. The metabolism of glucose proceeds by the glycolytic pathway (Embden–Meyerhof–Parnas or EMP pathway). There are 10 enzymes in this pathway, which oxidizes the six-carbon sugar to two molecules of pyruvate. In the process, there is a net gain of two molecules of ATP and two of NADH.

The pathway is illustrated in Figure 18–3. Glucose 6-phosphate, produced by the action of hexokinase on wort glucose, is converted to its isomer fructose 6-phosphate by the enzyme phosphoglucose isomerase. The fructose 6-phosphate formed (together with that produced by the action of hexokinase on wort fructose) is phosphorylated to give fructose 1,6-bisphosphate by the enzyme phosphofructokinase. Both phosphorylation reactions consume ATP. Fructose 1,6-bisphosphate is hydrolyzed by the enzyme aldolase to give two three-carbon phosphates (glyceraldehyde 3-phosphate and 3-phosphoglycerone (also known as dihydroxyacetone phosphate). The reaction favors the production of the latter, but this is isomerized to the former by the enzyme triosephosphate isomerase.

Glyceraldehyde 3-phosphate is oxidized in a reaction mediated by a dehydrogenase with NAD^+ as cofactor and involves the incorporation of phosphate into the substrate. The product is 1,3-biphosphoglycerate. Hydrolysis of the phosphoester linkage at the C–1 atom of this compound by phosphoglycerokinase to give 3-phosphoglycerate releases a large

WORT

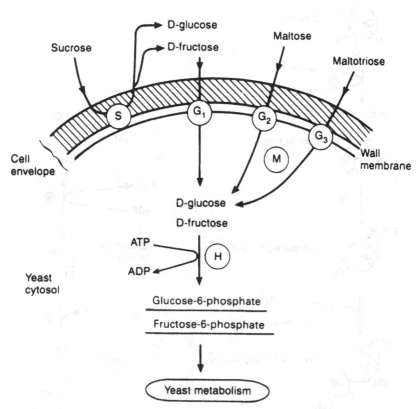

Figure 18–2 Diagrammatic representation of a simplified view of the uptake of wort fermentable sugars by yeast. The cell envelope is the wall (hatched area) and membrane (single line). G_1, glucose permease (in reality, there are at least 2 types of glucose transporters, one operating when glucose levels are high and the other at low glucose); G_2, maltose permease; G_3, putative maltotriose permease; S, invertase (sucrase) covalently attached within the cell wall; M, intracellular maltase (α-glucosidase); H, hexokinase; ATP, adenosine triphosphate; ADP, adenosine diphosphate.

amount of energy. This energy is used to synthesize ATP. Since there are two molecules of glyceraldahyde 3-phosphate metabolized for each molecule of glucose entering the pathway, then the two ATP consumed at the beginning of the pathway are now effectively restored. The enzyme phosphoglyceromutase converts 3-phosphoglycerate to 2-phosphoglycerate and the removal of water by enolase yields phosphoenol pyruvate. The hydrolysis of the carbon to phosphorus (C–P) bond of this compound by

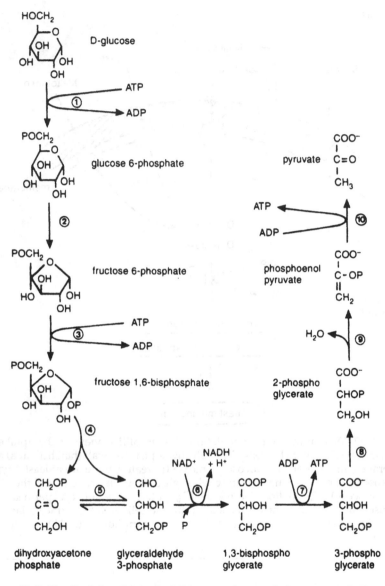

Figure 18–3 The Embden–Meyerhof–Parnas pathway of glucose metabolism (also known as the glycolytic pathway or glycolysis). ATP, adenosine triphosphate; ADP, adenosine diphosphate; NAD$^+$, nicotinamide adenine dinucleotide; NADH, reduced NAD^{1+}. The enzymes of the pathway are indicated by circled numbers. 1, hexokinase; 2, hexoseisomerase; 3, phosphofructokinase; 4, aldolase; 5, triosephosphate isomerase; 6, glyceraldehyde 3-phosphate dehydrogenase; 7, phosphoglycerokinase; 8, phosphoglyceromutase; 9, enolase; 10, pyruvate kinase.

pyruvate kinase also releases sufficient energy to synthesize ATP. The overall pathway, therefore, generates a net 2 ATP. This is the end of the glycolytic pathway. However, the cell has consumed NAD^+ and produced NADH. The supply of the oxidized form of the cofactor is finite and it must be regenerated from its reduced form. If this is not done, then the energy-generating reactions cease. In brewer's yeast, oxidation of NADH is achieved at the expense of further metabolism of pyruvate (Figure 18–4a). Two enzymes, pyruvate decarboxylase and alcohol dehydrogenase are used. The first releases CO_2 from the substrate and gives ethanal (acetaldehyde); the second uses ethanal as substrate, NADH as cofactor, and yields ethanol (alcohol) and regenerated NAD^+ in the process. These major end

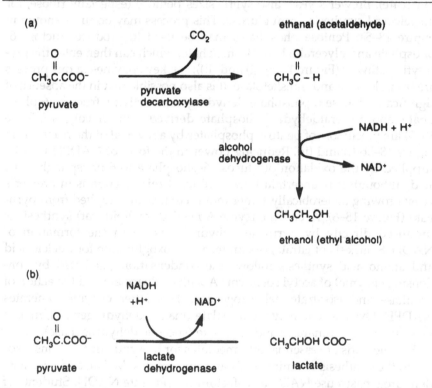

Figure 18–4 Final reaction of glycolysis in fermenting brewer's yeast. Regeneration of NAD^+ from NADH. The alcohol (ethanol) and most of the CO_2 of fermentation are produced in these reactions. Beer acetaldehyde (ethanal) also arises from this process. (b) Alternative enzymatic regeneration of NAD^+ used by some lactic acid bacteria and human skeletal muscle under exercise. NAD^+, nicotinamide adenine dinucleotide; NADH, reduced NAD^+

products of catabolism are of course major constituents of beer. The ethanol produced in the absence of oxygen gives yeast a competitive advantage in natural environments because it is inhibitory to many microbes.

The glycolytic pathway is found in most living organisms. Several types of lactic acid bacteria use it but regenerate NAD^+ by reducing pyruvate to lactic acid (Figure 18–4b). The enzyme used, lactate dehydrogenase, is also active in exercizing human skeletal muscle.

18.4.2 Provision of biosynthetic intermediates and NADPH

The glycolytic pathway is amphibolic and provides key intermediates for biosynthetic reactions (particularly, hexose and triose phosphates and pyruvate). Brewer's yeast also synthesizes pentose (especially ribose, for nucleic acid synthesis) from glucose. This process may occur as shown in Figure 18–5a. Pentose phosphates may be used to produce fructose 6-phosphate and glyceraldehyde 3-phosphate, which can then enter the glycolytic pathway (Figure 18–5b (i) and (ii)); the key enzymes in this process are transaldolase and transketolase. It is also possible that in the absence of significant glucose 6-phosphate dehydrogenase activity, fructose 6-phosphate, and glyceraldehyde 3-phosphate derived from glycolysis (Figure 18–3) are precursors of pentose phosphates by a reversal of the reactions in Figure 18–5b (i) and (ii). Reducing power in the form of NADPH may be supplied by the oxidation of glucose 6-phosphate to 6-phosphogluconic acid, although it is uncertain how significant this reaction is in brewer's yeast growing anaerobically. Other intermediates are supplied from pyruvate (Figure 18–6). Acetyl coenzyme A needed for lipid (fat) synthesis is produced directly by pyruvate dehydrogenase with the formation of NADH. Synthesis of citrate, isocitrate, and 2-oxoglutarate for nucleic acid and amino acid synthesis follows the condensation (mediated by condensing enzyme) of acetyl coenzyme A with oxaloacetate and the action of aconitase and isocitrate dehydrogenase. This latter enzyme generates NADPH. Oxaloacetate may yield malate (malate dehydrogenase); malate gives fumarate (fumarase) and succinate (succinic dehydrogenase). These carboxylic acids are essential intermediates for biosynthetic reactions leading to the synthesis of amino acids and nucleotides. Malate and succinate dehydrogenases use NAD^+ as cofactor and generate NADH. Students of biochemistry will recognize that Figure 18–6 is the tricarboxylic acid cycle (Krebs or citric acid cycle) except that one key enzyme—2-oxoglutarate dehydrogenase (which converts 2-oxoglutarate to succinate) is missing.

Clearly when pyruvate is used to generate biosynthetic intermediates, acetaldehyde is no longer produced. Consequently, the NADH produced,

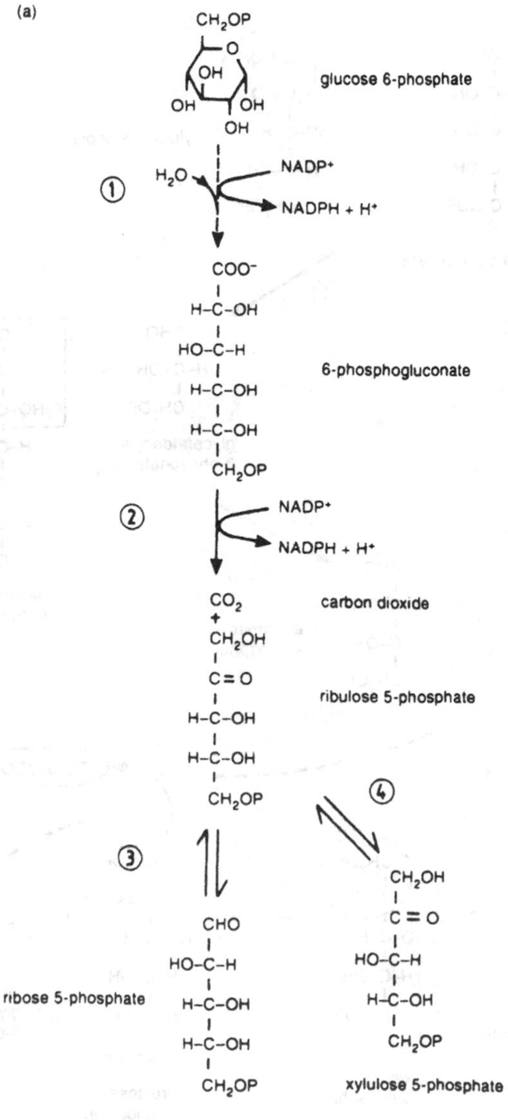

(a)

CH$_2$OP

glucose 6-phosphate

① H$_2$O — NADP$^+$

NADPH + H$^+$

COO$^-$
|
H–C–OH
|
HO–C–H
|
H–C–OH
|
H–C–OH
|
CH$_2$OP

6-phosphogluconate

② — NADP$^+$

NADPH + H$^+$

CO$_2$
+
CH$_2$OH
|
C = O
|
H–C–OH
|
H–C–OH
|
CH$_2$OP

carbon dioxide

ribulose 5-phosphate

③ ④

CHO
|
HO–C–H
|
H–C–OH
|
H–C–OH
|
CH$_2$OP

ribose 5-phosphate

CH$_2$OH
|
C = O
|
HO–C–H
|
H–C–OH
|
CH$_2$OP

xylulose 5-phosphate

Figure 18–5 Reactions of the hexone monophosphate pathway. (a) Formation of reduced nicotinamide adenine dinucleotide phosphate (NADPH) and synthesis of pentose sugars from glucose. Numbers refer to enzymes involved: 1, glucose 6-phosphate dehydrogenase; 2, 6-phosphogluconic acid dehydrogenase; 3, phosphoriboismerase; 4, phosphoketopentose epimerase. NADP$^+$ nicotinamide adenine dinucleotide phosphate, NADPH reduced NADP$^+$. (b) (i) Conversion of pentoses to hexose and tetraose. (ii) Formation of hexose and triose from tetraose and pentose. Enzymes: 1, transaldolase; 2, transketolase. The former transfers the

(b)

(i)

```
    CHO                    CH2OH
    |                      |
  H-C-OH                   C=O
    |                      |
  H-C-OH        +        HO-C-H          xylulose 5-phosphate
    |                      |
  H C-OH                 H-C-OH
    |                      |
   CH2OP                  CH2OP
```

ribose 5-phosphate

①

```
    CHO                    CH2OH
    |                      |
  H-C-OH         +         C=O
    |                      |
   CH2OP                 HO-C-H
                           |
glyceraldehyde           H-C-OH
3-phosphate               |
                        H-C-OH
                          |
                        H-C-OH
                          |
                         CH2OP

                    sedoheptulose
                    7-phosphate
```

②

```
   CH2OH
    |
    C=O               CHO
    |                  |
  HO-C-H             H-C-OH
    |         +        |             erythrose
  H-C-OH             H-C-OH          4-phosphate
    |                  |
  H-C-OH              CH2OP
    |
   CH2OP

fructose
6-phosphate
```

- - - - → (TO GLYCOLYSIS)

(ii)

```
    CHO              CH2OH                 CH2OH
    |                |                     |
  H-C-OH            C=O          ①        C=O                CHO
    |        +       |                     |                  |
  H-C-OH          HO-C-H         →       HO-C-H       +     H-C-OH
    |                |                     |                  |
   CH2OP          H-C-OH                 H-C-OH              CH2OP
                    |                     |
erythrose         CH2OP                 H-C-OH         glyceraldehyde
4-phosphate                               |            3-phosphate
                                        CH2OP
                  xylulose
                  5-phosphate           fructose
                                        6-phosphate
```

two-carbon unit (dashed circle) from a keto- sugar to the aldehyde group of aldo-sugar. Transketolase transfers the three-carbon fragment (dashed box) from a keto- sugar to the aldehyde group of an aldo- sugar. These reactions can generate one hexose and one triose plus one CO_2 from two molecules of pentose. Alternatively, because all the reactions are reversible, one hexose and one triose (from glycolysis) can be used to produce pentose (reactions (ii)). This pentose can generate ribose, as shown in Figure 18–5a.

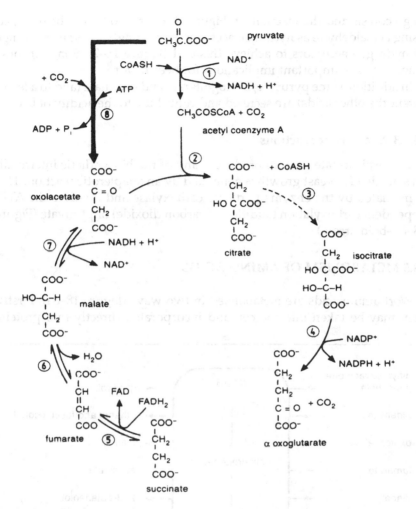

Figure 18–6 Production of tri- and di-carboxylic acids for biosynthetic reactions. NAD^+ nicotinamide adenine dinucleotide; NADH, reduced NAD^+; $NADP^1$ nicotinamide adenine dinucleotide phosphate; NADPH, reduced $NADP^+$; CoASH, coenzyme A; FAD, flavin adenine dinucleotide; $FADH_2$ reduced FAD; ATP, adenosine triphosphate; ADP, adenosine diphosphate; P_i, inorganic phosphate. The enzymes involved are numbered; 1, pyruvate dehydrogenase; 2, condensing enzyme (citrate synthase); 3, aconitase; 4, isocitrate dehydrogenase; 5, succinate dehydrogenase; 6, fumarase; 7, malate dehydrogenase; 8, pyruvate carboxylase. As well as providing NADPH and intermediates for biosynthesis, malate dehydrogenase (7) is used to regenerate NAD^+. Succinate and oxoglutarate are excreted in substantial amounts in beer. Pyruvate carboxylase (8, heavy line) mediates an anaplerotic reaction, topping up the oxaloacetate pool. The broken arrow (aconitase 3), indicates that more than one step is involved.

in glycolysis and the reactions in Figure 18–6, can no longer be oxidized using acetaldehyde as hydrogen acceptor. Accordingly, yeast uses a range of hydrogen acceptors to achieve this end (Figure 18–7). Some of these reactions have important implications for beer flavor.

In addition, some pyruvate, 2-oxoglutarate and succinate (and to a lesser extent the other acids) are secreted and contribute to the acidity of beer.

18.4.3 Anaplerotic reactions

The oxaloacetate used to produce many of the biosynthetic intermediates needed for yeast growth is generated by an anaplerotic reaction. This is mediated by the enzyme pyruvate carboxylase and involves the ATP-dependent carboxylation (addition of carbon dioxide) of pyruvate (Figure 18–6—bold arrow).

18.5 METABOLISM OF AMINO ACIDS

Wort amino acids are metabolized in two ways (Figure 18–8a). Firstly, they may be taken into the cell and incorporated directly into protein.

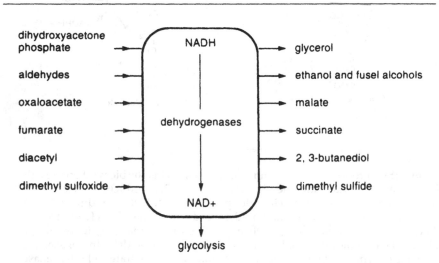

Figure 18–7 Regeneration of NAD^+ by fermenting yeast. All the compounds on the right are in beer. They influence flavor and taste. They are derived in and by yeast from the intermediates on the left using dehydrogenases and regenerating NAD^+. Diacetyl is undesirable in many beers; its reduction by yeast is highly desirable (2,3-butanediol has negligible flavor). Dimethyl sulfide may be formed by yeast using a dehydrogenase using $NADP^+$ rather than NAD^+.

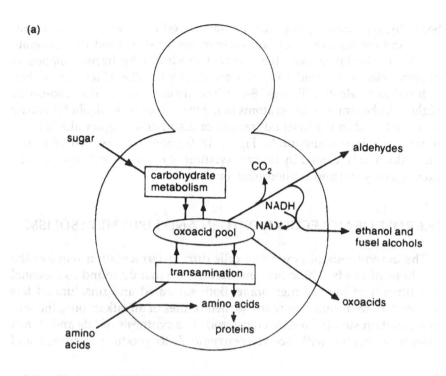

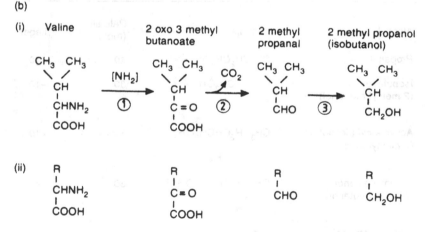

Figure 18–8 Formation of higher alcohols, aldehydes, and oxo-acids from amino acids and sugars. (a) Overview of main metabolic processes. (b) (i) Specific example, formation of isobutanol from valine. (ii) Generalized pathway R any amino acid side chain. [NH$_2$] indicates group transferred to transaminase enzyme. Enzymes involved: 1, transaminase; 2, decarboxylase; 3, dehydrogenase. NAD$^+$, nicotinamide adenine dinucleotide; NADH, reduced NAD$^+$.

Secondly, the amino group may be transferred to an enzyme (transaminase) and the remaining carbon skeleton secreted or used to regenerate NAD^+. In the latter case, the oxo-acid produced by transamination is decarboxylated to yield CO_2 and an aldehyde. The aldehyde is then reduced to an alcohol (Figure 18–8b). The alcohol produced is a so-called higher alcohol (more carbon atoms than ethanol) or fusel alcohol (because they are found in the fusel oil fraction of distillates). Higher alcohols are important flavor compounds. Figure 18–9 shows representative examples. Aldehydes formed by decarboxylation of two oxo-acids may also be excreted by yeast and influence flavor.

18.6 ROLE OF MOLECULAR OXYGEN AND LIPID METABOLISM

The biosynthesis of new yeast cells during fermentation requires the synthesis of lipids. These macromolecules are mainly found as essential constituents of the cell membrane. Both saturated and unsaturated fats are used. Also in common with the membranes of all eukaryotes, those of yeast contain sterols (mainly ergosterol). The synthesis of fats and sterols essentially begins with acetyl coenzyme A to produce saturated and

Alcohol	Formula	Pale ale (mg/l)	Lager
Propanol	$CH_3CH_2CH_2OH$	40	~10
Isobutanol (2 methyl propanol)	CH_3CHCH_2OH | CH_3	30	~10
Active amyl alcohol (2 methyl butanol)	$CH_3CH_2CHCH_2OH$ | CH_3	15	<10
Iso amyl alcohol (3 methyl butanol)	$CH_3CHCH_2CH_2OH$ | CH_3	50	40
2 phenyl ethanol (β phenyl ethanol)	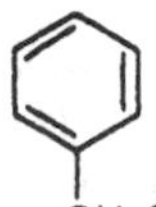 CH_2CH_2OH	40	30

Figure 18–9 Some representative fusel alcohols and their levels in beer.

unsaturated acyl CoA molecules and sterols (Figure 18–10). Biosynthesis of unsaturated molecules and sterols requires an oxidative step; this oxidation is achieved at the expense of molecular oxygen, forming water. The oxygen in wort is used for this purpose. Different yeasts require different amounts of oxygen and the requirement can be replaced in laboratory fermentations by supplying unsaturated fatty acids and sterols to the medium. Oxygen deficiency results in poor fermentation and probably results in an increased level of acetyl coenzyme A in cells. This in turn can produce elevated levels of esters in beer, thus markedly influencing flavor (see below). Supplying the correct amount of oxygen, is therefore, very important in ensuring consistent fermentations.

18.6.1 Formation of esters

The reaction between an alcohol and an acyl coenzyme A molecule produces an ester (Figure 18–11). Sources of acyl CoA molecules are acetyl CoA itself (which produces acetate esters) and intermediates in fat synthesis. Since the main alcohol in yeast is ethanol, the most common ester

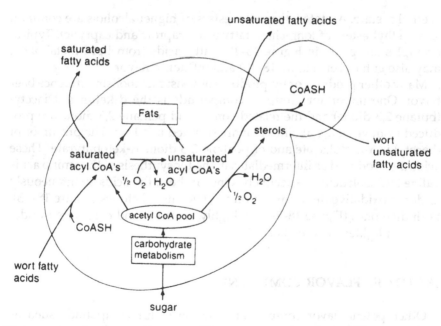

Figure 18–10 Overview of lipid metabolism and the role of molecular oxygen. CoASH, coenzyme A.

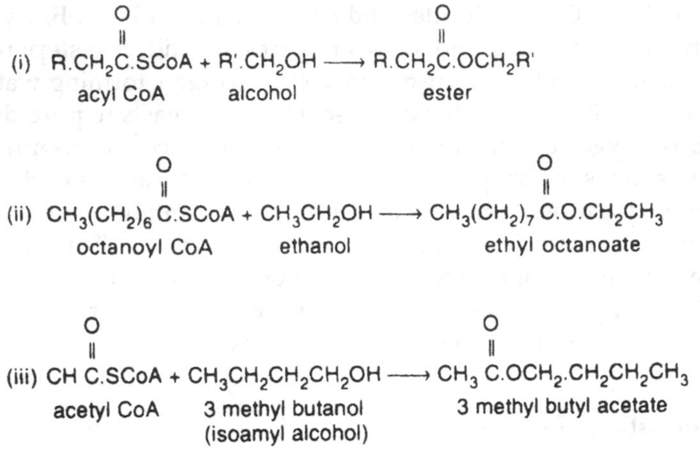

$$\text{(i)}\quad \underset{\text{acyl CoA}}{R.CH_2\overset{\overset{\displaystyle O}{\|}}{C}.SCoA} + \underset{\text{alcohol}}{R'.CH_2OH} \longrightarrow \underset{\text{ester}}{R.CH_2\overset{\overset{\displaystyle O}{\|}}{C}.OCH_2R'}$$

$$\text{(ii)}\quad \underset{\text{octanoyl CoA}}{CH_3(CH_2)_6\overset{\overset{\displaystyle O}{\|}}{C}.SCoA} + \underset{\text{ethanol}}{CH_3CH_2OH} \longrightarrow \underset{\text{ethyl octanoate}}{CH_3(CH_2)_7\overset{\overset{\displaystyle O}{\|}}{C}.O.CH_2CH_3}$$

$$\text{(iii)}\quad \underset{\underset{\text{(isoamyl alcohol)}}{\text{acetyl CoA}\quad\text{3 methyl butanol}}}{CH\overset{\overset{\displaystyle O}{\|}}{C}.SCoA + CH_3CH_2CH_2CH_2OH} \longrightarrow \underset{\text{3 methyl butyl acetate}}{CH_3\overset{\overset{\displaystyle O}{\|}}{C}.OCH_2.CH_2CH_2CH_3}$$

Figure 18–11 Formation of esters from alcohols and acyl coenzyme A. (i) Generalized formula where R is an aliphatic chain. (ii) Formation of ethyl octanoate. (iii) Formation of 3-methyl butyl acetate (isoamyl acetate). The most concentrated ester in beer is from acetyl CoA and ethanol (ethyl acetate; ethyl ethanoate).

---

is ethyl acetate. Acetate (ethanoate) esters of higher alcohols are common as are ethyl esters of long-chain fatty acid (caproic and caprylic). Typical examples are given in Figure 18–12. Fatty acids from lipid metabolism may also enter beer (Figure 18–10) and influence flavor.

Many other products or by-products of yeast metabolism influence beer flavor. One important group of compounds is the **diketones**. Diacetyl (butane 2,3 dione) and the related compound pentane 2,3 dione are produced from yeast metabolites that are secreted into beer. The precursor of diacetyl is α-acetolactate and of pentane 2,3 dione α-ketobutyrate. These acids are produced as intermediates in the biosynthesis of the amino acids valine and isoleucine respectively. Once in beer, the acids spontaneously undergo oxidative decarboxylation to yield the diketones (Figure 18–13). Both diketones (Figure 18–14) are highly aromatic and considered undesirable in lighter-flavored beers.

18.7 OTHER FLAVOR COMPOUNDS

Other potent flavor components include sulfur compounds such as hydrogen sulfide (odor of rotten eggs). The formation of these compounds is related to metabolism of sulfur-containing amino acids and sulfate ions

Name	Formula	Concentration (mg/l)	Flavor threshold (mg/l)
Esters			
Ethyl acetate	$CH_3CH_2\,O\,C\,CH_3$ $\overset{\|}{}$ O	8–70	33
Isoamyl acetate (3 methyl butyl acetate)	CH_3 $\diagdown$ $CH\,CH_2\,CH_2\,O\,C\,CH_3$ $CH_3\diagup\overset{\|}{}$ O	0.4–6	1.6
Ethyl octanoate (ethyl caprylate)	$CH_3CH_2\,O\,C\,(CH_2)_6\,CH_3$ $\overset{\|}{}$ O	1.5	0.9
Ethyl decanoate (ethyl caproate)	$CH_3\,CH_2\,O\,C\,(CH_2)_8\,CH_3$ $\overset{\|}{}$ O	0.2	1.5
Fatty acids			
Butyric	$CH_3\,(CH_2)_2\,COOH$	0.6	2
Octanoic	$CH_3\,(CH_2)_6\,COOH$	6	14
Isovaleric (3 methyl butyric)	CH_3 $\diagdown$ $CH\,CH_2\,COOH$ $CH_3\diagup$	1.3	1.5

Figure 18–12 Some representative esters and fatty acids and their levels and taste thresholds in beer.

in wort. Some (such as dimethyl sulfide) are mainly derived from malt constituents.

The range and amounts of various minor metabolites produced by yeast are dependent upon the yeast strain itself. This is largely why brewers use selected strains for fermentation and go to considerable lengths to maintain them. However, the same yeast on different worts or on the same wort under different fermentation conditions will produce different products. This is why the brewer aims to keep a consistent quality of raw materials and process.

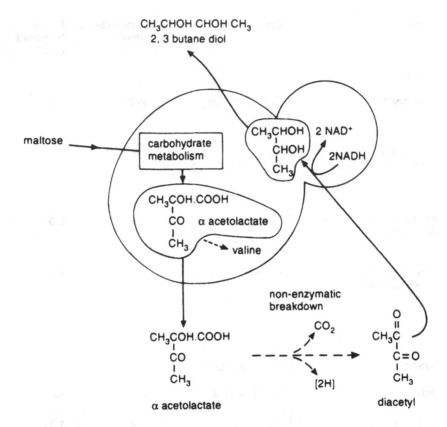

Figure 18–13 Overview of the non-enzymatic formation of diacetyl and its conversion by yeast to inocuous 2,3 butane diol. Similar reactions involving α-keto-butyrate, the non-enzymatic formation of 2,3 pentane dione and its reduction by yeast to 2,3 pentane diol also occur. NAD$^+$, nicotinamide, adenine dinucleotide; NADH, reduced NAD$^+$.

Factors having the most influence on the production of minor metabolites by a given yeast strain are: wort composition (especially the amino acid content and amount); availability of molecular oxygen; temperature of fermentation (higher temperatures give more metabolites in beer); whether or not the fermentation is mixed or agitated; and the speed at which yeast is separated from beer. Consistent beer quality can be achieved only if these factors are carefully controlled.

Diketone	Formula	Beer (mg/l)	Flavor threshold (mg/l)
Diacetyl (2,3 butanedione)	O O ‖ ‖ CH_3 C C CH_3	0.06	0.15
2,3 pentanedione	O O ‖ ‖ CH_3 CH_2 C C CH_3	0.01	0.90

Figure 18–14 Diketones in beer.

18.8 YEAST FOOD AND ENERGY RESERVES

When given access to a plentiful supply of nutrition, organisms use any excess over need to provide food stores in the form of carbohydrate or fat. For example, in many plants, starch is made in seeds and provides the energy for germination. In animals, glycogen is stored in the liver and fat in fatty tissue. Yeast is no exception and energy reserves are found particularly in the form of the non-reducing disaccharide trehalose and polysaccharide glycogen. Trehalose is two α D-glucose molecules linked together by an α-1–1 bond. Glycogen is a polymer of glucose very similar in structure to amylopectin (Figures 2–17b and 13–1b). It differs in that it is a larger molecule and has more α-1–6 branches and shorter side chains. These carbohydrates have normally accumulated in cells at the end of a brewery fermentation. There is evidence that yeast, both in storage and very early in fermentation, uses them as reserves to provide glucose for glycolysis. This is of importance to brewers for if all the reserves are expended during yeast storage when the yeast is pitched, fermentation may be slow to start.

18.9 DNA MICRO-ARRAY TECHNOLOGY—WHICH PATHWAYS / ENZYMES IS A YEAST USING?

The publication of the complete sequence of *Saccharomyces cerevisiae* genome (i.e., the genetic code for the entire chromosomal DNA of the yeast) has opened up the prospect of answering detailed questions about the metabolic status of yeast. These questions can be addressed to wider issues of overall metabolism as well as to specific details of particular enzyme processes. Thus, it is now feasible to compare the metabolic status of yeast grown under different environmental conditions (e.g., on worts produced using glucose adjuncts versus non-adjunct worts). This amazing advance should enable biochemists to make a detailed examination leading to an understanding at the molecular level of those characteristics that distinguish different strains of brewer's yeast.

The new technology uses the knowledge of the genome to separately synthesize (using the polymerase chain reaction [PCR]) all 6,400 known yeast genes and fix them as individual spots of DNA in an array on a microscope slide. Yeast is then grown under control and test conditions. When a gene is expressed, its message is present in the cells and can be extracted and amplified copies made (by a variation of pcr). In making the copies, fluorescent dyes are incorporated (different colors e.g., red and green for the control and test culture extracts, respectively). The extracts are mixed and then reacted with the microarray. The copies of the messages bind specifically to the gene from which they were originally derived. When the reacted array is viewed under a fluorescence microscope, genes expressed in the control only will be red, those in the test only, green and those in both cases, yellow mixture of red and green. This technique, therefore, shows the pattern of genes specifically turned on or off under the different conditions in which the yeast was grown. Modern confocal fluorescence microscopes also enable the investigator to make quantitative estimates of the amount of fluorescence, which is proportional to the amount of message expressed in the cells.

The process may also be used to characterize particular yeast strains by culturing them under identical conditions and examining the differences in their gene expression. Modern confocal fluorescence microscopes also enable the investigator to make quantitative estimates of the amount of fluorescence,which is proportional to the amount of message expressed in the cells.

The technology can be applied as described to look at "whole genome expression" or in a more selective way. So, for example, arrays could be made of genes known to be involved in flocculation or production of flavor compounds. It seems that application of the technique will be limited only by the imagination of the experimeter and the perceived commercial importance of the investigation.

CHAPTER 19

Finishing processes

D uring primary fermentation yeast catalyzes a massive transformation of sugar into alcohol and carbon dioxide (CO_2). A relatively small proportion of the sugar (perhaps 5–8%) is converted into yeast mass and a small, but vitally important, quantity of beer flavor compounds. The product is indubitably beer. This beer however is far from suitable for sale because it contains suspended particles and is therefore hazy, lacks sufficient carbonation, the flavor is not fully matured, it is physically and microbiologically unstable and its flavor/color may need to be adjusted. Finishing processes correct these shortcomings. **Maturation, carbonation, clarification** and **stabilization** comprise the objectives of finishing processes.

19.1 MATURATION and CARBONATION

Maturation of beer flavor requires the presence of yeast as a catalyst. There are many schemes of finishing that prolong the contact of beer with yeast after the primary fermentation is complete to achieve maturation. These can be divided into two general schemes called **secondary fermentation** or **storage/aging**.

Secondary fermentation is fermentation by the yeast after the primary fermentation subsides. Secondary fermentation is controlled by low temperature and low yeast count, and so is much slower than primary fermentation. Such processes include cask- or bottle-conditioned beers, krausen fermentations and lagering. In each case an amount of sugar is fermented which matures the beer and the CO_2 gas produced carbonates the product.

In storage/aging processes, in contrast, the amount of yeast or, more usually, the temperature is too low to permit significant fermentation. In such cases all the fermentable sugar is exhausted in primary fermentation and none is added in storage/aging. Such practices depend mainly on modification of primary fermentation for maturation (e.g. diacetyl rest or warm conditioning) and CO_2 injection for carbonation because the storage/aging conditions permit little or no maturation and carbonation from yeast action.

19.1.1 Secondary fermentations

Traditional cask-conditioned ale provides a useful model for illustrating secondary fermentation for finishing processes. Ale produced in small traditional fermenting squares entraps little carbonation from the primary fermentation. After substantial yeast removal by skimming and settlement, the beer with up to 2 million cells of yeast/ml is transferred to casks and necessary additions are made. These include priming sugar (usually hydrolyzed sucrose called invert sugar) for secondary fermentation (for maturation and carbonation) and finings, usually isinglass to accelerate clarification. Many beers might also be adjusted by adding hops or hop products for the fragrance of 'dry hop' aroma from the essential oils of the hop. Some additional bitterness might accrue. Color, as caramel or roasted malt extract, might be added and sometimes potassium metabisulfite (SO_2) to restrict bacteria. The casks are then sealed and held for a few days or a week to 'come into condition' before being served. At the end of this period the beer in the cask is clear with a deposit of (mainly) yeast and a floating mass of hops (if whole hops are used for dry-hopping). Thus all the objectives of finishing can be achieved in one simple and relatively short process. For a bottle-conditioned beer all the necessary additions must be made in bulk before bottling.

The **krausen process** of secondary fermentation is not unlike cask conditioning in that additional sugar must be added to provide the fermentable extract. In this case however the addition is called the **krausen** and comprises about 10% volume (up to 17% in the case of high-gravity brews) of freshly fermenting wort. The beer is held in large horizontal tanks (in one particularly large brewing company above a layer of beechwood chips in the 'chip' tanks), for 1–3 weeks at about 8°C (46°F). The tank is left open for the first day or so to allow the gas to purge out sulfury aromas and other particularly volatile materials and the tanks are then sealed to entrap CO_2 for carbonation of the product. During this prolonged cold secondary fermentation the beer flavor matures, the beer is carbonated by gas entrapment and substantially clarifies by sedimentation.

Lagering follows the same general practice as krausen fermentations but the sugar for the prolonged secondary fermentation is part of the extract of the original wort. That is, the beer is cooled to about 8°C (46°F) or below and moved to a secondary vessel before primary fermentation is complete. Secondary lager fermentations may last up to several weeks or even months. As in the other processes described, after lagering the beer is mature in flavor, fully fermented, carbonated and substantially clarified.

In all cases of secondary fermentation (except cask beer) the beer is finally reduced to very low temperature and transferred to another vessel for further treatment. The objective here is to achieve further yeast settlement and chill stabilization by allowing time for the chill haze (a protein–carbohydrate–tannin (polyphenols)–metal ion complex) to form before filtration. In most cases additional methods for chill stabilization are also used (see below).

19.1.2 Storage and aging

Storage or aging practices follow a somewhat prolonged primary fermentation which yields mature beer. This is achieved by lengthening the time at primary fermentation temperature in a stage called the "diacetyl rest". Maturation is monitored by measuring the decline in diacetyl and its precursor (α-aceto-lactate) during this period. Storage/aging of beer comprises relatively short-term cold storage of up to a week or 10 days at −2 to 4°C (28 to 39°F). The objective is to permit yeast to scavenge any oxygen incorporated during transfer to the aging tank, to encourage the formation of chill haze particles and to allow time for yeast sedimentation. Additional techniques may be used to achieve chill stability (see below).

Thus, whether beers undergo a secondary fermentation or storage/aging, they eventually arrive at the same point just before filtration, that is mature in flavor, reduced in yeast population, carbonated and very cold.

Not all microbreweries can afford multiple transfers of beer and prolonged processes. There is a tendency therefore to favor short processes which leads primarily to the use of *S. cerevisiae* and the production of ales. Cylindroconical unitanks are popular because these permit the primary and secondary fermentations, yeast removal, cold storage and flavor adjustment to take place in one vessel. In such a process after the primary fermentation subsides hops and possibly finings might be added and shortly thereafter the vessel sealed to entrap CO_2. After the vessel comes to pressure and the yeast exhausts the fermentable extract, the tank may be cooled slowly at first for maturation and then rapidly to about minus 1 to 3°C (30 to 37°F) for chill stabilization and clarification. Filtration follows. Throughout the process yeast is removed routinely from the cone.

19.2 CLARIFICATION

Sufficient time of storage at low temperature will eventually produce bright beer in accordance with **Stoke's Law**. This predicts that large dense particles will settle faster than small light ones, and any particle will settle best in a liquid of low density and low viscosity as follows:

$$u_g = \frac{d^2 (\partial_v - \partial_l) g}{9\mu}$$

where u_g is the terminal settling velocity of a particle under gravitational force; d the diameter of the particle; ∂_v the density of the particle; ∂_l the density of the liquid; g acceleration due to gravity; and μ the viscosity of the liquid.

19.2.1 Centrifugation

Brewers achieve substantial sedimentation of yeast and other suspended particles during cold storage, because aggregation of particles, yeast flocculation and fining agents will increase particle size. However, sedimentation to yield bright beer is almost always too long and too unreliable for modern practice.

Increasing gravitational force by using continuous centrifuges substantially accelerates clarification as predicted by the equation above. The main practical use of centrifuges (assuming their noise and cost is acceptable) is in preliminary reduction of suspended particles, primarily yeast, before the beer enters cold storage/aging. In such cases non-flocculent yeast could be used in the primary fermentation. Centrifuges are rarely used for final beer clarification because significant shear force would be required to achieve this and the beer would heat up too much.

19.2.2 Filtration

Filters allow rapid and dependable clarification. There are two main mechanisms by which particles may be filtered from flowing liquids (and gasses): **depth filtration** and **absolute filtration**. All brewery techniques of filtration, except membrane filtration, are depth filtration in principle although this may not be entirely obvious from practical operations.

Depth filters operate on the principle that a deep bed of filter material provides a long and complex route for the beer to pass through (commonly called a tortuous path). Although each part of the route may have physical dimensions sufficient to permit particles to pass, the particles simply cannot thread these tortuous paths and are entrapped in the filter mass, primarily near the surface or a short distance within the mass. Such filters are powder filters or sheet filters operated with or without powder dosing (see below).

If there is a significant quantity of suspended material to be removed then a filter with a continuously renewable surface must be employed so that the accumulation of particles filtered from the beer and entrapped in the surface layers of the depth filter does not clog or blind the filter. A powder filter (using e.g. diatomaceous earth (DE, also called kieselguhr) or perlite) does this best. Where there is very little suspended solids, for example when a bright beer is being 'polish' filtered, a renewable surface is not required and **sheet filters** or **cartridge filters** work well.

In the special case where the only suspended solids are a few micro-organisms the beer can be filtered sterile by proper choice of filter sheet or cartridge or a filter **membrane**. A membrane is an absolute filter. It is extremely thin and carries a defined number of holes of defined size per unit area. Each hole is dimensionally smaller than the particle size to be filtered, and the removal of one particle seals one hole. Pore sizes of 0.45 μm or less are commonly used for sterile filtration.

Powder (diatomaceous earth) filtration

Diatomaceous earth (DE) is the primary filter aid material used to clarify beer. The DE chosen must be low in beer-soluble iron. A suitable filter apparatus provides a micromesh screen of some sort, either as (a) a stainless steel wire mesh, (b) as a filter cloth or filter sheet or (c) in the form of small holes in a tube (candle filter). In all cases the mesh serves as a base upon which to build a bed of diatomaceous earth (Figure 19–1). The initial bed is called the **pre-coat** and most simply comprises a coarse grade of DE though mixtures of different DEs and DE plus cellulose are common. The pre-coat is built up by circulating the beer through the filter and heavily dosing it with DE until the exit beer is bright. This signals that the support mesh is fully bridged over or coated with the pre-coat material. A second pre-coat of finer DE is often applied. The beer can then be delivered from the storage tank to the receiving tank through the filter.

Of course, as the flow continues, particles from the beer accumulate on the filter surface (Figure 19–1). Yeast and protein particles can pack tightly and imperviously and blind the filter with surprising speed. The pressure required to force the beer through the filter bed then rises and/or the flow rate slows dramatically. This is reflected in the provisions of Darcy's Law. This equation relates the flow through the filter (Q, quantity of filtrate col-

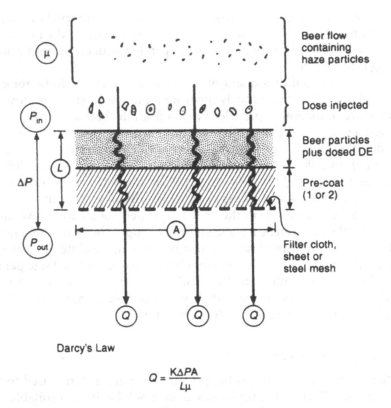

Darcy's Law

$$Q = \frac{K\Delta PA}{L\mu}$$

Figure 19–1 Diagrammatic representation of filtration through a bed of diatomaceous earth. μ, the viscosity of the beer containing haze particles; DE, diatomaceous earth (also called kieselguhr); P_{in} inlet pressure; P_{out} outlet pressure; ΔP ($P_{in} - P_{out}$) the pressure across the filter bed; L, depth of the filter bed; A, area of filter bed; Q, quantity of filtrate collected per unit time (flow through the filter); K, the permeability coefficient of the bed (related to the porosity of the filter bed).

lected per unit time) to the porosity of the filter bed (K, the permeability coefficient of the bed), the area of the filter surface (A), the pressure across the filter bed (ΔP), the viscosity of the beer (μ) and the depth of the bed (L):

$$Q = \frac{K\Delta PA}{L\mu}$$

Examination of Darcy's equation (Figure 19–1) confirms the inter-relations among flow rate and pressure across the bed and filter bed thickness for example. Similarly, blinding of the bed by the accumulation of particles, or its collapse under pressure, can be viewed as a reduction in the value of K (bed permeability). Also of importance is beer viscosity. If β-glucans survive mashing they can increase beer viscosity and can also increase suspended solids in beer and so reduce filtration efficiency. Some brewers view adding finings to reduce suspended solids in beer and adding β-glucanase to reduce beer viscosity as equivalent actions. Both serve to improve filtration efficiency but of course by different mechanisms.

To prolong the filter run brewers use two strategies (a) they settle the beer to remove as much suspended matter by sedimentation as possible before filtration and (b) during filtration they strive to keep the surface of the filter open by injecting DE into the beer just before it enters the filter. This is the 'body feed' or 'dose'. More hazy beer requires a greater dose rate, but typically some 80 to 100g/hl of DE is used. By this means the pressure required to force the beer through the filter bed rises quite slowly during the filter run. Nevertheless, the pressure does rise as the thickness of the bed increases, as it begins to compact and as it progressively fills the upstream capacity of the filter. In such cases 'breakthrough' of DE and beer particles can occur and downstream trap filters remove this material. Eventually the filter must be emptied of sludge, cleaned and the cycle begun again. All else being equal the length of a filter run depends on the amounts of suspended matter to be removed. Generally a coarser grade of DE is used to filter large particles (e.g. yeast).

Powder filters may be (a) vertical or horizontal leaf filters in which the support screen is a wire mesh, or (b) a plate and frame filter in which the support is a cloth or (c) candle filters which support DE on a long thin perforated rod. All work well as filters and since contemporary examples of each may be found in the industry none has established a clear dominance in the field in all applications. Filters are chosen because they are effective filters and also because they can be easily emptied, cleaned and put back on stream quickly and even automatically. The discharged sludge cannot be used again (at least under the present pricing structure for DE) and, because it is a most unsightly and easily identified liquid waste, dry discharge from the filter is highly prized for solid waste disposal. Some breweries recover DE from liquid waste streams by rotary vacuum filters and dispose of it as a solid.

Sheet filters

Sheet filters are mostly used where the beer is already bright and little suspended matter needs to be removed. Filter sheets comprise fibrous material (usually cellulose) that can be chopped and chemically frayed or shredded to markedly alter its filtration properties, and diatomaceous earth and resins to bind the whole together. The thick suspension is discharged onto a moving wire table where the water drains, leaving the formed sheet. This is then rolled and dried and cut to size. Sheets adsorb as well as filter. The grade of sheet is chosen for the work required. Suitable grades of sheet filters are used for polish filtration to render bright beer brilliant. Tighter filtration demands smaller pore size and consequently lower flow rate. Sheets can produce sterile beer for 'bottled draft' products, and can be back washed and steam sterilized in this application.

19.3 STABILIZATION

Beer that remains brilliantly clear when chilled to 4°C (40°F) or below is a necessary quality in most beer markets these days. Consumers have no difficulty seeing haze and, especially if they are accustomed to brilliant beer, view it as a negative quality. Preventing future haze formation in the trade is therefore of central concern during beer finishing. Beer haze is primarily a complex of proteins and polyphenols (tannins) with metal ions and usually some carbohydrate components.

19.3.1 Chill stability

Chill haze appears when the beer is chilled and re-dissolves when the beer warms up. Cyclic warming and cooling soon causes a permanent haze which will not disappear on warming. Brewers strive to remove the haze-forming material in the brewery. Prolonged cold storage at −1 or −2°C (28 to 30°F) precipitates the haze-forming complexes which can be removed by filtration of the beer while it is still very cold.

It is usual to represent the haze-forming reaction as the gross simplification

$$P + T = PT \text{ (precipitate)}$$

which nevertheless serves to illustrate the application of additional chill-proofing treatments which are commonly used. These agents attack the chill-haze forming components. Silica hydrogels and xerogels are the primary materials used to remove the proteinaceous portion of beer (P) that might engage in chill-haze formation. These are mostly derived from the hordein fraction of the malt (Chapter 11). It is also possible to use proteolytic enzymes (e.g. papain) or special preparations of tannic acid These treatments are declining in use however because of possible labeling requirements and, with tannic acid, high beer losses The polyphenol components (T) are partially removed by adsorption onto PVPP (polyvinyl-polypyrrolidone). Collagen finings and sometimes gelatin (a degraded form of collagen) can also promote chill-stability though they are not much used specifically for this purpose.

Silica gels and PVPP have the advantage of being insoluble in beer and so are "processing aids" not a beer "ingredient". They will not therefore be required to be identified on beer labels. They also act almost instantaneously and so can be conveniently injected into a beer stream on the way to the filter or even incorporated into filter sheet material. They are very effective.

19.3 CARBONATION

All beers have some degree of carbonation (sometimes also with nitrogenization) which makes important contributions to the mouth-feel of beer. Mouthfeel is sensed by touch and is not a flavor perception. Mouthfeel therefore is not restricted to perceptions of "body" or "fullness" but to the whole range of stimuli that affect touch which might also include (for example) temperature, astringency, acidity, viscosity, oiliness as well as tingle or burn associated with carbonation and mellowness associated with nitrogenization. Gas break-out also brings up and helps maintain the foam on poured beer. Longer storage under pressure seems to promote foam consistency and stability. Sufficient carbonation/nitrogenization are therefore important beer quality parameters.

19.3.1 Carbon dioxide in beer

Generally beer will gain about 1.5 volumes of CO_2 per volume of beer as a result of fermentation at atmospheric pressure (depending on temperature of fermentation) although much higher levels can accrue locally in deep tanks because of hydrostatic pressure. Carbonation therefore is used to increase the natural level of carbonation to that required in the marketplace which, in the UK, might be about 2.0 volumes of gas per volume of beer and, elsewhere, is some 2.6 to 2.8 volumes and possibly more. These are package levels. Most beers are overcarbonated by 10% or so to account for losses during packaging. One volume of CO_2 per volume of beer is 198 g/hl. (about 0.2% weight/weight depending on the specific gravity of the beer or about 0.5 lb. of gas per barrel)

If uncarbonated water or beer is in a closed container under an atmosphere of carbon dioxide *or* nitrogen the gas will dissolve in the beer until it is saturated. At saturation, which is the equilibrium point for the specific conditions of temperature and pressure, as much gas leaves the beer as enters it. The rate of carbonation (or nitrogenization) can be described by this formula:

$$V = K_L A (C_E - C)$$

In which V is the rate of gas solution

K_L is the liquid mass transfer coefficient
A is the area of contact between the gas and the beer
C_E is the concentration of gas that will be in the beer at equilibrium
C is the concentration of gas in the beer now

$C_E - C$ is therefore the driving force for gas solution (similar to the $T_1 - T_2$ term in heat transfer equations) and as C approaches C_E the rate of carbonation slows down. Carbonation is very slow from an atmosphere of carbon dioxide under quiescent conditions at the temperature and pressure (e.g. 15 psig at 43°F (1 bar at 6°C)) required for the correct amount of CO_2 in beer (e.g. 2.6 volumes/volume, 0.5 % weight/weight). Brewers therefore use lower temperature, higher pressure, and inject masses of very small gas bubbles into moving beer to substantially improve all the terms on the right of the above equation. This is best done in-line though gas can be sparged into beer in tanks. Practical carbonation conditions are therefore different from the equilibrium conditions required to maintain a prescribed amount of gas in solution and, under such forcing conditions, carbonation can be rapid. Overcarbonation should be avoided as this is difficult to correct.

However, when beer is confined in the BBT or a keg or a bottle, equilibrium conditions apply. The concentration of gas in the beer is then predicted by Henry's law:

$$P_{gas} = H_{gas} M_{gas}$$

P = absolute gas pressure (gauge pressure + 1 bar or psig + 14.7) above the beer at equilibrium
H = Henry's constant for the gas under equilibrium conditions (temperature and pressure)
M = the molar fraction of the gas in the beer.

In practice, the gas content of packaged beer is calculated from the pressure over the beer at a single test temperature (25°C, ASBC method), when a modified form of the equation above can be used:

$$CO_2 \text{ (\% by weight)} = (P - p) \times 0.00965$$

Where P = the absolute pressure over the beer (psia)

p = the correction for the headspace air (psia, if significant)

The factor 0.00965 = g CO_2/lb. of absolute pressure per 100g beer. The value is derived from Henry's law constant for solubility of CO_2 in water and adjusted empirically for beer.

CO_2% by weight can be converted to volumes simply:

CO_2 (volumes) = 22,414 × CO_2 (% by weight) × s.g./44.01 × 100

This reduces to:

CO_2 (volumes) = 5.093 × CO_2 (% by weight) × s.g.

where:

22,414 = the volume of one mole of gas at STP
44.01 = the molecular weight of CO_2
s.g. = the specific gravity of the beer

For the most part in brewing the gas is either CO_2 or N_2, sometimes both. In mixed gasses the total absolute pressure (P_{gas}, above) is apportioned between the gasses (their partial pressures) according to the % composition of the gas mixture (Dalton's Law). This means that if there is significant air (or nitrogen) in the gas of the headspace of a bottle, can or BBT the amount of CO_2 in solution can be overestimated. Henry's constant for nitrogen is about 100 times less than for CO_2 and so, under the same conditions, much less N_2 dissolves than CO_2. Therefore, to acquire significant nitrogenization and to maintain the gas in solution, much higher pressure and lower temperature is required than for CO_2. As a result, nitrogen easily breaks out of solution at quite low concentration and this quality is used in the manufacture of many ales and stouts for draft dispense in the UK and is the function of "widgets" (see page w) in bottled and canned products.

When nitrogen is used as an inert gas for beer transfer, sufficient partial pressure of CO_2 must be maintained to prevent loss of CO_2 from the beer. For the same reason, mixed gas (CO_2 *and* N_2) is used to dispense ordinary carbonated beer when a high overpressure is necessary to move the beer long distances or up several floors. If pure CO_2 were used for this, the beer would acquire excessive carbonation, with time, and would fob at dispense. However, by mixing nitrogen with the CO_2 the total pressure (motive force) can be increased as needed, without increasing the partial pressure of CO_2 needed for proper carbonation.

19.3.2 Recovery of carbon dioxide

Carbon dioxide for use about the brewery is recovered from fermentation which is a rich source of the gas. Theoretically each unit of glucose yields 0.489 units of CO_2 and so up to about 4 kg of CO_2 is evolved from each hectoliter of fermenting wort (roughly 10 lb. per barrel) depending on the content of fermentable sugar. Recovery then depends on the design of the particular fermenter-recovery system; large ones are much more efficient than small ones. About 20% of gas is vented to purge oxygen out of the fermenter headspace and recovery system, and an additional 20% might be lost in the inefficiencies of washing, drying and deodorizing the gas and compressing it to a liquid. Recovery therefore falls well short of theoretical but is sufficient to make CO_2 recovery economically beneficial in large breweries who are more or less self sufficient for the gas (although CO_2 production does not always match needs) and may even have excess CO_2 for sale. Up to 5–6 lb. of CO_2 might be needed per barrel of beer produced (2.0 to 2.4 kg per hl.). The gas is used for carbonation, counterpressure in storage vessels, deaeration of water, purging beer transfer lines and for packaging. Packaging, especially of cans, might consume up to half the total. CO_2 is a "greenhouse' gas.

Nitrogen is economical to purchase or it can be produced from air by burning a pure fuel in air to exhaust the oxygen or by membrane technology.

CHAPTER 20

Beer packaging and dispense

20.1 SMALL PACKAGES

Packaging of beer in kegs and casks assumes a draft beer drinking culture in which such products can be successfully marketed. Where this does not exist, or is emerging, beer is far more commonly packaged in bottles and cans than in kegs. The beer travels long distances and endures a long shelf-life (120 days or more) in these small containers and the disadvantageous surface-to-volume ratio demands extreme attention to the beer, the package and packaging if the consumer's expectation of a quality beer are to be met. Without such attention to detail, the beer suffers rapid loss of fresh beer flavor followed by further flavor breakdowns such as oxidized and light-struck (skunky) flavors and aromas. Eventually bottle beer haze and sediments form. Good packaging decisions can minimize these problems, but not eliminate them.

20.1.1 Packaging materials

Glass is chemically inert, non-tainting, impermeable to gases and resists cleaning and sanitizing agents. Flint or clear glass provides little protection of the product from light. Brown or amber glass usefully absorbs visible light below about 550 nm (Figure 20–1) providing some protection of the product from photolysis. This reaction forms isopentenyl mercaptan, the 'skunky' aroma from the side chain of iso-α-acids and hydrogen sulfide (H_2S) in beer (see figure 15.4). Modern thin-walled non-returnable brown bottles and green glass bottles (which absorb relatively weakly in the 400–550 nm range) are less protective than returnable ones.

Metal cans protect beer from light but could contribute flavors from lubricants, can linings and decorative coatings and by direct metal contact. With these problems under control, aluminum cans are an ubiquitous lightweight and highly successful beer container. Plastic (PET-poly-

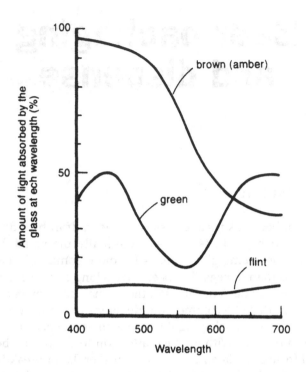

Figure 20–1 Absorption of light by different colored glass.

ethylene terephthalate) bottles have been used for soft drinks and beer in Britain and Europe. They have not yet found use in the USA, presumably because their resistance to gas permeation and light protection is currently insufficient for that market.

20.1.2 Packaging processes

Beer for packaging as it comes from the final filtration stage is brilliantly clear and very cold (about −1 or −2°C) and has acquired adequate carbonation by some mechanism, e.g. secondary fermentation and/or CO_2 injection. Typically the carbonation is somewhat higher (by up to 0.2 volumes) than that required in the finished package to allow for CO_2 losses in packaging. This beer is pumped to the bright beer or package release tanks either continuously (in which case these tanks act as surge tanks between the filter and filler) or in batches. The quality of beer is checked finally and released to package.

Minimizing exposure to oxygen

Beer processing and handling becomes more demanding and delicate as it approaches the final package. After yeast removal, entrance of oxygen can be devastating to package beer stability and this is stringently fought. Package beer pipes and tanks are cleaned as infrequently as possible (e.g. weekly) and perhaps with acid cleaners or just sanitary water to avoid interrupting the CO_2 atmosphere under which the beer is handled and held. Pipes and even tanks can be filled with de-aerated water to eliminate air; the water is then displaced with CO_2 or with beer flow. Similarly, beer can be chased from a system with water. Beer is filled into tanks without turbulence and below a CO_2 atmosphere and perhaps below beer remaining from a previous filler run to avoid air contact. This also avoids **fobbing** or **foaming**; collapsed and dried foam can cause suspended particles in the beer. Foaming also implies some loss of CO_2 which is undesirable at this stage. Package tanks are therefore maintained at equilibrium pressure to retain the CO_2 in solution; this is generally about 10–12 p.s.i.g. for beer at 0–1°C and 2.7 to 2.8 volumes of CO_2. When moving beer from one tank to another a constant overpressure is most easily attained by connecting the headspace of the tank being emptied to the one being filled. The three conditions mentioned above determine the main quality control criteria for package beer release: that the beer meet specification for CO_2, dissolved oxygen (0.2 mg/l or less) and clarity (and of course beer flavor). There may be a wide variety of other quality control parameters applied to the beer before release including, for example, color.

A packaging line (Figure 20–2) is a long and complex arrangement of machines linked by conveyors. The objective is to operate the line at high efficiency; 100% efficiency is the design capacity of the filler. To assure that the filler is never starved of empty packages nor choked by full ones, the filler is the slowest machine in the packaging line. Accumulator tables are provided at strategic points so that if one machine should temporarily halt, the rest of the line can continue to operate.

The capacity of the package cellar tanks must reasonably match the volume and mix of products to be packaged and to permit production of sufficient bright beer to keep up with the demand of the packaging plant. Flow of beer from the bright beer tanks to headers and hence to the filler bowl of the filler itself requires pressure provided by gravity or pumps. Beer flows down a line one way to the filler and all is well as long as the filler runs (Figure 20–3a). When it stops, flow in the beer line stops and, if the stoppage is a long one, the beer can warm up with serious consequences for beer quality, for beer recovery (i.e. re-cooling the beer) and for

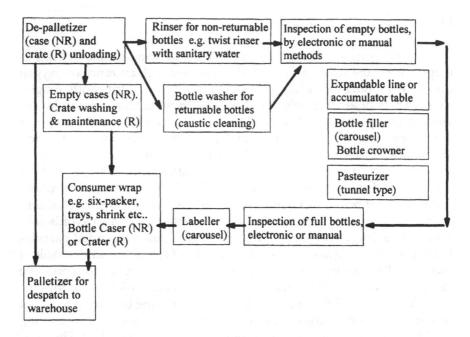

Figure 20–2 Diagram of a filling line. The filler and crowner are shown in the same box as these are always close together. The machines generally have slower operating speeds as they approach the filler (which is the slowest machine in the line) then operate progressively faster after the filler. There may be several accumulators in the line. The number of packages filled, relative to the design capacity of the filler, is the efficiency of the line

fill control. In such a system starting and stopping are complex procedures and requires skilled communication between the brewery and the bottle shop. However, the system itself is technologically simple and relatively low cost (unless automated). A **packaging loop**, which is a header in which beer is continuously circulated and cooled, solves many of these problems by maintaining a constant supply of cold beer in the header (Figure 20–3b). Automatic sensing of pressure and temperature of the beer in the loop is necessary to control the speed of the pump.

Filling systems

Beer flows into the filler bowl, located at the center of the machine or as a ring bowl around its periphery, where the appropriate bowl fill is maintained by monitoring the beer level or gas pressure. Most fillers are rotary (carousel type, Figure 20–4) machines, the filling capacity of which is determined by the number of heads or valves. A modern machine can easily approach (bottles) or exceed (cans) 2000 containers per minute. When filling a bottle or can (Figure 20–5) the concerns are to avoid introducing dissolved oxygen, to achieve an accurate fill level and to leave a headspace above the beer with minimum air. After the container makes contact with the filling valve of the machine (called a 'bell' in a bottle filler and a 'tulip' in a can filler) the filler commonly evacuates the container, sometimes twice and/or purges it with CO_2. Cans, being fragile, cannot be evacuated to remove air. The container is then counterpressured to match the pressure

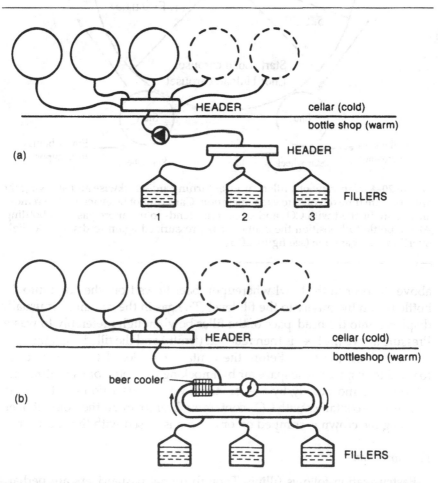

Figure 20–3 Diagrammatic representation of small pack filling. (a) Single line from header to fillers. (b) Improved system with a packaging loop between header and fillers.

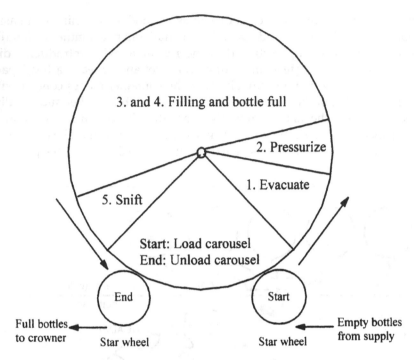

Figure 20–4 Plan view of a filler carousel turning anti-clockwise and showing the approximate time devoted to each operation. Cans cannot be evacuated to remove air but are flushed with CO_2 and so canning tends to use more gas than bottling. At the bottle full position the container is pressurized again to displace a slight overfill in the gas-tube (see figure 20.5).

above the beer in the bowl whereupon beer flows from the bowl into the bottle or can by gravity to the fill level. The gas in the container is usually displaced into the headspace of the filler bowl though preferebly to waste. Pressure above the beer is then relieved ('snift') and the filled container disconnects from the filler. Before the container is closed the beer is overfoamed to displace headspace air by knocking the bottle or can, ultrasonic vibration or, most easily, by a jet of sterile water. This action fills the headspace of the container with CO_2-containing beer foam. At the point of overfoaming the crown is crimped on or the can is closed with the end unit.

Pasteurization

Pasteurization follows filling. Though tunnel pasteurizers are perhaps the least ideal technology (e.g. in term of complexity, space and energy effi-

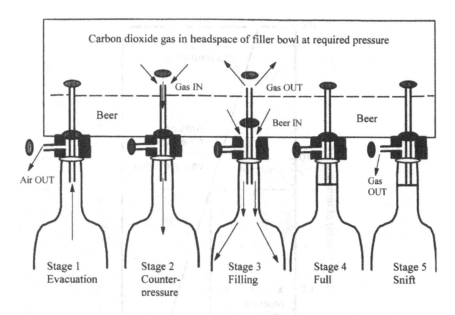

Figure 20–5 Diagram of the sequence of a bottle-filling operation with a "short-tube" filler. However, only gas, not beer travels through the tube. Beer flows down the wall of the bottle in stage 3 by gravity after the pressure in the bottle and bowl are equalized. Stage 2 might include a gas flush (especially for cans), and double evacuation is a feature of some machines (bottles only).

ciency) for microbiological stabilization of beer, they are the most common because they are reliable and effective. The great advantage is that the product is heat-treated in its final closed package. Other methods of microbiological stabilization such as 'flash' or bulk heat treatment or filtration require aseptic packaging and rigorous quality control standards to ensure that post-pasteurization contamination did not occur, e.g. during filling.

The objective of pasteurization is to reduce the chance that microorganisms survive in beer to the required low level. This depends on many factors including the kind of beer being pasteurized, the number and kind of microbes present, the size of the package, and the level of comfort brewers have with minimal pasteurization treatments. One Pasteurization Unit (PU) is defined a 1 minute at 60°C (140°F) or its equivalent based on the death rate curve of microorganisms found in beer. This logarithmic curve (Figure 20–6), established by empirical observations, relates time and temperature of heat treatment to the survival/death of such microorganisms. The line shown joins all the time–temperature combinations that

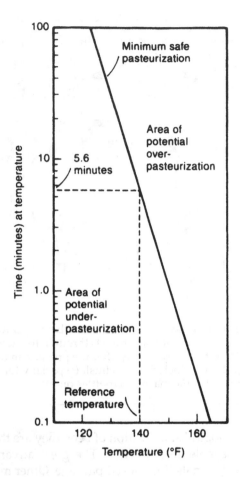

Figure 20–6 Pasteurization diagram for beer microbes. The continuous line running from top left to bottom right defines the time–temperature conditions needed to deliver minimum pasteurization units (PUs).

just killed the test suspension of beer-related organisms. These points all have the same heat-treatment value; the kill time required at the reference temperature of 60°C (140°F) is 5.6 minutes or 5.6 PU. This is the minimum safe heat treatment for beer. The slope of the line is the Z value (12.5°F or 7°C) which implies that for each change of 7°C (12.5°F) the time of heat treatment can be shortened or lengthened by a factor of 10, i.e. 60°C (140°F) at 10 minutes, 67°C (152.5°F) at 1 minute and 74°C (165°F) at 0.1 minute all have the same lethal value (10 PU).

Brewers typically operate in the range of 5–15 PU. This amount of heat however must be received at the center of the bottle or can and so beer closer to the container surface receives more heat. The beer in package might easily spend 40 minutes in a pasteurizer to be heated, held at the required temperature and cooled (Figure 20–7). Especially if the pasteurizer should stop, some beer might be heated enough to affect beer flavor. This is the reason that other techniques of microbiological stabilization are attractive to brewers. Bulk pasteurization heats beer in a thin flowing stream as it passes through a heat exchanger and heating up and cooling down can be very rapid and energy efficient. Relatively high temperatures keep the treatment time very short (a high-temperature/short-time

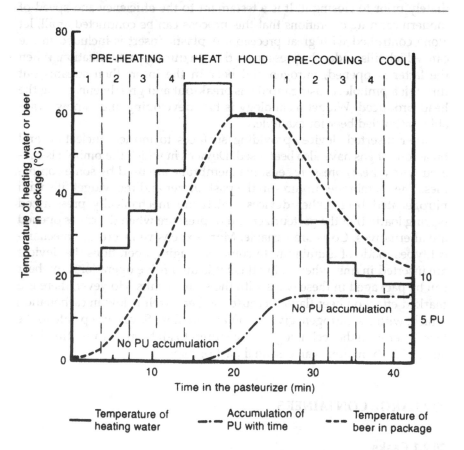

Figure 20–7 Temperature and pasteurization unit profiles of beer in small packs in a tunnel pasteurizer. PU, pasteurization unit.

strategy) so that overheating is minimized. Finally, since the beer is cellar beer (low in air) there is less likely to be any flavor effects of heat plus oxygen. Alternatively, beer can be sterilized without heat by filtration through sheets or membranes (above) or ceramic filters. In such cases aseptic filling must follow which is a demanding technology. The filter-sterilized beer can legally be called draft beer.

'Widgets'

Recently, canned beers have appeared in the market which form a most stable foam head when poured. These beers are packaged with nitrogen and CO_2. A very small amount of liquid nitrogen is jetted into a can immediately prior to closing it. It is a testament to the efficiency and speed of modern canning operations that this process can be conducted at all, let alone controlled with great precision. A plastic insert is included in the can, which fills under the pressure in the can during pasteurization. When the latter is opened, nitrogen and beer in the insert then streams out through a pinhole, causing rapid gas break-out and greatly enhancing the head produced. Widget technology is fast developing and is now available for bottled beer for example.

Other inserted devices providing surfaces to induce nucleation and break out of gas have also been used. Devices in which the pinhole is covered with a heat/pressure sensitive membrane are used by some companies. The membrane ruptures in the pasteurizer and the widget fills with nitrogenated beer. Other devices which are mechanically pressurized (spring loaded) in the pasteurizer and depressure when the can is opened are alternatives. Costs are variable. More expensive inserts are mechanical types made of aluminum. In some packaging operations the devices are inserted in line, whereas in others it is an off-line operation. Any beer can be packaged in these ways with the same results. However, there is a marked effect on beer flavor because the beer itself is low in carbonation which, with the nitrogen, gives a soft and mellow flavor and palate to the beer. Therefore the technology is not suited to all beers because the sharp sting of carbonation is an essential part of many.

20.2 LARGE CONTAINERS

20.2.1 Casks

Beer in bulk is packaged in the brewery into large containers of varying capacity (Table 20–1). With live (traditional draught) beers containing

Table 20–1 Capacities of bulk beer containers

Container	UK gallons	US gallons	Liters
Cellar tank	180 or 360	216 or 432	819 or 1637
English hogshead	108	130	491
English barrel	36	43	164
English kilderkin	18	22	82
English firkin	9	11	41
English pin	4.5	5.5	21
US barrel	26	31	117
Modern keg	22	26	100
	11	13	50

yeast the vessels were manufactured from oak and barrel shaped (referred to here as **casks**). Stainless steel and aluminum casks have largely replaced the wooden ones. In a traditional cask, one opening (approximately 25 mm) is located at the edge of one of the ends and the other (approximately 50 mm) in the center of the body. After rigorously cleaning the cask and visually inspecting the interior, a wooden plug (keystone) is placed in the edge port and the cask filled through the larger one. The filling operation is done manually by gravity feed from a reservoir (cask racker). In a traditional ale brewery, priming sugar, finings and dry hops may be added and then the barrel is sealed using another wooden plug (shive). The container is then held in the brewery for a few days before transfer to the retail outlet. Once received the cask is placed on a stand (stillage) and after 12 to 24 hours the cask is tapped by hammering a metal tap through the keystone. The day before the beer goes on sale a small quantity is withdrawn to relieve the pressure and a hard wooden peg (spile) driven through the center of the shive. Keystone and shive contain knockout sections which of course end up inside the cask. The peg is eased and beer sampled to check for its condition (mainly CO_2 content). The peg is always pushed in tight outside trading hours and loosened before drawing any beer off. When beer is in very 'high condition' (too much CO_2), a soft porous spile may be used to allow venting of excess gas.

In the retail outlet, beer in casks can be drawn off directly from the tap into a glass. It is of course more usual for the beer to be held in a cellar (at about 15°C or lower temperature). In this case some mechanism for bringing beer to the bar is needed. The earliest system was the beer engine (Figure 20–8). This is a simple lift pump comprising a cylinder and manually operated piston. Non-return valves are placed in the base of the piston and base of the cylinder (at its junction with the inlet). The cask is con-

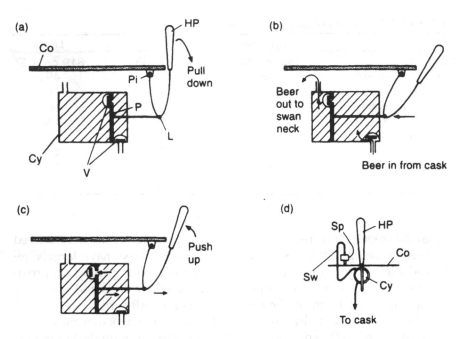

Figure 20–8 Diagrammatic representation of a beer engine. (a) At rest, primed both valves closed. HP, hand-pull; Co, bar (counter) surface; Pi, pivot; L, rotating link; P, piston seal; V, values; Cy, cylinder. (b) Beer dispenser: hand-pull drawn down, piston moves forward and the valve in the piston is forced shut so beer is forced out. At the same time, the valve in the inlet pipe is pulled open and beer is drawn in from the cask. (c) Return stroke: hand-pull is pushed up, pressure on beer behind piston causes inlet valve to close whereas the valve in the piston opens enabling the piston to return 'through' the beer. A second downward pull dispenses further beer. (d) Diagram of beer engine from behind the bar. Sw, swan neck dispense pipe (stainless steel); Sp, sparkler—used to adjust the amount of air mixing with the beer at dispense to control the degree of foaming. The cylinder may be mounted horizontally (as shown) or vertically using a slightly different arrangement (as described in the text). At the end of a day the lines are emptied so before beer can be dispensed the cylinder needs to be filled. This process, called 'priming', is done using several down/up strokes of the hand-pull.

nected to the beer engine with flexible piping. When the handle (the most prominent feature on the bar) is pulled down, the piston rises. The suction pressure causes the valve at the base of the cylinder to open and beer is drawn in beneath the piston. The differential pressure across the piston (greatest above) ensures that the piston valve remains closed. At the end of the down stroke, the handle is returned to the upright position. The pressure beneath the piston is now greatest so the cylinder valve closes and the piston valve opens. The piston passes down through the column of beer. At the end of this movement the handle is pulled again and the beer is now lifted up and passes out through the swan neck pipe. The whole process is referred to as 'pulling a pint'. The beer engine needs priming before it can operate effectively and this leads to considerable wastage. Modern beer engines are fabricated from glass and stainless steel and have hygienic seals and valves. The original ones were brass and leather!

With the beer engine, when beer is drawn, air passes into the cask. This increases the potential for microbiological spoilage by aerobic microbes (Chapter 15). Accordingly it is common to connect a cylinder of CO_2 to the cask so that beer withdrawn is replaced with gas retaining the anaerobic conditions. No significant pressure is applied, the intention being to maintain a blanket of CO_2 on top of the beer. Beer engines are easily replaced by electric pumps operated by a microswitch on the bar tap which may be manufactured to look like the hand pull of the beer engine!

20.2.2 Kegs

A major change in brewing technology was the filtration and pasteurization of the beer in the brewery and packaging under pressure of CO_2 into metal **kegs**. More recently the use of sterile filtration rather than pasteurization of beer before packaging carbonated beer into kegs has been introduced. These products, like their small pack counterparts, can be called draught products. Kegs are often of an overall cylindrical shape (straight-sided) and a typical capacity is 50 liters (Figure 20–9). In the USA 15.5 gallon (half-barrel) kegs are used. Kegs of similar capacities but of the traditional barrel shape are used by some companies. Straight-sided kegs are more easily handled automatically, whereas the barrel is often preferred for manual handling. The principal difference between a cask and keg is that the former has two apertures and the latter only one. Special kegs have been developed to allow live beer (beer containing yeast) packaging and dispense. Such kegs require special attention at the cleaning stage. In some countries (especially UK and Australia), large-capacity cellar tanks may be used to hold beer. These are 500–1000-liter capacity. In some situations they are filled at the brewery and demounted into the cellar at the outlet (demountable tanks) or maintained in the cellar and filled from beer carried in a road tanker. In the latter case the tanker driver may also be responsible for removing and replacing a sterile plastic bag (which holds the beer) in the tank prior to filling. Large serving vessels are used in most brew pubs in the USA.

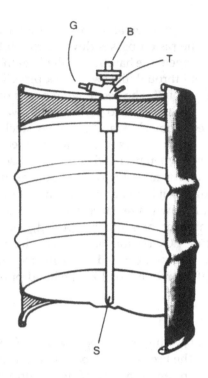

Figure 20–9 Diagram of a cut-through keg. S, spear, T, tapping head; B, connection for beer line; G, connection for gas line.

Sterilization and sanitation

The automated equipment used to fill kegs comprises a series of stations for washing, sterilizing and filling. The overall requirements for exclusion of oxygen, sanitation of equipment and maintenance of a counter-pressure of CO_2 are the same as those for bottling or canning plant. Where pasteurized beer is packaged, it has been treated by the 'flash' pasteurization process. A flash pasteurizer is similar in design to a plate heat exchanger used to cool wort (Chapter 7). Beer held in the unit at temperatures of 70–72°C for the time required to give the required amount of pasteurization (15–25 PU). Typically at least 10% more PU are given than applied to the same beer packaged in small packs. This increase takes into account that neither flow rate through the pasteurizer nor temperature control in the unit are absolute, but fluctuate.

The kegging operation requires an aseptic filling process. The design of plant and the cleaning and sanitation of it are vitally important in restricting access of microbes and atmospheric oxygen. Nonetheless, the shelf-life of kegged products is measured in weeks as opposed to the months typical of pasteurized small packs.

An empty stainless steel keg is automatically placed (usually inverted) onto a moving belt and passes to a washer. Here a pressure test occurs to check that there is no leak (if there is the keg is rejected). Residual beer is purged from the keg. A washing cycle of hot water, hot caustic soda (2%), hot water, acid (a nitric and phosphoric mixture) is used. Solutions are forcefully jetted into the kegs via the spear in the reverse direction to beer flow during dispense (see below). Between each stage, liquids are purged from the keg. Kegs are then sterilized using pressured steam, purged to remove condensate, cooled, and then counter-pressured with CO_2. The keg is then filled with finished beer, held cold in filtered form in bright beer tanks. Checks for over/underfilling are conducted and out-of-specification kegs rejected. Finally it is automatically stacked for removal to storage and subsequent transport to the retail outlet. Kegs typically spend about 1 minute at each stage of the process.

Aluminum kegs are cleaned in the same manner but only acid cleaners are employed.

Keg handling and dispense

The single opening of a keg (Figure 20–9) is fitted with a spear. The spear however operates as a double entry port allowing for example gas to pass into the top of the keg and displace beer up through the shaft of the spear (Figure 20–10). Connection to the spear is made with a keg tapping head which has inlet and outlet connections.

Carbonated beer in keg may be transferred to the bar using gas pressure. When CO_2 is used it is providing carbonation as well as propulsive force. This system can cause problems, especially in over-carbonating the beer and causing excessive fobbing. Mixed gas dispense systems using nitrogen and CO_2 allow the former to provide the energy for movement of the beer and the latter to give the right degree of carbonation. The use of nitrogen also influences beer flavor and appearance. Nitrogen forms smaller bubbles giving rise to a thicker, creamier head on beer.

Control of beer temperature in long runs of pipe is achieved using python systems (Figure 20–11). These contain the beer pipes and pipes carrying cooling fluid and are wrapped in insulation. The chemical com-

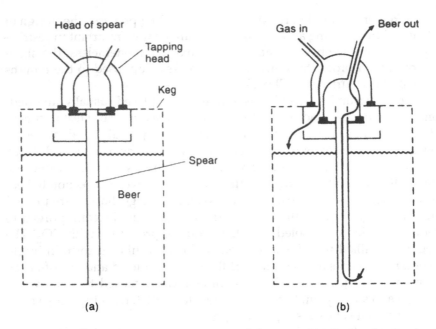

Figure 20–10 Diagrammatic representation of the operation of a keg (pressurized cask) spear. (a) Before locking the tapping head onto the keg. (b) After locking the tapping head in position. The spear creates a double port permitting gas pressure to displace beer up through the tube. Arrows show the direction of flow of fluids. In cleaning operations, the beer outlet port can be the cleaning agent inlet.

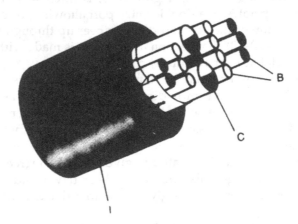

Figure 20–11 'Python' system for beer dispense lines. The large-diameter pipes (C) carry the cooling liquid. Smaller-diameter pipes (B) carry the beer. Beers dispensed at lower temperatures are piped closest to the cooling lines. The outer sheath (I) is a butyl rubber material which is highly insulating.

position of pipework is also rigidly controlled to ensure that no materials which may adversely affect beer flavor are able to leach out.

Where gas is not used to move the beer, electric pumps are installed and activated by switches in response to the operation of the bar tap (Figure 20–12).

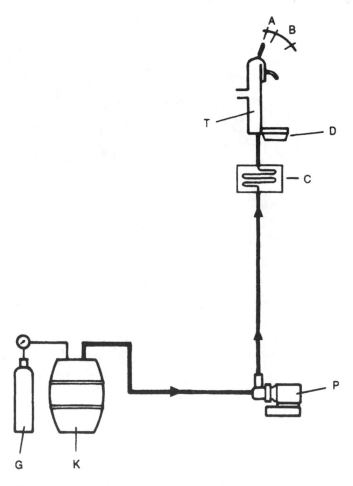

Figure 20–12 Beer dispense using an electrically driven pump. Moving the level at the top of the bar tap (T) from the vertical through position A activates an electrical switch and turns on the pump (P). At position B the tap is fully open and beer flows from the cask (K) through the tap. The gas cylinder (G) maintains a slight pressure to keep gas in solution and prevent the beer going flat. C is an in-line beer chiller and D a drip tray.

Control of beer foaming is an important consideration in dispensing. Non-turbulent flow in pipes is desired but a certain degree of turbulence desired at the extreme outlet of the tap. A restrictor (sparkler) is often incorporated at the end of the tap to allow adjustment of turbulence. This helps to create a foam (head) on the dispensed product.

All dispense lines need regular and rigorous cleaning. Detergent sanitizers and sterilants are usually circulated through the system and taps taken apart for special attention.

Additional reading

The following lists are intended primarily to be a guide for those wishing to gain more detailed knowledge than given in this book. Some texts, although no more (sometimes less) detailed than this one, present different approaches and are therefore of considerable educational value.

We have selected key texts from 1980 onwards. Some books from earlier years are included where we believe they offer important contributions to brewing science and technology.

Recent books on yeast molecular biology (primarily *Saccharomyces cerevisiae* and non-production laboratory strains) abound and are really for the specialist; accordingly we have excluded the majority of these but those included will also serve as pointers to others.

Periodicals: this list includes all the key journals which carry general and specific (state-of-the-art research) information on aspects of brewing science and technology. In addition two journals addressing the specific needs of the US micro- and home-brewers are included.

BOOKS

Brewing science and technology

Briggs, D.E., Hough, J.S., Stevens, R. and Young, T.W. (1981–1982). *Malting and Brewing Science*, 2nd edn, Chapman & Hall, London.

Broderick, H.M. (1982). Beer packaging: a manual for the brewing and beverage industries. Master Brewers Association of the Americas, Madison, Wis.

Broderick, H.M. (ed.) (1977). *The Practical Brewer: A Manual for the Brewing Industry*, 2nd edn, Master Brewers Association of the Americas, Madison, Wis.

Clerck, Jean de. (1957–58). *A Textbook of Brewing*. Chapman & Hall, London.

Findlay, W.P.K. (1971). *Modern Brewing Technology*. Macmillan Press, London.

Fix, G.J. (1989). *Principles of Brewing Science*. Brewers Publications, Boulder, Col.

Hardwick, W.A. (1995). *Handbook of Brewing. Marcel Dekker Inc., New York.*
Hough, J.S. (1985). *The Biotechnology of Malting and Brewing.* Cambridge University Press, Cambridge, New York.
Inoue, T., Tanaka, J-I., and Mitsui, S. (1992). *Recent Advances in Japanese Brewing Technology.* Gordon and Breach Science Publishers, Philadelphia.
Institute for Brewing Studies (1991). *Brewery Planner: A Guide to Opening your own Small Brewery.* Brewers Publications, Boulder, Col.
Jeffrey, E.J. (1956). *Brewing: Theory and Practice,* 3rd edn, N. Kaye, London.
Lloyd Hind, H. (1950). *Brewing: Science and Practice.* Chapman & Hall, London.
Pierce, J.S. (ed.) (1990). *Brewing Science and Technology,* Series II, vol. 2: *Malting, Wort Production and Fermentation,* vol. 3: *Quality;* vol. 4 *Engineering.* The Institute of Brewing, London.
Pollock, J.R.A. (1979). *Brewing Science* (3 volumes). Academic Press, London.
Preece, I. (1954). *The Biochemistry of Brewing.* Oliver & Boyd, Edinburgh.

Dictionaries and other source books

Downard, W.L. (1980). *Dictionary of the history of the American brewing and distilling industries.* Greenwood Press, Westport, Conn.
Forget, C. (1988). *The Association of Brewers' dictionary of beer and brewing.* The most complete collection of brewing terms written in English. Brewers Publications, Boulder, Col.
Gourvish, T.R. and Wilson, R.G. (1984). *The British brewing industry, 1830–1980.* Cambridge University Press, Cambridge, New York.
Heyse, K.U. (1989). A practical dictionary of brewing and bottling. [English–German, German–English]. H. Carl, Nurnberg.
Jackson, M. (1994). *The Simon & Schuster pocket guide to beer.* Simon & Schuster, New York.
Kiss, B. and Horvath, I. (1991). *Brewery dictionary.* OMIKK, Budapest.
Zentgraf, G. (1989). *A pictorial guide to modern brewery.* (With explanations of pictures in English). H. Carl, Nurnberg.
Zimmermann, C.E. (1981). English and foreign publications on hops. U.S. Dept. of Agriculture, Science and Education, Washington, DC.

Analytical methods

American Society of Brewing Chemists. *Methods of analysis of the American Society of Brewing Chemists* (1992). 8th revised edn, ASBC, St. Paul, Minn.

European Brewery Convention Congress. Enari, T.-M. (1987). *Analytica-EBC*, 4th edn, European Brewery Convention. Analysis Committee. Braurerei-und Getranke-Rundschau, Zurich, Switzerland.

Institute of Brewing. *Recommended methods of analysis*. (1986–). [Revised edn]. The Institute of Brewing, London.

General

Gump, B.H. (ed.) (1993). *Beer and Wine Production; Analysis, Characterization, and Technological Advances*. American Chemical Society, Washington, DC.

Linskens, H.F. and Jackson, J.F. (eds) (1988). *Beer Analysis*. Springer-Verlag, Berlin, New York.

Nykanen, L. and Suomalainen, H. (1983). *Aroma of Beer, Wine, and Distilled Alcohol Beverages*. D. Reidel, Dordrecht, Boston.

Barley, malt and hops

Briggs, D.E. (1978). *Barley*. Chapman & Hall, London.

Briggs, D.E. (1995). *Malts and Malting*. Kluwer Academic/Plenum.

Brown, J. (1983). *Steeped in tradition; the malting industry in England since the railway*. University of Reading, Institute of Agricultural History, Reading.

Filmer, R. (1982). *Hops and Hop Picking*. Shire Publications, Aylesbury.

MacGregor, A.W. and Bhatty, R.S. (eds) (1993). *Barley; Chemistry and Technology*. American Association of Cereal Chemists. St. Paul, Minn.

Rasmussen, D.C. (ed.) (1985). *Barley*. American Society of Agronomy, Madison, Wis.

Rybacek, V. (ed.) (1991). *Hop Production*. Elsevier, Amsterdam, New York.

Stevens, R. (ed.) (1987). *Brewing Science and Technology, Series II*, vol. 1: hops. The Institute of Brewing, London.

Tomanko, R.R. (1986). *Steiner's Guide to American Hops, Book III*. Hopsteiner, New York.

Tomlan, M.A. (1992). *Tinged with gold; hop culture in the United States*. University of Georgia Press, Athens, Georgia.

Verzele, M. and De Keukeleire, D. (1991). *Chemistry and Analysis of Hop and Beer Bitter Acids*. Elsevier, Amsterdam, New York.

Yeast and microbiology

Barnett, J.A., Payne, R.W. and Yarrow, D. (1990). *Yeasts; Characteristics and Identification*. Cambridge University Press, Cambridge, New York.

Berry, D.R. (1982). *Biology of Yeast*. Edward Arnold, London.
Berry, D.R., Russel, I. and Stewart, G.G. (eds.) (1987). *Yeast Biotechnology*. Allen & Unwin, London, Boston.
Carr, J.G. (1968). *Biological Principles in Fermentation*. Heinemann Educational, London.
Kirsop, B.E. and C.P. Kurtzman (eds), in collaboration with Nakase, T. and Yarrow, D. (1988). *Yeasts*. Cambridge University Press, Cambridge, New York.
Kirsop, B.E. and Snell, J.J.S. (1984). *Maintenance of Microorganisms*. Academic Press, London.
Kreger-van Rij, N.J.W. (ed.) (1984). *The Yeasts; a Taxonomic Study*, 3rd revision, Elsevier Science Publishers, Amsterdam.
Panchal, C.J. (1990). *Yeast Strain Selection*. Marcel Dekker, New York.
Priest, F.G. and Campbell, I. (eds) (1987). *Brewing Microbiology*. Elsevier Applied Science, London, New York.
Reed, G. and Nagodawithana, T.W. (1991). *Yeast Technology*. Van Nostrand Reinhold, New York.
Rose, A.H. and Harrison, J.S. (eds) (1987). *The Yeasts*, 2nd edn, 4 vols. Academic Press, London, San Diego.
Skinner, F.A., Passmore, S.M. and Davenport, R.R. (1980). *Biology and Activities of Yeasts*. Academic Press, London, New York.
Verachtert, H. and De Mot, R. (eds) (1990). *Yeast; Biotechnology and Biocatalysis*. Marcel Dekker, New York.

STATISTICS, PROCESS CONTROL AND QUALITY ASSURANCE

Oakland, J.S. (1996). *Statistical Process Control*. Butterworth Heinemann.
Stewart, J.R., Mauch, P. and Straka, F. (1996). *The 90-Day ISO 9000, Implementation Guide The Basics*. St. Lucie Press.

PERIODICALS

Brauwelt; Zeitschrift fuer das gesamte Brauwesen und die Getraenkewirtschaft. Nuernberg, Germany: Verlag hans Carl GmbH, [1861–].
The Brewer. London, Brewers' Guild Publications, Ltd. [1910–].
Brewers' Almanac. Washington, DC, Beer Institute.
Brewers Digest. Chicago, IL: Siebel Publishing Co., Inc. [1926–].
Brewers Digest Annual Buyers Guide and Brewery Directory. Chicago, IL: Siebel Publishing Co., Inc., [1926–].

Brewers' Guardian. Hampton, England: Hampton Publishing Ltd., [1871–].
Brewers Guild Directory. London: Brewers' Guild Publications Ltd., [1923–].
Brewing and Beverage Industry International. Mindelheim, Germany: Verlag W. Sachon, [1989–].
Brewing and Distilling International. Burton upon Trent, England: Brewery Traders Publications Ltd., [1865–].
Brewing and Malting Barley Research Institute. Annual Report. Winnipeg, Canada: Brewing and Malting Barley Research Institute, [1952–].
Brewing Industry News. Riverdale, Il.
Brewing Research Convention. Current Awareness Monthly. Zoeterwoude, Netherlands: European Brewery Convention, [1952–].
Brygmesteren; Scandinavian Brewers' Review. Hellerup, Denmark: Dansk Bladforlag ApS, [1944–].
European Brewery Convention. Proceedings of the Congress. Zoeterwoude, Netherlands, European Brewery Convention, [1950–].
Ferment. London: Institute of Brewing, [1988–].
Journal of the American Society of Brewing Chemists. [St. Paul, Minn.] American Society of Brewing Chemists, [1942–].
Journal of the Institute of Brewing. London, Institute of Brewing, [1904–].
Kirin Brewery Company Technical Report of Kirin Tokyo, Japan: Kirin Brewery Company, [1956–].
Louvain Brewing Letters. Place Croix du Sud, Belgium: Universite Catholique de Louvain; Unite de Brasserie et des Industries Alimentaires, [1988–].
Master Brewers Association of the Americas. Communications. Chicago: Master Brewers Association of the Americas, [1940–].
Master Brewers Association of the Americas. Technical quarterly—Master Brewers Association of the Americas. Madison, Wis.: Master Brewers Association of the Americas.
The New Brewer. Boulder, Co: The Institute for Brewing and Fermentation Studies, [1983–].
Research Institute of Brewing. Tokyo, Japan: Research Institute of Brewing, [1905–].
Zymurgy. Boulder, Co: American Homebrewers Association Inc., [1978–].

Index